AF361869

Global Understanding of Accretion and Ejection around Black Holes

Global Understanding of Accretion and Ejection around Black Holes

Editor

Santanu Mondal

MDPI • Basel • Beijing • Wuhan • Barcelona • Belgrade • Manchester • Tokyo • Cluj • Tianjin

Editor
Santanu Mondal
Theoretical Astrophysics
Indian Institute of Astrophysics
Bangalore
India

Editorial Office
MDPI
St. Alban-Anlage 66
4052 Basel, Switzerland

This is a reprint of articles from the Special Issue published online in the open access journal *Galaxies* (ISSN 2075-4434) (available at: www.mdpi.com/journal/galaxies/special_issues/Accretion_and_Ejection_around_Black_Holes).

For citation purposes, cite each article independently as indicated on the article page online and as indicated below:

LastName, A.A.; LastName, B.B.; LastName, C.C. Article Title. *Journal Name* **Year**, *Volume Number*, Page Range.

ISBN 978-3-0365-5610-9 (Hbk)
ISBN 978-3-0365-5609-3 (PDF)

Cover image courtesy of International Centre for Radio Astronomy Research

Contents

About the Editor

Santanu Mondal

Santanu Mondal is a Ramanujan Fellow at the Indian Institute of Astrophysics. He completed his Ph.D. in 2015 from Indian Centre for Space Physics, India. He was a Kreitman Fellow at the Ben-Gurion University of the Negev, Israel and FONDECYT Fellow at the Universidad de Valparaiso, Chile. He is also a life member of Astronomical Society of India. He was a visiting scientist of NASA/GSFC in 2014 and MPIA, Garching in 2017. His works include theory and observation of accretion-ejection flows around black holes, numerical simulation of the Fermi Bubbles, and the imaging of galaxy clusters.

Preface to "Global Understanding of Accretion and Ejection around Black Holes"

Black holes are one of the most interesting and fascinating objects in the universe, not because of their name, but because of their activity. They produce enormous amount of radiation from the accretion disks surrounding them and also a large amount of mechanical energy as jets. These make them important objects in the universe. However, their importance is recognized only recently in the long history of astronomy and astrophysics. In the early 1960s, the discovery of quasars was a trigger for the start of studying accretion disks. Accretion disks are now known in various compact objects, low mass black holes, ultraluminous X-ray sources, including active galactic nuclei, etc. These objects also produces jets occasionally in the relativistic and sub-relativistic limit.

The content of this book includes different aspects of accretion and ejection flows around black holes of all scales. The radiation coming from the disk and jet varies in a broad range from radio to gamma rays. However, their origin, spectral behaviour, linking with timing properties, and (anti)correlation between different bands are still lacking proper physical explanation. Therefore, we put the focus of discussion of this book related to the theory of accretion-ejection flows and its application to explain observations from different ground- and space-based observatories.

We hope the articles in this book will serve as a reference for astronomers in this field and interested astronomers from some other fields.

I thank all the contributing authors for making this Special Issue possible. I gratefully acknowledge the MDPI/Galaxies editorial and publication team for their constant support and effort for making this issue successful. My special thanks go to Ms. Queens Cui, the Managing Editor of this journal. Finally, I thank my colleagues, family, and my wife Suparna Mandal for their constant support during the COVID pandemic period, during which the majority of this issue was completed.

Santanu Mondal
Editor

 *galaxies*

MDPI

Editorial

Global Understanding of Accretion and Ejection around Black Holes

Santanu Mondal 

Indian Institute of Astrophysics, II Block Koramangala, Bangalore 560034, India; santanu.mondal@iiap.res.in

Citation: Mondal, S. Global Understanding of Accretion and Ejection around Black Holes. *Galaxies* **2022**, *10*, 64. https://doi.org/10.3390/galaxies10030064

Received: 13 April 2022
Accepted: 20 April 2022
Published: 27 April 2022

Publisher's Note: MDPI stays neutral with regard to jurisdictional claims in published maps and institutional affiliations.

Accretion and ejection around compact objects, mainly around black holes, both in low mass, and supermassive, is rich and has been studied exhaustively. However, the subject is expanding and growing rapidly after the launch of different space-based satellites and ground-based telescopes in multiwavelength bands, leaving a range of questions on accretion and ejection mechanisms. The proper understanding of the underlying physical mechanisms responsible for observational evidence is still lacking for several reasons. For instance, different phenomena are studied as a separate system, the hydrodynamics of one component ignores the other, spectral properties ignore timing properties or vice versa, and there is not enough instrumental resolution to decipher small scale physics.

With the advent of high-resolution satellite observations, it is possible to look at the problems globally as a complete package in a more consistent way. Recently, many new low mass black hole candidates have been discovered; however, very little is known about those systems, e.g., mass, spin parameter, orbital period. The study in the spectrotemporal domain also needs proper understanding of spectral state change, quasi periodic oscillation (QPO) frequency evolution, hardness intensity diagram, and line emissions. Therefore, it is a good time to take a break, recapitulate what has been carried out, what is ongoing, and what can be carried out in the future with upcoming missions, both from a theoretical and observational perspective, bringing together experts across the electromagnetic spectrum to gain new insights into the physical process of accretion.

This special issue covers both theoretical and observational aspects of accretion and ejection around both low mass and supermassive black holes. The papers published in this Special Issue provide novel, interesting phenomenological and theoretical contributions to this mainline topic. There is a range of questions related to the underlying physical processes which are responsible for the observed variabilities. This Issue puts some light on those basic questions and provide a roadmap for the resolution. For example, the physical picture of luminous active galactic nuclei (AGN) with powerful jets has been several decades-old problems in high-energy astrophysics. One of the fundamental problems is that to produce such powerful jets strong magnetic fields are to be transported from the accretion disk to the black hole horizon. This gives rise to greater disk thickness, which in turn decreases the radiative efficiency of the disk. Ref. [1] pointed out that recent numerical simulations have been partially successful to address the issue and much work is needed to single out the physical processes. They provided a road map for a resolution that counterrotating black holes with thin disks should be further explored in new simulations.

In addition, for a class of AGNs where the radiation is dominated by the accretion disk and its corona are known as Seyfert galaxies. This class of AGNs shows variability in different timescales in their observed lightcurve and emitted spectra, however, the physical origin of which is highly illusive, most importantly when the question comes on the origin of the inner hot corona and its geometrical and thermodynamical properties. Ref. [2] studied a well-known source Mrk 335 using several models including both physical and phenomenological and inferred the origin of variability from the basic accretion flow parameters. Authors suggested that the cooling and viscosity are important to make the corona variable, similar to low mass black hole candidates (LMBHCs, ref. [3]), which actually depends on the variation of the mass accretion rate (ref. [4,5]). The matter which is falling onto a black hole can be isothermal or adiabatic depending on the radiation mechanisms taken into account. Ref. [6] studied the Bondi accretion onto a supermassive black hole for the variable adiabatic index of the flow using Paczyński–Wiita potential. Earlier, a detailed study of accretion processes with varying the polytropic index was also done by ref. ([7,8] and others).

The other class of AGNs of which the radiation is mainly dominated by the jets is called blazars and radio-loud AGN. In their high activity state jet is the dominating contributor to the observed features, however, the disk can also contribute when the jet activity is low or moderate ref. [9]. Therefore, distinguishing the central engine responsible for the variabilities in these objects is one of the front line topics. Refs. [10–12] studied these issues and related variability properties in this special issue. Apart from continuum modeling, line emission is crucial to understand the gaseous outflows from these systems. This indeed helps us to understand the role of the Eddington luminosity, mass accretion rates, and the geometry of the flow in originating variable line profiles (see ref. [13] in this context and references therein), the wind/outflow velocities, and also from which region these lines are originating. Ref. [14] addressed some of these issues for quasars.

If we look at the black holes of small scales, they too show variability in their spectro-temporal domain, including QPOs, jet/outflows, the signature of magnetic field etc. Refs. [15,16] studied low LMBHCs, XTE J1908+094 and V404 Cygni, and determined their mass accretion rates, corona radii etc. Ref. [16] estimated the equipartition magnetic field strength of V404 Cygni source during the outburst phase and estimated the value of 900 Gauss. The authors did not find any QPO frequency in their power density spectra. They suggested that the absence of QPOs is due to the non-satisfaction of resonance condition (see ref. [17]).

It is to be noted that the mass of the central compact objects is hardly known, therefore, estimating its value is important to estimate different length scales of the flows correctly. Refs. [2,15] estimated the masses of the AGN and LMBHC from their spectral fitting, which agree with the dynamical mass measurement of the sources.

In addition to the fluid dynamics, one can also find the geodesic orbit of particles in black string spacetime geometry. Ref. [18] studied the same and found that the presence of the compact extra dimension leads to an increase in the number of the isofrequency pairing of geodesic orbits.

To summarize, the papers compiled in this Special Issue presented interesting results and gave useful insights onto the accretion-ejection processes around black holes. Furthermore, the articles on this Issue also put light on what can be carried out using future observations.

Funding: S.M. acknowledges Ramanujan Fellowship research grant #RJF/2020/000113 from SERB-DST, Govt. of India.

Institutional Review Board Statement: Not Applicable.

Informed Consent Statement: Not Applicable.

Data Availability Statement: This paper does not include any data as it is a review article on the papers published in the Special Issue titled the same as the title of the article.

Conflicts of Interest: The author declares no conflict of interest.

References

1. Singh, C.; Garofalo, D.; Lang, B. Powerful Jets from Radiatively Efficient Disks, a Decades-Old Unresolved Problem in High Energy Astrophysics. *Galaxies* **2021**, *9*, 10. [CrossRef]
2. Mondal, S.; Stalin, C. Changing Accretion Geometry of Seyfert 1 Mrk 335 with NuSTAR: A Comparative Study. *Galaxies* **2021**, *9*, 21. [CrossRef]
3. Mondal, S.; Chakrabarti, S.K.; Nagarkoti, S.; Arévalo, P. Possible Range of Viscosity Parameters to Trigger Black Hole Candidates to Exhibit Different States of Outbursts. *Astrophys. J.* **2017**, *850*, 47. [CrossRef]

4. Chakrabarti, S.; Titarchuk, L.G. Spectral Properties of Accretion Disks around Galactic and Extragalactic Black Holes. *Astrophys. J.* **1995**, *455*, 623. [CrossRef]
5. Mondal, S.; Chakrabarti, S.K. Spectral properties of two-component advective flows with standing shocks in the presence of Comptonization. *Mon. Not. R. Astron. Soc.* **2013**, *431*, 2716. [CrossRef]
6. Ramírez-Velásquez, J.; Sigalotti, L.; Gabbasov, R.; Klapp, J.; Contreras, E. The Radiative Newtonian $1 < \gamma \leq 1.66$ and the Paczyński–Wiita $\gamma = 5/3$ Regime of Non-Isothermal Bondi Accretion onto a Massive Black Hole with an Accretion Disc. *Galaxies* **2021**, *9*, 55. [CrossRef]
7. Chakrabarti, S.K.; Das, S. Model dependence of transonic properties of accretion flows around black holes. *Mon. Not. R. Astron. Soc.* **2001**, *327*, 808–812. [CrossRef]
8. Chattopadhyay, I.; Ryu, D. Effects of fluid composition on spherical flows around black holes. *Astrophys. J.* **2009**, *694*, 492. [CrossRef]
9. Mondal, S.; Rani, P.; Stalin, C.S.; Chakrabarti, S.K.; Rakshit, S. Flux and spectral variability of Mrk 421 during its moderate activity state using *NuSTAR*: Possible accretion disc contribution? *arXiv* **2022**, arXiv:2204.02132. [CrossRef]
10. Rajput, B.; Pandey, A. γ-ray Flux and Spectral Variability of Blazar Ton 599 during Its 2021 Flare. *Galaxies* **2021**, *9*, 118. [CrossRef]
11. Valtonen, M.; Dey, L.; Gopakumar, A.; Zola, S.; Komossa, S.; Pursimo, T.; Gomez, J.; Hudec, R.; Jermak, H.; Berdyugin, A. Promise of Persistent Multi-Messenger Astronomy with the Blazar OJ 287. *Galaxies* **2022**, *10*, 1. [CrossRef]
12. Fedorova, E.; Hnatyk, B.; Del Popolo, A.; Vasylenko, A.; Voitsekhovskyi, V. Non-Thermal Emission from Radio-Loud AGN Jets: Radio vs. X-rays. *Galaxies* **2022**, *10*, 6. [CrossRef]
13. Mondal, S.; Adhikari, T.P.; Singh, C.B. Emission lines from X-ray illuminated accretion disc in black hole binaries. *Mon. Not. R. Astron. Soc.* **2021**, *505*, 1071. [CrossRef]
14. Marziani, P.; Deconto-Machado, A.; Del Olmo, A. Isolating an Outflow Component in Single-Epoch Spectra of Quasars. *Galaxies* **2022**, *10*, 54. [CrossRef]
15. Chatterjee, D.; Jana, A.; Chatterjee, K.; Bhowmick, R.; Nath, S.; Chakrabarti, S.; Mangalam, A.; Debnath, D. Properties of Faint X-ray Activity of XTE J1908+094 in 2019. *Galaxies* **2021**, *9*, 25. [CrossRef]
16. Jana, A.; Shang, J.; Debnath, D.; Chakrabarti, S.; Chatterjee, D.; Chang, H. Study of Accretion Flow Dynamics of V404 Cygni during Its 2015 Outburst. *Galaxies* **2021**, *9*, 39. [CrossRef]
17. Chakrabarti, S.K.; Mondal, S.; Debnath, D. Resonance condition and low-frequency quasi-periodic oscillations of the outbursting source H1743-322. *Mon. Not. R. Astron. Soc.* **2015**, *452*, 3451. [CrossRef]
18. Shaymatov, S.; Atamurotov, F. Geodesic Circular Orbits Sharing the Same Orbital Frequencies in the Black String Spacetime. *Galaxies* **2021**, *9*, 40. [CrossRef]

Article

Isolating an Outflow Component in Single-Epoch Spectra of Quasars

Paola Marziani [1,*] , Alice Deconto-Machado [2] and Ascension Del Olmo [2]

1 National Institute for Astrophysics (INAF), Astronomical Observatory of Padova, IT-35122 Padova, Italy
2 Instituto de Astrofisíca de Andalucía, IAA-CSIC, Glorieta de la Astronomia s/n, E-18008 Granada, Spain; adeconto@iaa.es (A.D.-M.); chony@iaa.es (A.D.O.)
* Correspondence: paola.marziani@inaf.it; Tel.: +39-0498293415

Abstract: Gaseous outflows appear to be a universal property of type-1 and type-2 active galactic nuclei (AGN). The main diagnostic is provided by emission features shifted to higher frequencies via the Doppler effect, implying that the emitting gas is moving toward the observer. However, beyond the presence of blueshift, the observational signatures of the outflows are often unclear, and no established criteria exist to isolate the outflow contribution in the integrated, single-epoch spectra of type-1 AGN. The emission spectrum collected the typical apertures of long-slit spectroscopy or of fiber optics sample contributions over a broad range of spatial scales, making it difficult to analyze the line profiles in terms of different kinematical components. Nevertheless, hundred of thousands of quasars spectra collected at moderate resolution demand a proper analysis of the line profiles for proper dynamical modeling of the emitting regions. In this small contribution, we analyze several profiles of the H I Balmer line Hβ from composite and individual spectra of sources radiating at moderate Eddington ratio (Population B). Features and profile shapes that might be traced to outflow due to narrow-line region gas are detected over a wide range of luminosity.

Keywords: active galactic nuclei; optical spectroscopy; ionized gas; broad line region

Citation: Marziani, P.; Deconto-Machado, A.; Del Olmo, A. Isolating an Outflow Component in Single-Epoch Spectra of Quasars. *Galaxies* **2022**, *10*, 54. https://doi.org/10.3390/galaxies10020054

Academic Editor: Santanu Mondal

Received: 15 February 2022
Accepted: 17 March 2022
Published: 24 March 2022

1. Introduction

Type-1 active galactic nuclei (AGN) are characterized by the presence of broad and narrow optical and UV lines (for introductions, see, e.g., [1–6]). Spectra show a mind-boggling variety of broad emission line profiles not only among different objects but also among different lines in the spectrum of the same object. Sulentic [7] carried out measurements of spectral shifts and asymmetries exhibited by the broad lines relative to the narrow ones, proposing an empirical classification scheme for the broad H I Balmer line Hβ. Among the classes identified by Sulentic [7], two stand out: AR,R and AR,B, where AR means red-ward asymmetric, and the letter after the comma indicates either a shift of the line peak toward red or blue.

Fast forward more than 30 years, type-1 quasars are now being contextualized on the basis of the main sequence (MS) trends (e.g., [1,8,9]). Type-1 AGNs have been grouped into two main populations, Population A and B, defined on the basis of Balmer line widths (more specifically of Hβ: FWHM H$\beta \lesssim 4000$ km s^{-1} for Population A; FWHM H$\beta \gtrsim 4000$ km s^{-1} for Population B [1,10] at low and moderate luminosity $\log L \lesssim 46$ (erg s^{-1})). The classification of the quasar population along MS has its main physical foundation on systematic differences in Eddington ratio [11]: Population A sources typically have $L/L_\mathrm{Edd} \gtrsim 0.2$, with extreme Population A sources reaching $L/L_\mathrm{Edd} \gtrsim 1$ [12], values close to the expected theoretical limit for super-Eddington accretion rate [13–15]. Usually, Pop. B sources present lower values of the Eddington ratio when compared with the ones of Pop. A. The governing parameter of the MS itself appears to be an Eddington ratio convolved with the effect of orientation (e.g., [16,17]).[1]

Sources showing prominent Hβ red asymmetries (i.e., AR,R according to Sulentic [7]) are classified as belonging to Population B [10,11]. The red asymmetry itself can be considered as a defining feature of Population B sources, hinting at the presence of a "very broad component" (VBC) at the line base [18–23]. The physical properties of the region associated with VBC are largely undetermined (e.g., [24]) but the general consensus is that the region is located at the innermost radii of the broad line region (BLR), closest to the central continuum source. This inference follows from the deduction of a velocity field dominated by virial motions, at least for several population B sources [25,26]. The dynamical conditions of the "very broad line region" (VBLR) are the subject of current debate [27]. Two main alternatives have been proposed: infall and obscuration [28], or gravitational and transverse redshift [27,29–33]. Both mechanisms are, however, still consistent with a virial velocity field as the main broadening factor.

Gaseous outflows appear to be ubiquitous in type-1 AGN, although their traceability and their kinetic power varies greatly along the main sequence [34,35]. The signature of outflows in the optical and UV spectra is provided by the blueshift of emission lines with respect to the rest frame, under the assumption that the shift is due to a Doppler effect on the wavelength of lines emitted by gas moving toward us and that the receding side of the flow is mainly hidden from view (e.g., [36]). While there is unambiguous evidence of outflows from the emitting regions of quasars radiating at high Eddington ratios, the situation is by far less clear for Pop. B where the accretion rate is modest, as implied by a Eddington ratio $\lesssim 0.2$. High-resolution X-ray and ultraviolet (UV) observations of the prototypical Population B source NGC 5548 reveal a persistent ionized outflow traced by UV and X-ray absorption and emission lines [37]. However, the C$_{IV}$ emission line profile lacks strong evidence of such an outflow also because of the prominent red line wing merging with He$_{II}\lambda$1640 [38].

In this short note, we address the very specific issue of the origin of sources showing a blueshift at the peak of the Hβ emission line, e.g., of the AR,B classification. The focus is on the Hβ line because the line is a singlet, and its peak is isolated from other contaminants, offering a clear view of its broad and narrow components. The [O$_{III}]\lambda\lambda$4959,5007 lines recorded along with Hβ help assess the nature of the Hβ line profile. In addition, the narrow, high-ionization [O$_{III}]\lambda$5007 emission lines are known to be affected by outflows, as indicated by the frequent blueward asymmetries and even systematic shifts [39–43]. Section 2 presents the data used in this work, which comprise a set of composite spectra covering a wide range in luminosity and redshift, for which Hβ and [O$_{III}]\lambda\lambda$4959,5007 emissions have been covered with optical and IR spectroscopic observations. Details on how the spectral analysis was performed are shown in Section 3. The main results come from the profile comparison of Hβ and [O$_{III}]\lambda$5007 (Section 4) and are briefly analyzed in terms of the physical conditions of the line emitting gas, as well as of the dynamical parameters of the outflow (Section 5).

2. Data

The data analyzed in this paper refer to the most widely populated spectral type of Population B, B1, defined by FWHM Hβ in the range of 4000–8000 km s^{-1} [44]. Median composite spectra covering the Hβ range were computed over spectral type B1 sources belonging to two samples of low-to-moderate redshift and luminosity [44,45], and one sample of intermediate z and high luminosity [46]. The [44] composites are based on the individual observations of Marziani et al. [47] that involved 97 B1 spectra. The [45] composites are SDSS spectra in the redshift range of 0.4–0.7, covering both Mg$_{II}\lambda$2800 and Hβ. The radio-quiet B1 composite was computed over 179 spectra, while CD and FR-II composites involved 16 and 23 spectra, respectively. The [46] B1 composite included 22 high-luminosity, Hamburg ESO (HE) quasars. Median composites were constructed from continuum-normalized (at 5100 Å) spectra, after a determination of the heliocentric redshift based on [O$_{II}]\lambda$3727 or a narrow component of Hβ, two low-ionization narrow line that provide the best estimators of the systemic redshift of the host galaxy [48]. The accurate redshift correction allowed for the preservation of the spectral resolution of the individual spectra. The [45] composites should, therefore, have a resolving power $\lambda/\delta\lambda \sim 2000$. The resolving power is only slightly lower for [44], $\lambda/\delta\lambda \sim 1000$. The HE ISAAC near-IR observations were all collected with a narrow slit (0.6 arcsec) that yielded $\lambda/\delta\lambda \sim 1000$, comparable to the spectra of the samples observed with optical spectrometers. The main physical properties associated with the composite spectra are summarized in Table 1, where the first column lists an identification code, and the following columns list the redshift range and the median values of bolometric luminosity, black hole mass M_{BH}, and the Eddington ratio L/L_{Edd}. In addition to the composite spectra, the spectra of two quasars of extreme luminosity at intermediate redshift (Deconto-Machado et al. 2022, in preparation) provide examples of two opposite cases: One where a prominent outflow signature is detected (Q0029+079), and one in which there is no obvious evidence of outflow (HE0001-2340). The last two lines of Table 1 consider composites for core-dominated (CD) and Fanaroff–Riley (FR) sources belonging to spectral type B1 from the [45] sample. These two composite were defined to address the somewhat controversial issue of the mild-ionized outflow presence among radio-loud, jetted AGN.[2] The data of Table 1 confirm that the empirical selection of spectral type B1 corresponds to the selection of modest L/L_{Edd} radiators. At the higher redshift and luminosity, L/L_{Edd} appears somewhat higher ($L/L_{Edd} \approx 0.3$) because of the preferential selection of higher L/L_{Edd} for a fixed black hole mass in flux limited surveys [51].

Table 1. Physical parameters.

Spectrum	z	$\log L$ (erg s^{-1})	$\log M_{BH}$ [a] (M$_\odot$)	$\log L/L_{Edd}$
		Composite spectra		
B1 [44]	0–0.7	45.63 [b]	8.52	−1.07
B1 [45]	0.4–0.7	46.31 [b]	9.19	−1.06
B1 [46]	0.9–2.6	47.29 [c]	9.63	−0.51
		Individual, high-L quasars		
HE0001−2340	2.2651	47.09 [c]	9.78	−0.86
Q0029+079	3.2798	47.43 [c]	9.95	−0.70
		Composite spectra, jetted		
B1 [45] CD	0.4–0.7	46.51 [b]	9.39	−1.05
B1 [45] FRII	0.4–0.7	46.62 [b]	9.44	−1.00

[a]: Black hole mass computed from the Hβ scaling law provided by Vestergaard and Peterson [52], using the Hβ full profile FWHM. Applying the average correction suggested for spectral type B1 would lower the mass by a factor 0.64 and increase the L/L_{Edd} ratio by the inverse of this factor. [b] Bolometric correction assumed a factor 10; [c] Bolometric correction assumed a factor 4, as appropriate for very high luminosity sources following Netzer [53].

3. Analysis

The non-linear multicomponent fits were performed using the SPECFIT routine from IRAF [54]. This routine allows a simultaneous minimum-χ^2 fit of the continuum approximated by a power law and the spectral line components yielding FWHM, peak wavelength, and intensity of all line components. In the optical range, we fit the Hβ profile as well as the [OIII]$\lambda\lambda$4959,5007 emission lines and the FeII multiplets accounted for by a scaled and broadened template [55]. The details of multi-component analysis has been provided in several previous papers (e.g., [56]) and will not be repeated here. Suffice to say that the broad profiles of Pop. B sources can be successfully modeled with two Gaussians: (1) one narrower, unshifted or slightly shifted to the red; and (2) one broader, with FWHM $\sim 10{,}000$ km s^{-1}, and shifted by few thousands km s^{-1} to the red [57]. This model accounts for the AR,R profile type. In addition to the model decomposition, we measured several parameters on the full broad profile [58]. The definitions of the centroids and of the asymmetry index *A.I.* are reported below for convenience:

$$c\left(\frac{i}{4}\right) = \frac{v_{r,B}\left(\frac{i}{4}\right) + v_{r,R}\left(\frac{i}{4}\right)}{2}, \, i = 1, 2, 3; \, \frac{i}{4} = 0.9,$$
(1)

where the radial velocities are measured with respect to the rest frame at fractional intensities $\frac{i}{4}$ for each value of the index i on the blue and red side of the line with respect to the rest frame.

$$A.I.\left(\frac{1}{4}\right) = \frac{v_{r,B}\left(\frac{1}{4}\right) + v_R\left(\frac{1}{4}\right) - 2v_{r,P}}{v_{r,R}\left(\frac{1}{4}\right) - v_{r,B}\left(\frac{1}{4}\right)}.$$
(2)

Note that the *A.I.*, unlike the centroids, is defined as a shift with respect to the line peak radial velocity $v_{r,P}$ ($v_{r,P}$ is measured with respect to rest frame; a suitable proxy is provided by $c(0.9)$).

4. Results

4.1. Broad Hβ

Figure 1 shows the continuum-subtracted spectra and their models for the [44–46] composite spectra (top, middle and bottom panel, respectively). The measurements of the broad Hβ line parameters are reported in Table 2. For each spectrum, Table 2 lists the normalized flux F of the Hβ full broad profile (Hβ_{BC} + Hβ_{VBC}+ Hβ_{BLUE}), its equivalent width W Hβ in Å, and the normalized fluxes of Hβ_{BC} and Hβ_{VBC} separately. The following columns report several parameters for the Hβ blue-shifted excess with respect to the standard Population B decomposition involving only Hβ_{BC} and Hβ_{VBC}: normalized flux, equivalent width, peak shift, FWHM, and skew. The last columns yield the normalized flux and the equivalent width of the FeIIλ4570 emission blend, as defined by Boroson and Green [55]. The equivalent width values correspond roughly to the normalized flux so that they are reported only for the main features. The normalized fluxes can be approximately converted into luminosities by multiplying them by the luminosity values reported in Table 1 divided by the bolometric correction and by 5100, i.e., by the wavelength in Å at which the continuum was normalized. Table 3 reports FWHM, A.I., and centroids as defined in Section 3 for the *broad* Hβ profile (Hβ_{BC} + Hβ_{VBC}+ Hβ_{BLUE} i.e., without considering the narrow (Hβ_{NC}) and semi-broad (Hβ_{SBC}) components associated with narrow-line region emission). Only at the highest L blueshifted emission with a broad profile (Hβ_{BLUE}) is detected in the Hβ profile: In this case, the Hβ_{BLUE} contribution is $\lesssim 5\%$ of the total line luminosity for the [46] composite and reaches about 1/3 of the total line luminosity in the admittedly extreme Q0029 case. In no case, however, was Hβ_{BLUE} able to create a significant shift to the blue close to the line base: red asymmetry dominates and even the Q0029 Hβ broad profile is "symmeterized" toward the line base, with a centroid at $\frac{1}{4}$ peak intensity close to 0 km s^{-1}.

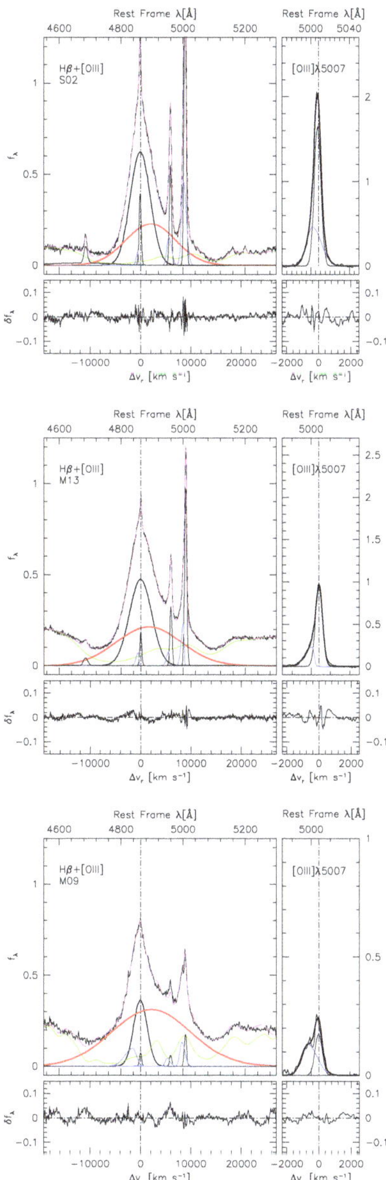

Figure 1. Analysis of the Hβ + [Oiii]$\lambda\lambda$4959,5007 region for [44] (**top**), [45] (**middle**), and [46] (**bottom**) B1 composite spectra. Continuum subtracted spectra are shown in the rest frame, over the range 4550–5300 Å (left panel), with an expansion around [Oiii]λ5007 (right panels). Thin solid lines: continuum-subtracted spectrum; dashed magenta line: model spectrum; thick black line: Hβ broad component; red thick line, Hβ very broad component; thin smooth black lines: narrow components of Hβ and [Oiii]λ5007; blue lines: blue shifted components. Green lines trace the scaled and broadened Feii emission template. The lower panel show the observed minus model residuals in radial velocity scale.

Table 2. Broadline properties measurements.

Spectrum	$H\beta$		$H\beta_{BC}$	$H\beta_{VBC}$		$H\beta_{BLUE}$			$FeII\lambda4570$	
	F	W [a]	F	F	F	Shift [b]	FWHM [b]	Skew [c]	F	W
Composite spectra										
B1 [44]	95.3	86.7	49.8	45.4	...	...	...	...	18.1	14.6
B1 [45]	122.3	126.5	52.5	69.8	...	...	...	...	47.8	43.0
B1 [46]	123.3	129.1	19.6	99.1	4.6	−1535	3611	0.5	39.2	34.1
Individual, high-L quasars										
HE0001	99.3	95.1	26.7	72.7	...	...	...	...	20.6	16.3
Q0029	69.8	66.4	13.9	32.2	23.7	−2097	4711	1.2	25.8	21.4
Composite spectra, jetted										
B1 [45] CD	113.8	118.5	42.9	70.9	...	...	...	...	35.6	32.7
B1 [45] FRII	129.8	131.1	57.5	72.3	...	...	...	...	24.6	21.9

[a]: In units of Å; [b]: in units of km s^{-1}. [c] skew as reported by the SPECFIT routine; it is equal to the conventional definition of skew [59] + 1.

Table 3. $H\beta$ profile properties measurements.

Spectrum	FWHM [a]	AI	c(1/4) [a]	c(1/2) [a]	c(3/4) [a]	c(0.9) [a]
Composite spectra						
B1 [44]	5560 ± 170	0.12 ± 0.03	680 ± 230	250 ± 80	160 ± 70	130 ± 50
B1 [45]	6540 ± 210	0.12 ± 0.06	740 ± 340	150 ± 110	50 ± 90	40 ± 60
B1 [46]	6010 ± 450	0.28 ± 0.06	2120 ± 490	−50 ± 220	−230 ± 70	−270 ± 50
Individual, high-L quasars						
HE0001	6510 ± 690	0.29 ± 0.09	2700 ± 560	1310 ± 340	900 ± 170	830 ± 110
Q0029	6200 ± 380	0.18 ± 0.10	430 ± 500	−380 ± 190	−500 ± 160	−500 ± 110
Composite spectra, jetted						
B1 [45] CD	6880 ± 240	0.23 ± 0.06	1520 ± 380	270 ± 120	70 ± 90	20 ± 60
B1 [45] FRII	6790 ± 220	0.10 ± 0.06	820 ± 330	320 ± 110	240 ± 90	230 ± 60

[a] In units of km s^{-1}.

4.2. [OIII]$\lambda5007$ and $H\beta$ Narrow-Line Emission

Table 4 summarizes the measurements of the components associated with the narrow-line region (NLR) emission, i.e., narrow and semi-broad components of $H\beta$ and [OIII]$\lambda5007$ ($H\beta_{NC}$, [OIII]$\lambda5007_{NC}$, and $H\beta_{SBC}$ and [OIII]$\lambda5007_{SBC}$), for which normalized flux, equivalent width, shift, and FWHM are reported. The skew parameter is reported only for semi-broad components, as the narrow components are assumed to be symmetric Gaussian, within a few tens km s^{-1} from the rest frame [48].

The [44] composite shows a broad + narrow component profile that is very well represented by three Gaussians: the symmetric unshifted $H\beta_{NC}$, unshifted $H\beta_{BC}$ and the $H\beta_{VBC}$ with a significant shift to red. A small blue-shifted excess appears at the interface between $H\beta_{NC}$ and $H\beta_{BC}$ and has been modeled by an additional Gaussian. Its intensity is so low that a very good fit with no significant worsening in the χ^2 can be achieved also without it. Most notably, the [OIII]$\lambda5007$ profile (enlarged in the right panel) is also fairly symmetric: A small centroid blueshift ~ -50 km s^{-1} is detected only at $\frac{1}{4}$ peak intensity (Table 5, where the [OIII]$\lambda5007$ full profile parameters are reported as in Table 3 for $H\beta$). The relatively large shift reported for [OIII]$\lambda5007_{SBC}$ is compensated by a red-ward skew (Figure 1, right panel on top row). In this case, the decomposition [OIII]$\lambda5007_{NC}$-[OIII]$\lambda5007_{SBC}$ is especially uncertain, and a more reliable measurement is provided by the centroid.

Table 4. Narrow line measurements.

Spectrum	Hβ_{NC}			FWHM^b	F	W^a	Hβ_{SBC}	FWHM^b	Skew^c	[OIII]λ5007$_{NC}$			FWHM^b	F	W^a	[OIII]λ5007$_{SBC}$	FWHM^b	Skew^c
	F	W^a	Shift^b				Shift^b			F	W^a	Shift^b				Shift^b		
		Composite spectra																
B1 [44]	3.33	3.00	−9	492	0.59	0.53	−349	881	…	14.6	14.0	12	499	8.86	8.45	−307	881	1.47
B1 [45]	1.45	1.45	−8	450	0.83	0.85	−480	1054	0.26	8.9	9.7	11	513	2.76	3.00	−417	1054	0.26
B1 [46]	0.65	0.70	−25	508	0.06	0.07	−752	932	1.46	1.6	1.8	−15	528	2.68	2.97	−688	932	1.46
		Individual, high-L quasars																
HE0001	0.11	0.10	−25	2202	1.05	0.99	−133	1301	0.41	2.6	2.7	239	592	4.39	4.43	−69	1301	0.41
Q0029	1.00	0.95	−6	1181	0.00	0.00	…	…	…	4.9	4.9	−728	1181	3.36	3.30	−2051	1340	1.12
		Composite spectra, jetted																
B1 [45] CD	1.92	1.96	−25	662	0.00	0.00	…	…	…	6.8	7.4	20	497	5.17	5.62	39	1439	0.10
B1 [45] FRII	0.94	0.98	−25	300	0.00	0.00	…	…	…	13.5	13.0	11	360	7.65	8.11	−279	585	2.12

^{a}: In units of Å; ^{b}: in units of km s^{-1}. ^{c} skew as reported by the SPECFIT routine; it is equal to the conventional definition of skew [59] + 1.

Table 5. [OIII]λ5007 profile measurement.

Spectrum	FWHM [a]	AI	c(1/4) [a]	c(1/2) [a]	c(3/4) [a]	c(0.9) [a]
			Composite spectra			
B1S02	580 ± 30	-0.10 ± 0.08	-40 ± 40	-10 ± 20	0 ± 20	0 ± 10
B1M13	560 ± 40	-0.23 ± 0.11	-80 ± 50	-20 ± 20	10 ± 10	30 ± 10
B1M09	1100 ± 120	-0.43 ± 0.05	-380 ± 40	-280 ± 60	-60 ± 20	-40 ± 10
			Individual, high$-L$ quasars			
HE0001	900 ± 70	-0.26 ± 0.07	-100 ± 40	0 ± 30	70 ± 30	80 ± 10
Q0029	2120 ± 140	-0.37 ± 0.04	-1360 ± 60	-1240 ± 70	-860 ± 50	-830 ± 30
			Composite spectra, jetted			
B1M13CD	490 ± 40	-0.18 ± 0.12	-90 ± 50	-70 ± 20	-10 ± 20	-10 ± 10
B1M13FRII	440 ± 30	-0.10 ± 0.09	-20 ± 30	10 ± 10	10 ± 10	10 ± 10

[a]: In units of km s^{-1}.

The [45] composite spectrum appears as a "goiter" at the top of the H$\beta_{\rm BC}$ broad profile. The [OIII]λ5007 profile is also fairly asymmetric, and it can be modeled by a narrower, almost unshifted component and a skewed Gaussian displaced to the blue by ≈ -500 km s^{-1}. The top of the Hβ profile is well fit by assuming two components with the same shift, width, and asymmetry of the model components' [OIII]λ5007 line. The consistency between the model of H$\beta_{\rm SBC}$ and [OIII]λ5007$_{\rm SBC}$ provides evidence that the Hβ blueshifted, and the skewed component is associated with a NLR outflow. The [46] composite can be equally modeled with the same skewed and blueshifted component for Hβ and [OIII]λ5007. However, this model would require an implausibly strong [OIII]$\lambda\lambda$4959,5007 emission. The fit shown in the bottom panel of Figure 1 assumes a broader component for Hβ emission.

At very high luminosity (Figure 2), a prominent outflow is apparently absent in one Pop. B Hβ profile (HE0001) but very prominent in another (Q0029). If the classification of Q0029 as a Population B source is correct, the model of the blue "goiter" at the side of the Hβ profile implies a strong contribution of blueshifted emission with a broad profile. The [OIII]λ5007 profiles are also different: the equivalent width W is higher and the shift is lower in the case of HE0001, where no significant Hβ outflow is detected. By all means, the properties of Q0029 appear more extreme. We predict that this source will show extreme CIV blueshift, with an amplitude of several thousands km s^{-1}.

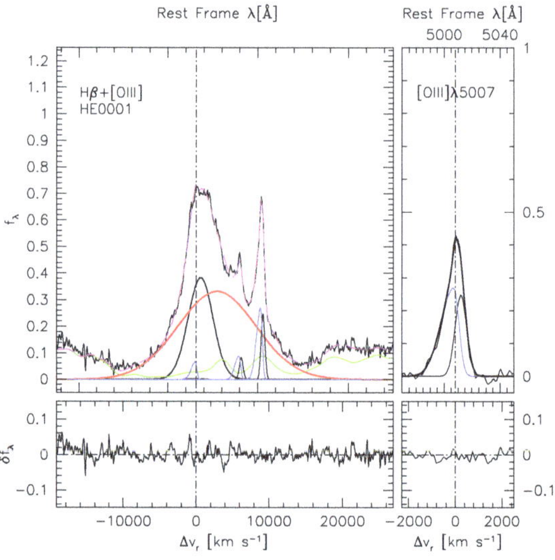

Figure 2. *Cont.*

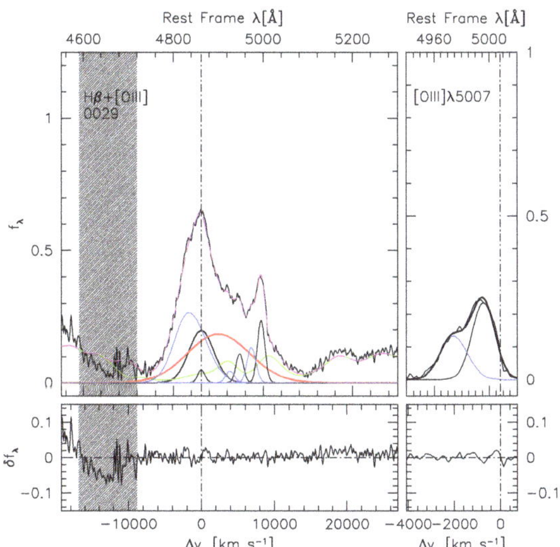

Figure 2. Analysis of the Hβ + [OIII]$\lambda\lambda$4959,5007 region for two high-luminosity, high-z quasars belonging to the B1 spectral type. The top one, HE0001-234, shows no appreciable evidence of blueshift, while the bottom one (HB89) 0029+073 requires a stronger blue shifted excess for [OIII]$\lambda\lambda$4959,5007 and an even stronger and broader one to fit Hβ. Color coding of the components is the same as in the previous Figure. The shaded area identifies a spectral region affected by atmospheric absorptions

The [OIII]λ5007 shift and the A.I. become more negative, and the equivalent width decreases with increasing luminosity. This is a pure luminosity effect that proceeds in the same sense of the effect of an increasing Eddington ratio in sample covering the full span of $L/L_{\mathrm{Edd}} \sim 10^{-2} - 1$, and it can be interpreted as a result of NLR evolution with redshift [35].

4.3. Jetted Sources

The CD and FR-II composites from the [45] sample (Figure 3) show that the [OIII]λ5007 blueshifted and skewed component is not detected in Hβ, implying that the intensity ratio is [OIII]λ5007/H$\beta \gg 1$ for this component. In addition, the [OIII]λ5007 profile for the FR-II composite spectrum is much more symmetric than that of the CD composite, for which its A.I. and centroid shifts are more consistent with the RQ composite of the same sample. This systematic difference may arise because of the different viewing angles expected for CD (seen almost pole on) and FR-II sources (seen at a viewing angle $\approx$ 40–60 [60]).

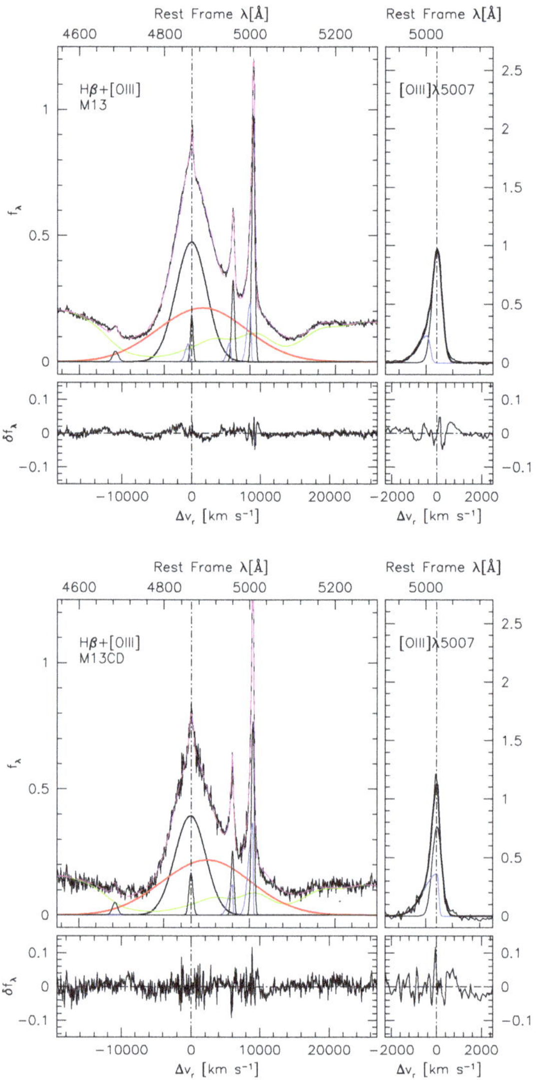

Figure 3. *Cont.*

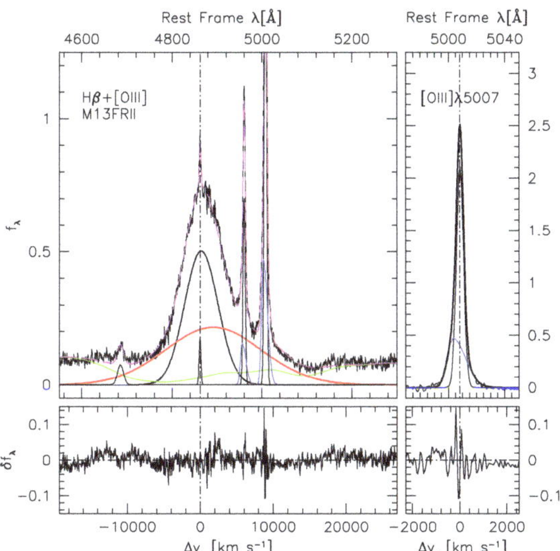

Figure 3. Analysis of the Hβ + [OIII]$\lambda\lambda$4959,5007 region for the RQ composite spectrum of Marziani et al. [45] (**top**), and for the CD and FR-II composite spectra (**middle** and **bottom**, respectively). Color coding of the components is the same as in the previous Figures.

5. Discussion

The analysis performed above has been focused on sources radiating at relatively modest L/L_{Edd} (Population B) but covering a wide range of redshifts ($0 \lesssim z \lesssim 3$) and luminosities. Significant outflow features have been detected in NLR, as traced by Hβ and [OIII]$\lambda\lambda$4959,5007 blue shifted components. At high luminosity, significant blueshifts are found not only in the [OIII]$\lambda\lambda$4959,5007 lines but also with a broader profile, hinting at an association with BLR emission.

5.1. How Important Is the Outflow Component?

The present analysis relies on the important assumption that the Population B profile at Hβ low-z and luminosity is not significantly affected by any outflowing gas. Reverberation mapping campaigns in the early 2000s provided evidence that the main broadening mechanism is indeed provided by a virial velocity field of gas orbiting around a point-like mass. More recent works point toward a more complex situation ([61–63] Bao et al., 2022, in preparation), although the main inference from velocity-resolved reverberation mapping studies for the sources with the red Hβ asymmetry is that the velocity field is predominantly virial, with the frequent detection of infall motions. The detection of infall is based on the shorter time delay of the red wing and not on the response of the line core.

5.2. Identifying an Outflow Component

The Hβ profile of Population B presents a clear inflection between Hβ_{BC} and Hβ_{NC} that can be explained on the basis of the expected radial emissivity of Hβ [64]. The identification of an outflow component may be achieved by considering the following options:

- No significant centroid blueshift in the broad profile of Hβ and symmetric appearance at the interface between Hβ_{NC} and Hβ_{BC}, with the peak of the broad profile showing no shift or a slight redshift: no evidence of outflow.
- No significant centroid blueshift in the broad profile of Hβ and "goiter" appearance at the interface between Hβ_{NC} and Hβ_{BC}: If the [OIII]λ5007 line shows a significant blueward asymmetry and a model of the [OIII]λ5007 line profile with a core and

semi-broad component is applicable to the Hβ profile, then it is likely that the outflow is mainly associated with NLR emission.

- Even modest centroid blueshift in the broad profile of Hβ at fractional intensity $\frac{3}{4}$ or 0.9, the outflow might involve BLR emission. In this case, H$\beta_{\rm BLUE}$ corresponds to the prominent blueshifted emission of the C\,IV line observed at high luminosity [56]. The detection of H$\beta_{\rm BLUE}$ is made more difficult by the C\,IV/Hβ ratio expected to be $\gg 1$.

5.3. Location and Physical Nature of the Outflow

Even in case of modest accretion rate, the outflow can be radiatively driven [65]. The ratio between the radiation and gravitation force can be written as $a_{\rm rad}/a_{\rm grav} \approx 7.2$ $L/L_{\rm Edd}\, N_{\rm c,23}^{-1}$ where $N_{\rm c,23}$ is the Hydrogen column density in units of 10^{23} cm^{-2} (e.g., [66]). For $L/L_{\rm Edd} \sim 0.1$, the gas of moderate common density $N_{\rm c,23} \sim 0.1$ could be accelerated to $a_{\rm rad}/a_{\rm grav} \sim 10$ (c.f. Equation (6) of Netzer and Marziani [65]) [67]. The first underlying assumption is that all of the photon's momentum in the ionizing continuum is transferred to the line-emitting gas. The second assumption is that the gas is optically thick relative to the ionizing continuum, and this condition is more easily verified if the ionization parameter is low, implying that the low column density gas located farther out from the AGN continuum source might be preferentially accelerated. This might explain why we see a signature due to a semi-broad component in Hβ, H$\beta_{\rm SBC}$, which is in turn associated with the [O\,III]$\lambda5007$ semi-broad component, likely at the inner edge of NLR, and it may be the main signature of outflow in low $L/L_{\rm Edd}$ sources.

Regarding BLR, at low luminosity, there is no signature of outflow, if our interpretation of the profile is correct. For Population B sources, however, the observed spectrum can be explained by the locally optimized cloud (LOC) scheme, in which a range of ionization parameters, density, and column density is assumed, and the emerging spectrum is set by the parameters at which lines are emitted most efficiently [68,69]. This is to say that there might be always gas as "light" as needed for an outflow; however, the outflow may not produce a significant signature in the emission line spectrum. A powerful outflow at a modest Eddington ratio may become possible only at high luminosity (e.g., [70–72]), as predicted from wind theory and confirmed by observations [56,73–75].

5.4. The Fate of the Outflowing Gas: No Feedback Effects at Low L

The mass outflow rate at a distance r can be written as follows if the flow is confined to a solid angle of Ω of volume $\frac{4}{3}\pi r^3 \frac{\Omega}{4\pi}$: $\dot{M}_{\rm o}^{\rm ion} = \rho\, \Omega r^2 v_{\rm o} = \frac{M_{\rm o}^{\rm ion}}{V}\Omega r^2 v_{\rm o} \propto L v_{\rm o} r^{-1}$ [76]; it implies $\dot{M}^{\rm ion} \sim 30 L_{44} v_{\rm o,1000} r_{\rm 1kpc}^{-1} \left(\frac{Z}{5Z_\odot}\right)^{-1} n_3^{-1}$, where the mass of ionized gas can be directly estimated from the line luminosity: $M_{\rm ion} \sim 1 \cdot 10^7\, L_{44}\left(\frac{Z}{5Z_\odot}\right)^{-1} n_3^{-1}$.^{3} The low-luminosity cases [44,45] imply that the outflow velocity is $v_{\rm o,1000} \sim 1$ from the peak shift of [O\,III]$\lambda5007_{\rm SBC}$, and the [O\,III]$\lambda5007_{\rm SBC}$ luminosity is $\log L_{\rm [OIII]} \sim 42$. Assuming $Z \approx 1Z_\odot$ as appropriate for Population B sources [27], $M_{\rm ion} \sim 5 \cdot 10^5 n_3^{-1}$ and $\dot{M}^{\rm ion} \sim 0.15 r_{\rm 1kpc}^{-1} n_3^{-1}$. By the same token, the thrust and kinetic power can be written as $\dot{M}v \sim 1.9 \cdot 10^{35} L_{44} v_{\rm o,1000}^2 r_{\rm 1kpc}^{-1}\left(\frac{Z}{5Z_\odot}\right)^{-1} n_3^{-1}$ and $\dot{\epsilon} \sim 10^{43} L_{44} v_{\rm o,1000}^3 r_{\rm 1kpc}^{-1}\left(\frac{Z}{5Z_\odot}\right)^{-1} n_3^{-1}$, which become $\dot{M}v \sim 1 \cdot 10^{34} r_{\rm 1kpc}^{-1} n_3^{-1}$ and $\dot{\epsilon} \sim 5 \cdot 10^{41} r_{\rm 1kpc}^{-1} n_3^{-1}$. Even assuming that we are observing a flow at $r \sim 10$ pc, the kinetic power is $\dot{\epsilon} \lesssim 10^{44}$ erg s^{-1}, a factor ≈ 100 below the bolometric luminosity and ≈ 1000 the Eddington luminosity of the [45] case. The emitting gas might be beyond or at the limit of the black hole sphere of influence given by $r \approx GM/\sigma_\star^2 \approx 8 \cdot 10^{19} M_{9,\odot}/\sigma_{\star,400}^2$ cm, where $\sigma_\star$ is the velocity dispersion associated with the bulge of the host galaxy in units of 400 km s^{-1}. At this radius, the escape velocity is expected to be $v_{\rm esc} \sim 500$ km s^{-1} for a 10^9 M$_\odot$ black hole. It is, therefore, doubtful whether the outflowing gas might be even able to escape from the sphere of influence of the black hole. Even less likely, the outflowing gas might "wreak havoc" galaxy-wide in the bulge and the disk of the host due to the small amount of gas masses involved in the outflow, and due to the escape velocity that can be as high as $v_{esc} \gtrsim 1000$ km s^{-1} in the inner regions of a massive spheroid or in a giant spiral such as the Milky Way [77].

The scenario might radically change at high luminosity: considering the [46] composite, the velocity of [OIII]$\lambda 5007_{SBC}$ is higher by a factor ≈ 2, and the line luminosity is higher by a factor ~ 10, implying a 20-, 40-, and ~ 100-fold increase over the [45] case in mass flow, thrust, and kinetic power, respectively. In the [46] case, the kinetic power would be comparable to the Eddington luminosity. An even more powerful outflow is expected for Q0029.

6. Summary and Conclusions

The analysis of outflow signatures carried out in the present paper has been focused on three samples of type-1 AGN covering a wide range of luminosity.

The detection of different kinematic components in single epoch profiles is a complicated issue. The apertures and slit widths used in ground-based observation add up the emission from the AGN continuum, BLR, NLR, and host galaxy, which are associated with widely different spatial scales. The case of Population B sources of spectral type B1 is particularly well-suited to analyze the presence of an outflow component in the Balmer Hβ line for sources that are radiating at modest Eddington ratios.

Generally speaking, the detection of significant systematic blueshifts in the centroid measurements can be taken as a signature of outflow. If the blueshift/blue asymmetry is confined at the top of the Hβ line and the Hβ narrow emission can be modeled as [OIII]$\lambda 5007$ assuming a semi-broad and a narrow component with a similar parameter, then the evidence of the outflow (the "goiter" in the line profile) remains confined to the NLR. However, if the Hβ centroid at $\frac{3}{4}$ or at lower fractional intensity is also blue shifted, it is likely that a BLR outflow is being detected. Low column density gas can be driven into an outflow by radiation forces. Blueshifts in the line core can be, therefore, straightforwardly interpreted by an outflow component, without invoking binary BLR, in turn pointing toward sub-parsec binary black holes. Other spectral types along the MS have been identified as frequently involving binary black hole candidates [78,79].

The estimates of mass flow, thrust, and kinetic power are highly uncertain because of the lack of spatially resolved data. This situation might be changing soon with the development of integral-field spectrographs. Nonetheless, even when maximizing the coarse estimates reported above, it is unlikely that the thrust and the kinetic power (only $\sim 10^{-2}$ the Eddington luminosity as derived for the [44,45] samples) might have a strong impact on the host galaxy's evolution and not to mention the possibility of driving the black hole mass—bulge correlation (e.g., and references therein [80]). Even if the [OIII]$\lambda 5007$ samples only emission from mildly ionized gas and the mass flow might be dominated by the higher-ionization gas, for low luminosity AGNs such as the prototypical Population B Seyfert-1 NGC 5548, kinetic luminosity remains a very small fraction of the Eddington luminosity [37,81]. The situation is expected to change at the "cosmic noon" at redshifts in the range of 1–2, when the most luminous quasars are observed, and of which the [46] composite provides a representative spectrum.

Author Contributions: P.M. wrote most of the paper; A.D.-M. contributed to the analysis of spectra. A.D.-M. and A.D.O. Both contributed with suggestions and critical reading. All authors have read and agreed to the published version of the manuscript.

Funding: Funding for the Sloan Digital Sky Survey has been provided by the Alfred P. Sloan Foundation and the U.S. Department of Energy Office of Science. The SDSS web site is http://www.sdss.org. (accessed on 14 February 2022). SDSS-III is managed by the Astrophysical Research Consortium for the Participating Institutions of the SDSS-III Collaboration including the University of Arizona, the Brazilian Participation Group, Brookhaven National Laboratory, Carnegie Mellon University, University of Florida, the French Participation Group, the German Participation Group, Harvard University, the Instituto de Astrofisica de Canarias, the Michigan State/Notre Dame/JINA Participation Group, Johns Hopkins University, Lawrence Berkeley National Laboratory, Max Planck Institute for Astrophysics, Max Planck Institute for Extraterrestrial Physics, New Mexico State University, University of Portsmouth, Princeton University, the Spanish Participation Group, University of Tokyo, University of Utah, Vanderbilt University, University of Virginia, University of Washington, and Yale University.

Institutional Review Board Statement: Not applicable.

Informed Consent Statement: Not applicable.

Data Availability Statement: Data could be made t available upon request.

Acknowledgments: A.D.M. and A.D.O. acknowledge financial support from the State Agency for Research of the Spanish MCIU through the project PID2019-106027GB-C41 and the "Center of Excellence Severo Ochoa" award to the Instituto de Astrofísica de Andalucía (SEV-2017-0709). A.D.M. acknowledges the support of the INPhINIT fellowship from the "la Caixa" Foundation (ID 100010434). The fellowship code is LCF/BQ/DI19/11730018.

Conflicts of Interest: The authors declare no conflict of interest.

Abbreviations

The following abbreviations are used in this manuscript:

AGN	Active Galactic Nucleus/i;
BLR	Broad Line Region;
BC	Broad Component;
CD	Core Dominated;
FR-II	Fanaroff-Riley II;
FWHM	Full Width Half-Maximum;
HE	Hamburg-ESO;
ISAAC	Infrared Spectrometer And Array Camera;
IR	Infrared;
LOC	Locally Optimized Cloud;
MDPI	Multidisciplinary Digital Publishing Institute;
MS	Main Sequence;
NC	Narrow Component;
NGC	New General Catalogue;
NLR	Narrow Line Region;
RL	Radio loud;
RQ	Radio quiet;
SBC	Semi-Broad Component;
SDSS	Sloan Digital Sky Survey;
UV	Ultra-violet;
VBC	Very Broad Component;
VBLR	Very Broad Line Region.

Notes

[1]   In flux limited samples, Pop. A and B may have similar luminosity distributions. If this is the case, Pop. B sources are expected to host more massive black holes, considering the systematic differences in the Eddington ratio.

[2]   We consider the attribute "radio-loud" as a synonym of relativistically jetted [49,50].

[3]   Note that the filling factor is not appearing explicitly because, by using line luminosity, we already are considering the volume of the line-emitting gas. The fraction of volume that is actually occupied by the line emitting gas then depends on its density.

References

1. Sulentic, J.W.; Marziani, P.; Dultzin-Hacyan, D. Phenomenology of Broad Emission Lines in Active Galactic Nuclei. *Annu. Rev. Astron. Astrophys.* **2000**, *38*, 521–571. [CrossRef]
2. Osterbrock, D.E.; Mathews, W.G. Emission-line regions of active galaxies and QSOs. *Annu. Rev. Astron. Astrophys.* **1986**, *24*, 171–203. [CrossRef]
3. Netzer, H. AGN emission lines. In *Active Galactic Nuclei*; Blandford, R.D., Netzer, H., Woltjer, L., Courvoisier, T.J.-L., Mayor, M., Ed.; Springer: Berlin/Heidelberg, Germany, 1990; pp. 57–160.
4. Peterson, B.M. *An Introduction to Active Galactic Nuclei*; Cambridge University Press: Cambridge, UK, 1997.

5. Osterbrock, D.E.; Ferland, G.J. *Astrophysics of Gaseous Nebulae and Active Galactic Nuclei*; University Science Books: Mill Valley, CA, USA, 2006.

6. Marziani, P.; Dultzin-Hacyan, D.; Sulentic, J.W. Accretion onto Supermassive Black Holes in Quasars: Learning from Optical/UV Observations. In *New Developments in Black Hole Research*; Kreitler, P.V., Ed.; Nova Press: New York, NY, USA, 2006; p. 123.

7. Sulentic, J.W. Toward a classification scheme for broad-line profiles in active galactic nuclei. *Astrophys. J.* **1989**, *343*, 54–65. [CrossRef]

8. Shen, Y.; Ho, L.C. The diversity of quasars unified by accretion and orientation. *Nature* **2014**, *513*, 210–213. [CrossRef] [PubMed]

9. Panda, S.; Czerny, B.; Adhikari, T.P.; Hryniewicz, K.; Wildy, C.; Kuraszkiewicz, J.; Śniegowska, M. Modeling of the Quasar Main Sequence in the Optical Plane. *Astrophys. J.* **2018**, *866*, 115. [CrossRef]

10. Sulentic, J.; Marziani, P.; Zamfir, S. The Case for Two Quasar Populations. *Balt. Astron.* **2011**, *20*, 427–434.

11. Marziani, P.; Zamanov, R.K.; Sulentic, J.W.; Calvani, M. Searching for the physical drivers of eigenvector 1: Influence of black hole mass and Eddington ratio. *Mon. Not. R. Astron. Soc.* **2003**, *345*, 1133–1144. [CrossRef]

12. Marziani, P.; Sulentic, J.W. Highly accreting quasars: Sample definition and possible cosmological implications. *Mon. Not. R. Astron. Soc.* **2014**, *442*, 1211–1229. [CrossRef]

13. Abramowicz, M.A.; Czerny, B.; Lasota, J.P.; Szuszkiewicz, E. Slim accretion disks. *Astrophys. J.* **1988**, *332*, 646–658. [CrossRef]

14. Mineshige, S.; Kawaguchi, T.; Takeuchi, M.; Hayashida, K. Slim-Disk Model for Soft X-Ray Excess and Variability of Narrow-Line Seyfert 1 Galaxies. *Publ. Astron. Soc. Jpn.* **2000**, *52*, 499–508.

15. Sadowski, A. Slim accretion disks around black holes. *arXiv* **2011**, arXiv:1108.0396.

16. Sun, J.; Shen, Y. Dissecting the Quasar Main Sequence: Insight from Host Galaxy Properties. *Astrophys. J. Lett.* **2015**, *804*, L15. [CrossRef]

17. Panda, S.; Marziani, P.; Czerny, B. The Quasar Main Sequence Explained by the Combination of Eddington Ratio, Metallicity, and Orientation. *Astrophys. J.* **2019**, *882*, 79. [CrossRef]

18. Peterson, B.M.; Ferland, G.J. An accretion event in the Seyfert galaxy NGC 5548. *Nature* **1986**, *324*, 345–347. [CrossRef]

19. Marziani, P.; Sulentic, J.W. Evidence for a very broad line region in PG 1138+222. *Astrophys. J.* **1993**, *409*, 612–616. [CrossRef]

20. Sulentic, J.W.; Zwitter, T.; Marziani, P.; Dultzin-Hacyan, D. Eigenvector 1: An Optimal Correlation Space for Active Galactic Nuclei. *Astrophys. J.* **2000**, *536*, L5–L9. [CrossRef]

21. Wang, J.; Li, Y. Strong Response of the Very Broad Hβ Emission Line in the Luminous Radio-quiet Quasar PG 1416-129. *Astrophys. J. Lett.* **2011**, *742*, L12. [CrossRef]

22. Punsly, B. Multi-epoch Observations of the Red Wing Excess in the Spectrum of 3C 279. *Astrophys. J. Lett.* **2013**, *762*, L25. [CrossRef]

23. Wolf, J.; Salvato, M.; Coffey, D.; Merloni, A.; Buchner, J.; Arcodia, R.; Baron, D.; Carrera, F.J.; Comparat, J.; Schneider, D.P.; et al. Exploring the diversity of Type 1 active galactic nuclei identified in SDSS-IV/SPIDERS. *Mon. Not. R. Astron. Soc.* **2020**, *492*, 3580–3601. [CrossRef]

24. Snedden, S.A.; Gaskell, C.M. The Case for Optically Thick High-Velocity Broad-Line Region Gas in Active Galactic Nuclei. *Astrophys. J.* **2007**, *669*, 126–134. [CrossRef]

25. Peterson, B.M.; Wandel, A. Keplerian Motion of Broad-Line Region Gas as Evidence for Supermassive Black Holes in Active Galactic Nuclei. *Astrophys. J.* **1999**, *521*, L95–L98. [CrossRef]

26. Peterson, B.M.; Ferrarese, L.; Gilbert, K.M.; Kaspi, S.; Malkan, M.A.; Maoz, D.; Merritt, D.; Netzer, H.; Onken, C.A.; Pogge, R.W.; et al. Central Masses and Broad-Line Region Sizes of Active Galactic Nuclei. II. A Homogeneous Analysis of a Large Reverberation-Mapping Database. *Astrophys. J.* **2004**, *613*, 682–699. [CrossRef]

27. Punsly, B.; Marziani, P.; Berton, M.; Kharb, P. The Extreme Red Excess in Blazar Ultraviolet Broad Emission Lines. *Astrophys. J.* **2020**, *903*, 44. [CrossRef]

28. Wang, J.M.; Du, P.; Brotherton, M.S.; Hu, C.; Songsheng, Y.Y.; Li, Y.R.; Shi, Y.; Zhang, Z.X. Tidally disrupted dusty clumps as the origin of broad emission lines in active galactic nuclei. *Nat. Astron.* **2017**, *1*, 775–783. [CrossRef]

29. Gaskell, C.M. Direct evidence for gravitational domination of the motion of gas within one light-week of the central object in NGC 4151 and the determination of the mass of the probable black hole. *Astrophys. J.* **1988**, *325*, 114–118. [CrossRef]

30. Corbin, M.R. QSO Broad Emission Line Asymmetries: Evidence of Gravitational Redshift? *Astrophys. J.* **1995**, *447*, 496. [CrossRef]

31. Popovic, L.C.; Vince, I.; Atanackovic-Vukmanovic, O.; Kubicela, A. Contribution of gravitational redshift to spectral line profiles of Seyfert galaxies and quasars. *Astron. Astrophys.* **1995**, *293*, 309–314.

32. Gavrilović, N.; Popović, L.Č.; Kollatschny, W. The gravitational redshift in the broad line region of the active galactic nucleus Mrk 110. In *IAU Symposium*; Karas, V., Matt, G., Eds.; Cambridge University Press: Cambridge, UK, 2007; Volume 238, pp. 369–370. [CrossRef]

33. Bon, N.; Bon, E.; Marziani, P.; Jovanović, P. Gravitational redshift of emission lines in the AGN spectra. *Astrophys. Space Sci.* **2015**, *360*, 7. [CrossRef]

34. Marziani, P.; Sulentic, J.W. Quasar Outflows in the 4D Eigenvector 1 Context. *Astron. Rev.* **2012**, *7*, 33–57.

35. Marziani, P.; Sulentic, J.W.; Stirpe, G.M.; Dultzin, D.; Del Olmo, A.; Martínez-Carballo, M.A. Blue outliers among intermediate redshift quasars. *Astrophys. Space Sci.* **2016**, *361*, 3. [CrossRef]

36. Leighly, K.M.; Moore, J.R. Hubble Space Telescope STIS Ultraviolet Spectral Evidence of Outflow in Extreme Narrow-Line Seyfert 1 Galaxies. I. Data and Analysis. *Astrophys. J.* **2004**, *611*, 107–124. [CrossRef]

37. Kaastra, J.S.; Kriss, G.A.; Cappi, M.; Mehdipour, M.; Petrucci, P.O.; Steenbrugge, K.C.; Arav, N.; Behar, E.; Bianchi, S.; Boissay, R.; et al. A fast and long-lived outflow from the supermassive black hole in NGC 5548. *Science* **2014**, *345*, 64–68. [CrossRef]

38. Fine, S.; Croom, S.M.; Bland-Hawthorn, J.; Pimbblet, K.A.; Ross, N.P.; Schneider, D.P.; Shanks, T. The CIV linewidth distribution for quasars and its implications for broad-line region dynamics and virial mass estimation. *Mon. Not. R. Astron. Soc.* **2010**, *409*, 591–610. [CrossRef]

39. Whittle, M. The Narrowline Region of Active Galaxies-Part Two-Relations Between OIII Profile Shape and Other Properties. *Mon. Not. R. Astron. Soc.* **1985**, *213*, 33.

40. Bennert, N.; Falcke, H.; Schulz, H.; Wilson, A.S.; Wills, B.J. Size and Structure of the Narrow-Line Region of Quasars. *Astrophys. J.* **2002**, *574*, L105–L109. [CrossRef]

41. Komossa, S.; Xu, D.; Zhou, H.; Storchi-Bergmann, T.; Binette, L. On the Nature of Seyfert Galaxies with High [O III] λ5007 Blueshifts. *Astrophys. J.* **2008**, *680*, 926–938. [CrossRef]

42. Zamanov, R.; Marziani, P.; Sulentic, J.W.; Calvani, M.; Dultzin-Hacyan, D.; Bachev, R. Kinematic Linkage between the Broad- and Narrow-Line-emitting Gas in Active Galactic Nuclei. *Astrophys. J.* **2002**, *576*, L9–L13. [CrossRef]

43. Marziani, P.; Martínez Carballo, M.A.; Sulentic, J.W.; Del Olmo, A.; Stirpe, G.M.; Dultzin, D. The most powerful quasar outflows as revealed by the Civ λ1549 resonance line. *Astrophys. Space Sci.* **2016**, *361*, 29. [CrossRef]

44. Sulentic, J.W.; Marziani, P.; Zamanov, R.; Bachev, R.; Calvani, M.; Dultzin-Hacyan, D. Average Quasar Spectra in the Context of Eigenvector 1. *Astrophys. J.* **2002**, *566*, L71–L75. [CrossRef]

45. Marziani, P.; Sulentic, J.W.; Plauchu-Frayn, I.; del Olmo, A. Is Mg II 2800 a Reliable Virial Broadening Estimator for Quasars? *Astron. Astrophys.* **2013**, *555*, 89.

46. Marziani, P.; Sulentic, J.W.; Stirpe, G.M.; Zamfir, S.; Calvani, M. VLT/ISAAC spectra of the Hβ region in intermediate-redshift quasars. III. Hβ broad-line profile analysis and inferences about BLR structure. *Astron. Astrophys.* **2009**, *495*, 83–112. [CrossRef]

47. Marziani, P.; Sulentic, J.W.; Zamanov, R.; Calvani, M.; Dultzin-Hacyan, D.; Bachev, R.; Zwitter, T. An Optical Spectroscopic Atlas of Low-Redshift Active Galactic Nuclei. *Astrophys. J. Suppl. Ser.* **2003**, *145*, 199–211. [CrossRef]

48. Bon, N.; Marziani, P.; Bon, E.; Negrete, C.A.; Dultzin, D.; del Olmo, A.; D'Onofrio, M.; Martínez-Aldama, M.L. Selection of highly-accreting quasars. Spectral properties of Fe II$_{opt}$ emitters not belonging to extreme Population A. *Astron. Astrophys.* **2020**, *635*, A151. [CrossRef]

49. Padovani, P. The faint radio sky: Radio astronomy becomes mainstream. *Astron. Astrophys. Rev.* **2016**, *24*, 13. [CrossRef]

50. Padovani, P. Active Galactic Nuclei at All Wavelengths and from All Angles. *Front. Astron. Space Sci.* **2017**, *4*, 35. [CrossRef]

51. Sulentic, J.W.; Marziani, P.; del Olmo, A.; Dultzin, D.; Perea, J.; Alenka Negrete, C. GTC spectra of z $\approx$ 2.3 quasars: Comparison with local luminosity analogs. *Astron. Astrophys.* **2014**, *570*, A96. [CrossRef]

52. Vestergaard, M.; Peterson, B.M. Determining Central Black Hole Masses in Distant Active Galaxies and Quasars. II. Improved Optical and UV Scaling Relationships. *Astrophys. J.* **2006**, *641*, 689–709. [CrossRef]

53. Netzer, H. Bolometric correction factors for active galactic nuclei. *Mon. Not. R. Astron. Soc.* **2019**, *488*, 5185–5191. [CrossRef]

54. Kriss, G. Fitting Models to UV and Optical Spectral Data. *Astron. Data Anal. Softw. Syst. III* **1994**, *61*, 437.

55. Boroson, T.A.; Green, R.F. The Emission-Line Properties of Low-Redshift Quasi-stellar Objects. *Astrophys. J. Suppl. Ser.* **1992**, *80*, 109. [CrossRef]

56. Sulentic, J.W.; del Olmo, A.; Marziani, P.; Martínez-Carballo, M.A.; D'Onofrio, M.; Dultzin, D.; Perea, J.; Martínez-Aldama, M.L.; Negrete, C.A.; Stirpe, G.M.; et al. What does CIVλ1549 tell us about the physical driver of the Eigenvector quasar sequence? *Astron. Astrophys.* **2017**, *608*, A122. [CrossRef]

57. Marziani, P.; Dultzin-Hacyan, D.; D'Onofrio, M.; Sulentic, J.W. Arp 194: Evidence of Tidal Stripping of Gas and Cross-Fueling. *Astron. J.* **2003**, *125*, 1897–1907. [CrossRef]

58. Zamfir, S.; Sulentic, J.W.; Marziani, P.; Dultzin, D. Detailed characterization of Hβ emission line profile in low-z SDSS quasars. *Mon. Not. R. Astron. Soc.* **2010**, *403*, 1759. [CrossRef]

59. Azzalini, A.; Regoli, G. Some properties of skew-symmetric distributions. *Ann. Inst. Statist. Math.* **2012**, *64*, 857–879. [CrossRef]

60. Urry, C.M.; Padovani, P. Unified Schemes for Radio-Loud Active Galactic Nuclei. *Publ. Astron. Soc. Pac.* **1995**, *107*, 803. [CrossRef]

61. Denney, K.D.; Peterson, B.M.; Pogge, R.W.; Adair, A.; Atlee, D.W.; Au-Yong, K.; Bentz, M.C.; Bird, J.C.; Brokofsky, D.J.; Chisholm, E.; et al. Diverse Kinematic Signatures from Reverberation Mapping of the Broad-Line Region in AGNs. *Astrophys. J.* **2009**, *704*, L80–L84. [CrossRef]

62. Du, P.; Brotherton, M.S.; Wang, K.; Huang, Z.P.; Hu, C.; Kasper, D.H.; Chick, W.T.; Nguyen, M.L.; Maithil, J.; Hand, D.; et al. Monitoring AGNs with Hβ Asymmetry. I. First Results: Velocity-resolved Reverberation Mapping. *Astrophys. J.* **2018**, *869*, 142. [CrossRef]

63. Barth, A.J.; Vogler, H.A.; Guo, H.; Treu, T.; Bennert, V.N.; Canalizo, G.; Filippenko, A.V.; Gates, E.; Hamann, F.; Joner, M.D.; et al. The Lick AGN Monitoring Project 2016: Velocity-resolved Hβ Lags in Luminous Seyfert Galaxies. *Astrophys. J.* **2022**, *925*, 52. [CrossRef]

64. Sulentic, J.W.; Marziani, P. The Intermediate-Line Region in Active Galactic Nuclei: A Region "Præter Necessitatem"? *Astrophys. J.* **1999**, *518*, L9–L12. [CrossRef]

65. Netzer, H.; Marziani, P. The Effect of Radiation Pressure on Emission-line Profiles and Black Hole Mass Determination in Active Galactic Nuclei. *Astrophys. J.* **2010**, *724*, 318–328. [CrossRef]

66. Ferland, G.J.; Hu, C.; Wang, J.; Baldwin, J.A.; Porter, R.L.; van Hoof, P.A.M.; Williams, R.J.R. Implications of Infalling Fe II-Emitting Clouds in Active Galactic Nuclei: Anisotropic Properties. *Astrophys. J.* **2009**, *707*, L82–L86. [CrossRef]
67. Marziani, P.; Sulentic, J.W.; Negrete, C.A.; Dultzin, D.; Zamfir, S.; Bachev, R. Broad-line region physical conditions along the quasar eigenvector 1 sequence. *Mon. Not. R. Astron. Soc.* **2010**, *409*, 1033–1048. [CrossRef]
68. Baldwin, J.; Ferland, G.; Korista, K.; Verner, D. Locally Optimally Emitting Clouds and the Origin of Quasar Emission Lines. *Astrophys. J.* **1995**, *455*, L119. [CrossRef]
69. Korista, K.; Baldwin, J.; Ferland, G.; Verner, D. An Atlas of Computed Equivalent Widths of Quasar Broad Emission Lines. *Astrophys. J. Suppl. Ser.* **1997**, *108*, 401. [CrossRef]
70. Murray, N.; Chiang, J. Disk Winds and Disk Emission Lines. *Astrophys. J.* **1997**, *474*, 91. [CrossRef]
71. Proga, D.; Stone, J.M.; Drew, J.E. Radiation-driven winds from luminous accretion discs. *Mon. Not. R. Astron. Soc.* **1998**, *295*, 595–617. [CrossRef]
72. Laor, A.; Brandt, W.N. The Luminosity Dependence of Ultraviolet Absorption in Active Galactic Nuclei. *Astrophys. J.* **2002**, *569*, 641–654. [CrossRef]
73. Bischetti, M.; Piconcelli, E.; Vietri, G.; Bongiorno, A.; Fiore, F.; Sani, E.; Marconi, A.; Duras, F.; Zappacosta, L.; Brusa, M.; et al. The WISSH quasars project. I. Powerful ionised outflows in hyper-luminous quasars. *Astron. Astrophys.* **2017**, *598*, A122. [CrossRef]
74. Vietri, G. The LBT/WISSH quasar survey: Revealing powerful winds in the most luminous AGN. American Astronomical Society Meeting Abstracts#. 2017; Volume 229. Available online: https://ui.adsabs.harvard.edu/abs/2017AAS...22930206V/abstract (accessed on 14 February 2022).
75. Vietri, G.; Piconcelli, E.; Bischetti, M.; Duras, F.; Martocchia, S.; Bongiorno, A.; Marconi, A.; Zappacosta, L.; Bisogni, S.; Bruni, G.; et al. The WISSH quasars project. IV. Broad line region versus kiloparsec-scale winds. *Astron. Astrophys.* **2018**, *617*, A81. [CrossRef]
76. Cano-Díaz, M.; Maiolino, R.; Marconi, A.; Netzer, H.; Shemmer, O.; Cresci, G. Observational evidence of quasar feedback quenching star formation at high redshift. *Astron. Astrophys.* **2012**, *537*, L8. [CrossRef]
77. Monari, G.; Famaey, B.; Carrillo, I.; Piffl, T.; Steinmetz, M.; Wyse, R.F.G.; Anders, F.; Chiappini, C.; Janßen, K. The escape speed curve of the Galaxy obtained from Gaia DR2 implies a heavy Milky Way. *Astron. Astrophys.* **2018**, *616*, L9. [CrossRef]
78. Ganci, V.; Marziani, P.; D'Onofrio, M.; del Olmo, A.; Bon, E.; Bon, N.; Negrete, C.A. Radio loudness along the quasar main sequence. *Astron. Astrophys.* **2019**, *630*, A110. [CrossRef]
79. del Olmo, A.; Marziani, P.; Ganci, V.; D'Onofrio, M.; Bon, E.; Bon, N.; Negrete, A.C. Optical spectral properties of radio loud quasars along the main sequence. In *Nuclear Activity in Galaxies Across Cosmic Time*; Ović, M., Marziani, P., Masegosa, J., Netzer, H., Negu, S.H., Tessema, S.B., Eds.; Cambridge University Press: Cambridge, UK, 2021; Volume 356, pp. 310–313. [CrossRef]
80. D'Onofrio, M.; Marziani, P.; Chiosi, C. Past, present and Future of the Scaling Relations of Galaxies and Active Galactic Nuclei. *arXiv* **2021**, arXiv:2109.06301.
81. Kriss, G.A. Coordinated UV and X-ray Observations of AGN Outflows. American Astronomical Society Meeting Abstracts#. 2017; Volume 229. Available online: https://ui.adsabs.harvard.edu/abs/2017AAS...22920901K/abstract (accessed on 14 February 2022).

Article

Non-Thermal Emission from Radio-Loud AGN Jets: Radio vs. X-rays

Elena Fedorova [1,*], Bohdan Hnatyk [1], Antonino Del Popolo [2,3] and Anatoliy Vasylenko [4] and Vadym Voitsekhovskyi [1]

[1] Astronomical Observatory, Taras Shevchenko National University of Kyiv, Observatorna Str. 3-b, 04053 Kyiv, Ukraine; bohdan_hnatyk@ukr.net (B.H.); v.v.voitsekhovskyi@gmail.com (V.V.)

[2] INAF-Osseravatorio Astrofisico di Catania, Universita di Catania, 95123 Catania, Italy; adelpopolo@oact.inaf.it

[3] Institute of Astronomy, Russian Academy of Sciences, 119017 Moscow, Russia

[4] Main Astronomical Observatory of National Academy of Sciences of Ukraine, 27 Akademika Zabolotnoho St., 03143 Kyiv, Ukraine; kvazarren@ukr.net

* Correspondence: efedorova@ukr.net

Abstract: We consider the sample of 55 blazars and Seyferts cross-correlated from the Planck all-sky survey based on the Early Release Compact Source Catalog (ERCSC) and Swift BAT 105-Month Hard X-ray Survey. The radio Planck spectra vs. X-ray Swift/XRT+BAT spectra of the active galactic nuclei (AGN) sample were fitted with the simple and broken power law (for the X-ray spectra taking into account also the Galactic neutral absorption) to test the dependencies between the photon indices of synchrotron emission (in radio range) and synchrotron self-Compton (SSC) or inverse-Compton emission (in X-rays). We show that for the major part of the AGN in our sample there is a correspondence between synchrotron and SSC photon indices (one of two for broken power-law model) compatible within the error levels. For such objects, this can give a good perspective for the task of distinguishing between the jet base counterpart from that one emitted in the disk+corona AGN "central engine".

Keywords: blazars; active galactic nuclei; X-rays; jets; synchrotron emission; inverse-Compton emission

Citation: Fedorova, E.; Hnatyk, B.; Popolo, A.D.; Vasylenko, A.; Voitsekhovskyi, V. Non-Thermal Emission from Radio-Loud AGN Jets: Radio vs. X-rays. *Galaxies* **2022**, *10*, 6. https://doi.org/10.3390/galaxies10010006

Academic Editor: Santanu Mondal

Received: 29 November 2021
Accepted: 30 December 2021
Published: 4 January 2022

Publisher's Note: MDPI stays neutral with regard to jurisdictional claims in published maps and institutional affiliations.

1. Introduction

The blazars are the radio-loud AGN with one jet directed along the line-of-sight to the observer (i.e., AGN type 0). Due to the jet alignment, an observer can see near pure jet base emission at the AGN center, as the emission of the disk+corona «central engine» is usually significantly weaker than the jet base one, and almost not visible in the blazar's X-ray spectrum. This makes blazars especially interesting when we intend to investigate the properties of the jet base emission. Here, we consider the dependence between the direct synchrotron emission of a jet base in the radio wave range and its inverse Compton (IC) or synchrotron self-Compton (SSC) emission in X-rays. Taking into account that non-blazar radio loud (RL) AGN have the same non-thermal component in their spectra, this dependence can open some possibilities to distinguish between nuclear (disk+corona) and jet base components in their high energy spectra extrapolating some parameters of the jet base synchrotron radio spectra to IC/SSC ones.

The jet base is of the order of a percent of pc in size. It is composed of ultra-relativistic plasma, either electron-positron (usually referred to as leptonic) or electron-proton (referred to as hadronic); the composition of this plasma is still being debated. Plasma particles are ejected at relativistic bulk velocity from an AGN central region and being accelerated at shock fronts inside the jet, emit synchrotron radiation visible at a wide range of wavelengths from radio to ultraviolet or even higher.

Depending on the peak frequency ν_S of the Synchrotron power (of peak intensity) $\nu_S F(\nu_S)$ two classes of BL Lacs were introduced in [1]: low-frequency peaked BL Lacs (LBL) and high-frequency peaked BL Lacs (HBL). In [2], similar classification was introduced to the sample of all blazars: low Synchrotron peaked blazars (LSP, $\nu_S \leq 10^{14}$ Hz), intermediate Synchrotron peaked blazars (ISP, $10^{14} \leq \nu_S \leq 10^{15}$ Hz), and high Synchrotron peaked blazars (HSP, $\nu_S > 10^{15}$ Hz) (see also discussion in [3–5]).

In addition to the dominant non-thermal jet-driven emission, accretion-driven thermal disk radiation, Comptonised by a hot $\sim$100 keV corona to power-law X-ray emission, can also make a significant contribution to the overall X-ray flux in some cases (3C120, 3C273, etc.) [2,6,7]. Signatures of such a disk+corona contribution in X-ray spectra are expected from LSP blazars in which IC/SSC jet components are dominant. Some ISP blazars with the intersection of falling synchrotron and rising IC/SSC spectra in the X-ray band (similarly to IBL S5 0716+714) are also promising candidates for the investigation of a jet-disk+corona interplay in X-ray band [2,8].

As it was shown in [9], the jet synchrotron and X-ray IC/SSC spectra are interconnected because they are produced by the same leptonic cosmic ray population. In particular, there is a relationship between photon indices in radio-band Γ_R and X-ray band Γ_X.

In our work, we investigate how this dependency manifest itself in radio and X-ray spectra of the sample of radio loud AGN created from the cross-correlation of the Swift BAT 105-Month Hard X-ray Survey (https://swift.gsfc.nasa.gov/results/bs105mon/ Accessed on 10 June 2020) and the Planck Early Release Compact Source Catalog (ERCSC, [10]) (http://www.sciops.esa.int/index.php?project=planck&page=Planck_Legacy_Archive Accessed on 15 June 2020). For this purpose, we use the 24–240 GHz Planck spectra and the Swift/XRT+BAT spectra of the same objects in X-rays and compare the photon indices.

The organisation of this paper is as follows. In the Section 2, we analyse the non-thermal MWL emission of AGN jets. In Section 3, we describe our AGN sample and spectral fitting procedure. In Section 4, we discuss an interconnection of radio and X-ray spectra of considered AGN and in Section 5 we draw out our conclusions.

2. Spectral Energy Distribution of Radio Loud AGN

Accelerated particles—cosmic-ray electrons and positrons (hereafter electrons)—of the jet in the most cases are distributed over the energies $E > E_m$ (Lorentz factor $\gamma = E/m_e c^2 > \gamma_m$) following a single power-law dependency $N(E) \propto E^{-p} \exp(-E/E_{cut})$ of the injected electron spectrum with the spectral index p and an exponential cut-off at $E \sim E_{cut}$ [9,11]. Such distribution of ultra-relativistic electrons ($\gamma_m \gg 1$) generates a two-hump spectrum consisting of the low-frequency synchrotron component and the high-frequency component due to inverse-Compton scattering (IC) of synchrotron photons (synchrotron-self-Compton SSC) or external low energy photons (CMB and other background radiation) by the same electron population [12]. In Figure 1, the two-hump spectra of two LSPs (3C279 and 3C273) and two HSPs (Mrk 421 ans Mrk 501) blazars are presented. In 3C273 case, the thermal contribution of the accretion disk in the optical-UV band is also visible.

The fast radiation cooling of the high energy relativistic electrons (if present) results in the broken power law spectrum of the final emitting electrons with the spectral index $p+1$ for $\gamma > \gamma_c$. For such a two-segment power law electron spectrum the corresponding synchrotron spectrum for the slow cooling case ($\gamma_c > \gamma_m$) is the four-segment power law $F_\nu \propto \nu^{-\alpha}$ with the spectral indices α_s (the photon index $\Gamma_s = \alpha_s + 1$) equal to -2.0, $-1/3$, $(p-1)/2$, and $p/2$ for $\nu < \nu_a, \nu_a < \nu < \nu_m, \nu_m < \nu < \nu_c$, and $\nu > \nu_c$, correspondingly. Here ν_a is the self-absorption frequency, ν_m is the frequency of synchrotron emission of electrons with minimum Lorenz factor $\gamma = \gamma_m$, ν_c is the same for electrons at the edge of fast cooling with $\gamma = \gamma_c$ and the typical case $\nu_a < \nu_m < \nu_c$ with the weak self-absorption regime $\nu_a < \nu_c$ is considered [9]. Synchrotron frequencies ν_i ($i = a, m, c$) are radiated by electrons with γ_i via γ^2-scattering of virtual photons with gyrofrequency $\nu_g = \omega_g/2\pi = eB/(2\pi m_e c)$ in magnetic field B: $\nu_i \sim \gamma_i^2 \nu_g$.

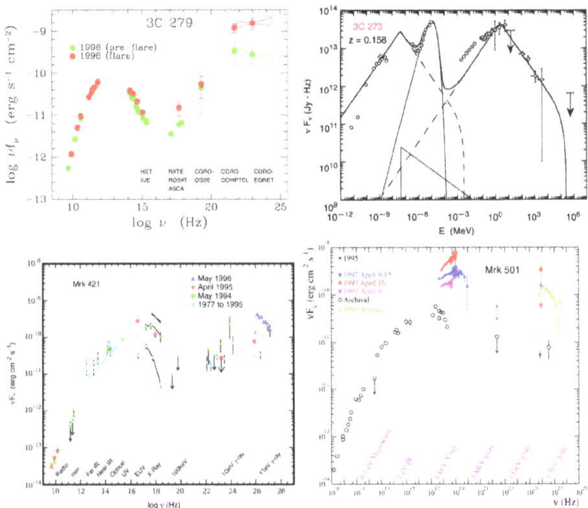

Figure 1. Two-hump multi-wavelength spectra of LSP (3C279 and 3C273) and HSP (Mrk 421 ans Mrk 501) blazars. In 3C273 case the thermal contribution of disk in optical-UV band is also visible. Credit: A.E. Wehrle/M.A. Catanese/J.H. Buckley/Whipple Collaboration.

In the case of IC/SSC emission, the real photons of background (IC) or newly generated synchrotron (SSC) radiation are γ^2-scattered by the relativistic electrons and characteristic frequencies of the SSC spectra are dependent on ones of synchrotron spectra $\nu_{ij}^{IC} \sim \gamma_i^2 \nu_j$. Therefore, the profiles of the SSC spectra are similar to the synchrotron ones and for the considered above case $\nu_a < \nu_m < \nu_c$ the four-segment power law spectrum has the spectral indices α_{IC} (the photon index $\Gamma_{IC} = \alpha_{IC} + 1$) equal to -1.0, $-1/3$, $(p-1)/2$, and $p/2$ for $\nu < \nu_{ma}^{IC}$, $\nu_{ma}^{IC} < \nu < \nu_{mm}^{IC}$, $\nu_{mm}^{IC} < \nu < \nu_{cc}^{IC}$, and $\nu > \nu_{cc}^{IC}$, correspondingly [9].

The multi-wavelength spectral energy distributions of blazars (BL Lacs and FSRQs) clearly demonstrates the two-hump shape (Figure 2, taken from [4]). As it is discussed in detail in [4], at low frequencies $\nu < \nu_t$ up to the self-absorption frequency $10^{11} \leq \nu_t/Hz \leq 10^{12}$ in the radio band a reasonable approximation of all observable spectra corresponds to power law with $\alpha_R = -0.1$ or photon index $\Gamma_R = 0.9$. In $\log(\nu L_\nu) - \log \nu$ representation where L_ν is the spectral luminosity, the peak frequency ν_S of the synchrotron emission determines the mentioned above classes (LSP, ISP, and HSP) of blazars. At frequency $\nu_{cut,S}$ the transition to the IC-dominated emission takes place. In a similar way, the peak frequency ν_C determines Compton-dominated hump and at $\nu_{cut,C}$ Compton-dominated emission decays. Between limiting frequencies ν_t and $\nu_{cut,C}$, the two-segment power law approximations for synchrotron humps (photon indices Γ_1 and Γ_2) and for IC/SSC humps (photon indices Γ_3 and $\Gamma_4 = \Gamma_2$) are presented and analysed in [4]. In Figure 2, the hump parameters are indicated for the FSRQs with $44 < \log(L_\gamma/erg\,s^{-1}) < 45$. From Table 1, in [4], it follows that interconnection between synchrotron radio-spectra at frequencies $\nu_t < \nu < \nu_S$ and IC/SSC X-ray spectra at frequencies $\nu > \nu_{cut,S}$ as signatures of emission produced by common cosmic-ray electron population is visible in some cases. Namely, in the cases of FSRQs with $44 < \log(L_\gamma/erg\,s^{-1}) < 47$ the photon indices $\Gamma_1 \approx 1.5$, while $1.4 < \Gamma_3 < 1.75$ and IC/SSC X-ray emission dominates for $\nu > \nu_{cut,S} \approx 10^{16}$ Hz. There are also promising sources among BL Lacs with $46 < \log(L_\gamma/erg\,s^{-1}) < 48$ for which $1.5 < \Gamma_1 < 1.65$ while $1.45 < \Gamma_3 < 1.62$ and IC/SSC X-ray emission dominates for $\nu > \nu_{cut,S} \approx 6 \times 10^{15}$ Hz.

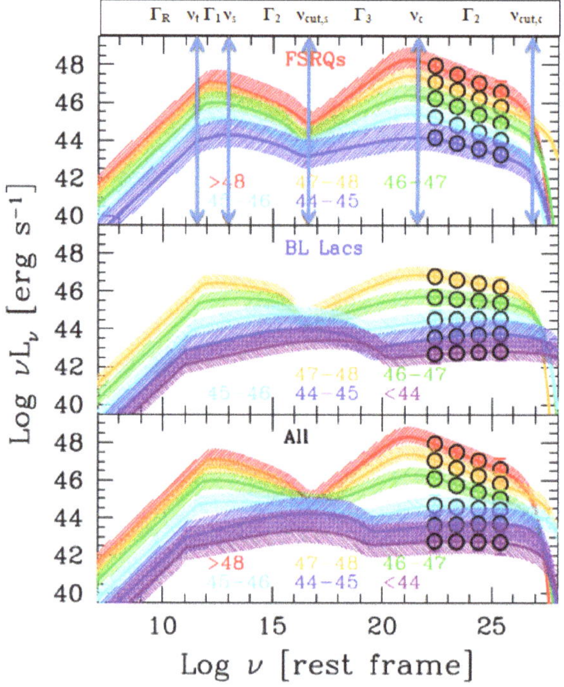

Figure 2. Spectral parameters of the two-hump multi-wavelength spectra for the new phenomenological Fermi blazar sequence, substantiated in [4]. Upper, middle and lower parts of the figure show the spectra of FSRQs, BL Lacs, and all sources, correspondingly. Different colours indicate different bins of the $\log(L_\gamma/erg\,s^{-1})$. Characteristic frequencies and photon indices are presented for FSRQs case with $44 < \log(L_\gamma/erg\,s^{-1}) < 45$. See detail in the text. Figure taken from [4].

3. The AGN Sample and Spectral Fitting

This research includes all the Swift/XRT+BAT datasets available in public data archive HEASARC for the objects of the sample identified as beamed AGN[1].

However, other sources identified these objects as BL Lacs (Table 1), FSRQs (Table 2), and Seyferts 1 (Table 3). In these Tables we show the source type, coordinates, and redshift following the SIMBAD data[2] and the neutral hydrogen absorbing column in the Galaxy following [13].

Table 1. Sub-sample of BL Lac type blazars.

Object	Type	RA [h, m, s]	Dec [°, ′, ″]	z	N_H [cm^{-2}]
PKS 0426-380	LSP	04 28 40.42	−37 56 19.58	1.11	2.3×10^{20}
PKS 0521-36	LSP	05 22 57.98	−36 27 30.84	0.05	4.1×10^{20}
PKS 0537-441	LSP	05 38 50.36	−44 05 08.93	0.89	3.5×10^{20}
S5 0716+714	ISP	07 21 53.44	+71 20 36.36	0.3	3.5×10^{20}
Mrk 501	HSP	16 53 52.21	+39 45 36.6	0.03	3.5×10^{20}
S5 1803+784	LSP	18 00 45.68	+78 28 04.01	0.68	4.1×10^{20}
PKS 2005-489	HSP	20 09 25.39	−48 49 53.72	0.07	4.1×10^{20}
PKS 2331-240	ISP	23 33 55.23	−23 43 40.65	0.05	1.7×10^{20}

Table 2. Sub-sample of FSRQ type blazars.

Object	Type	RA [h, m, s]	Dec [°, ′, ″]	z	N_H [cm^{-2}]
S5 0212+735	FSRQ-LSP	02 17 30.81	+73 49 32.61	2.36	3.9×10^{21}
PKS 0312-770	FSRQ	03 11 55.25	−76 51 50.84	0.22	8.0×10^{20}
4C +32.14	FSRQ	03 36 30.1	+32 18 29.34	1.26	2.7×10^{21}
4C +50.11	FSRQ	03 59 29.74	+50 57 50.16	1.52	2.7×10^{21}
PKS 0402-362	FSRQ	04 03 53.74	−36 05 01.91	1.42	6.1×10^{19}
PMN J0525-2338	FSRQ	05 25 06.50	−23 38 10.8	3.1	2.4×10^{20}
PKS 0528+134	FSRQ-LSP	05 30 56.46	+13 31 55.14	2.07	2.4×10^{20}
PKS 0537-286	FSRQ	05 39 54.28	−28 39 55.94	3.1	2.4×10^{20}
B2 0552+39A	FSRQ	05 55 30.80	+39 48 49.16	2.37	2×10^{21}
PMN J0623-6436	FSRQ	06 23 07.69	−64 36 20.71	0.12	4.7×10^{20}
PKS 0723-008	FSRQ	07 25 50.63	-00 54 56.54	0.12	1.7×10^{21}
4C +71.078	FSRQ	08 41 24.35	+70 53 42.28	2.21	3.1×10^{20}
S5 1039+81	FSRQ	10 44 23.06	+80 54 39.44	1.26	2.7×10^{20}
PKS 1127-14	FSRQ-LSP	11 30 07.05	−14 49 27.38	1.18	3.8×10^{20}
4C +49.22	FSRQ-LSP	11 53 24.46	+49 31 08.83	0.33	3.8×10^{20}
FBQS J1159+2914	FSRQ-LSP	11 59 31.83	+29 14 43.82	0.72	3.8×10^{20}
PG 1222+216	FSRQ	12 24 54.45	+21 22 46.38	0.43	2.3×10^{20}
3C 273	FSRQ-LSP	12 29 06.69	+02 03 08.59	0.15	2.3×10^{20}
3C 279	FSRQ-LSP	12 56 11.16	−05 47 21.53	0.53	2.3×10^{20}
PKS 1329-049	FSRQ	13 32 04.46	−05 09 43.3	2.15	2.4×10^{20}
PKS 1335-127	FSRQ	13 37 39.78	−12 57 24.69	0.54	6.5×10^{20}
PMN J1508-4953	FSRQ	15 08 38.94	−49 53 02.32	-	3.4×10^{21}
PKS 1510-08	FSRQ	15 12 50.53	−09 05 59.82	0.36	9.3×10^{20}
PKS 1622-29	FSRQ	16 26 06	−29 51 26.97	0.81	2.6×10^{21}
3C 345	FSRQ-LSP	16 42 58.81	+39 48 36.99	0.59	2.6×10^{21}
PKS 1830-21	FSRQ-LSP	18 33 39.92	−21 03 39	2.5	2.6×10^{21}
B1921-293	FSRQ	19 24 51.05	−29 14 30.12	0.35	2×10^{21}
4C +73.18	FSRQ	19 27 48.49	+73 58 01.57	0.3	1.1×10^{21}
PKS 2008-159	FSRQ	20 11 15.71	−15 46 40.25	1.18	2×10^{21}
QSO B2013+370	FSRQ	20 15 28.72	+37 10 59.51	0.85	1.3×10^{22}
B2 2023+33	FSRQ	20 25 10.84	+33 43	0.22	7.7×10^{21}
PKS 2052-47	FSRQ	20 56 16.35	−47 14 47.62	1.49	3.2×10^{20}
[HB 89] 2142-758	FSRQ	21 48	−75.575	1.13	7.7×10^{21}
PKS 2145+06	FSRQ	21 48 05.45	+06 57 38.6	1	5.6×10^{20}
PKS 2149-306	FSRQ-LSP	21 51 55.52	−30 27 53.69	2.35	1.8×10^{20}
4C +31.63	FSRQ-LSP	22 03 14.97	+31 45 38.26	0.29	1.2×10^{21}
II Zw 171	FSRQ	22 11 53.88	+18 41 49.85	0.06	5.1×10^{20}
[HB 89] 2230+114	FSRQ	22 31 48	+11.721	1.03	5.1×10^{20}
PKS 2227-088	FSRQ-LSP	22 29 40.08	−08 32 54.43	1.55	5.0×10^{20}
3C 454.3	FSRQ	22 53 57	+16 08 53	0.85	8.6×10^{20}

The Planck spectra within the frequency range 24 to 240 GHz were taken from the Early Release Compact Source Catalog (ERCSC) of compact sources observed by Planck during the 2009–2010 whole sky coverage. These Planck spectra are publicly available on the HEAVENS webpage[3].

The Swift/XRT spectra were obtained using the observations made by UK Swift Science Data Centre (UKSSDC) at the University of Leicester[4]. We have used single-pass centroid with the maximum of 10 attempts and 6 arcmin search radius.

The Swift/BAT spectra were obtained from the Swift BAT 105-Month Hard X-ray Survey official webpage[5].

To perform the spectra fitting the XSPEC package of the NASA HeaSoft version 6.27.2 software for astronomical data processing and analysis[6] was applied.

Table 3. Sub-sample of Seyfert 1 type AGN.

Object	Type	RA [h, m, s]	Dec [°, ′, ″]	z	N_H [cm^{-2}]
Mrk 1501	Seyfert 1	00 10 31.00	+10 58 29.5	0.09	7.1×10^{20}
QSO B0309+411	Seyfert 1	03 13 01.96	+41 20 01.18	0.13	1.7×10^{21}
PKS 0405-12	Seyfert 1	04 07 48.43	−12 11 36.66	0.57	4.2×10^{20}
PKS 1143-696	Seyfert 1	11 45 53.62	−69 54 01.79	0.24	2.8×10^{21}
3C 309.1	Seyfert 1	14 59 07.58	+71 40 19.86	0.9	2.4×10^{20}
3C 380	Seyfert 1	18 29 31.78	+48 44 46.15	0.69	2.4×10^{20}
8C 1849+670	Seyfert 1	18 49 16.07	+67 05 41.68	0.66	6.0×10^{20}
NGC 7213	Seyfert 1	22 09 16.21	−47 10 00.08	0.005	1.1×10^{20}

To fit the spectra in the both radio and X-ray range (here we include also the neutral hydrogen absorption in our model) we use the simple or broken power-law model taking into account that the flattering frequency ν_t due to the self-absorption may be located in the Planck range and the dip frequency $\nu_{cut,S}$ of the transition from synchrotron to IC component may be located in the Swift range or higher (Figure 2). So, if the self-absorption frequency ν_t falls into the Planck range 24 GHz $< \nu_t <$ 240 GHz, the two-segment power law approximation results in two photon indices: $\Gamma_{1,l} < 1$ for $\nu < \nu_t$ and $\Gamma_{1,h} > 1$ for $\nu > \nu_t$.

In a similar way, simple power law X-ray Swift spectra with $2 > \Gamma_3 \approx \Gamma_1 > 1$ are expected for the LSP and some IPS blazars with $\nu_{cut,S} \leq 10^{17}$ Hz, i.e., below the Swift range. Otherwise, the two-segment power law approximation of X-ray spectra will results in two photon indices: $\Gamma_{3,l} > 2$ for $\nu < \nu_{cut,S}$ and $2 > \Gamma_{3,h} > 1$ for $\nu > \nu_{cut,S}$. If $h\nu_{cut,S}$ exceeds the Swift/BAT range (175 keV) two photon indices: $\Gamma_{3,l} \geq 2$ for $E < E_{br}$ and $\Gamma_{3,h} \geq \Gamma_{3,l}$ for $E > E_{br}$ describe the falling part of the synchrotron spectra of HPS blazars.

In Table 4, we show the best-fit model parameters to the Planck and Swift/XRT+BAT spectra of the sample of 55 blazars and Seyferts. The models used to fit them were:

- *po* (simple power-law, for the Planck spectra);
- *bknpo* (broken power-law, for the Planck spectra);
- *po*tbabs* (absorbed power-law, for the Swift/XRT+BAT spectra);
- *bknpo*tbabs* (absorbed broken power-law, for the Swift/XRT+BAT spectra).

We explain the details of our Planck spectra fitting and data statistics in the Appendix A.

In columns 2–4 of the Table 4, we show the best-fit photon indices and break energies for the Planck spectra of the objects shown in the first column; in the last three columns we show the photon indices and the break energies for the X-ray (Swift/XRT+BAT) spectra. If the difference between the simple power-law and broken power-law is statistically significant (i.e., the null-hypothesis probability $P_{null} < 10\%$ where the null-hypothesis corresponds to the simple power-law model) we show the best-fit broken power-law model parameters. If there is not such difference between these two models in sense of χ-statistics and values of parameters, we show the simple power-law photon index only.

Table 4. The best-fit model parameters for the Planck and Swift/XRT+BAT spectra.

Object	Radio			X-rays		
Parameter ->	$\Gamma_{1,l}$	$E_b, 10^{-4}$ eV	$\Gamma_{1,h}$	$\Gamma_{3,l}$	E_b, keV	$\Gamma_{3,h}$
Mrk 1501	0.4 ± 0.8	1.6 ± 0.4	1.47 ± 0.09	1.62 ± 0.03	74 ± 69	2.91 ± 0.16
S5 0212+735	1.57 ± 0.08	4.0 ± 0.6	2.0 ± 0.3	1.41 ± 0.08	-	-
PKS 0312-770	0.4 ± 0.3	2.0 ± 0.2	1.95 ± 0.15	2.15 ± 0.08	2.7 ± 0.3	1.41 ± 0.09
QSO B0309+411	1.0 ± 0.2	-	-	2.7 ± 0.6	1.2 ± 0.2	1.8 ± 0.07
4C +32.14	1.62 ± 0.14	-	-	1.47 ± 0.07	25 ± 4	1.8 ± 0.3
4C +50.11	1.60 ± 0.05	-	-	1.66 ± 0.07	-	-
PKS 0402-362	1.13 ± 0.08	6.0 ± 0.6	1.45 ± 0.05	1.58 ± 0.04	9.5 ± 9.3	2.0 ± 0.5
PKS 0405-12	1.64 ± 0.09	-	-	1.79 ± 0.15	21 ± 4	2.7 ± 0.3
PKS 0426-380	0.7 ± 0.4	2.0 ± 0.6	1.57 ± 0.08	1.66 ± 0.07	4.1 ± 1.9	1.35 ± 0.07
PKS 0521-36	1.11 ± 0.09	4.1 ± 0.7	1.32 ± 0.03	1.58 ± 0.03	4.8 ± 1.8	1.86 ± 0.06
PMN J0525-2338	0.6 ± 0.6	2.5 ± 0.3	2.5 ± 0.5	1.5 ± 0.8	-	-
PKS 0528+134	1.9 ± 0.1	-	-	0.9 ± 0.3	1.5 ± 0.9	1.49 ± 0.07
PKS 0537-441	2.0 ± 0.4	-	-	2.0 ± 0.2	0.9 ± 0.2	1.68 ± 0.03
PKS 0537-286	1.10 ± 0.08	4.1 ± 0.8	1.43 ± 0.03	1.20 ± 0.06	10.9 ± 5.2	1.36 ± 0.04
B2 0552+39A	2.5 ± 0.3	1.9 ± 0.5	1.78 ± 0.10	1.45 ± 0.11	-	-
PMN J0623-6436	$0.87^{+0.17}_{-0.23}$	2.1 ± 0.1	1.59 ± 0.08	1.63 ± 0.11	-	-
S5 0716+714	0.75 ± 0.29	3.4 ± 0.6	1.28 ± 0.07	2.02 ± 0.09	6.1 ± 0.4	1.13 ± 0.06
PKS 0723-008	0.85 ± 0.06	3.7 ± 0.6	1.48 ± 0.06	1.62 ± 0.05	-	-
S5 1039+81	1.38 ± 0.15	-	-	1.25 ± 0.35	2.7 ± 1.8	1.65 ± 0.12
PKS 1127-14	0.6 ± 0.4	2.9 ± 0.6	1.70 ± 0.08	1.49 ± 0.06	18.4 ± 9.9	2.7 ± 1.6
PKS 1143-696	1.12 ± 0.35	3.3 ± 1.5	1.85 ± 0.27	2.5 ± 1.4	unc.	1.79 ± 0.14
4C +49.22	1.29 ± 0.05	-	-	1.68 ± 0.05	5.4 ± 2.8	1.84 ± 0.04
FBQS J1159+2914	1.17 ± 0.06	-	-	1.53 ± 0.04	47^{+90}_{-19}	>1.6
PG 1222+216	0.28 ± 0.22	2.4 ± 0.6	1.62 ± 0.12	2.16 ± 0.17	1.1 ± 0.1	1.43 ± 0.03
3C 273	0.96 ± 0.01	3.8 ± 0.2	1.76 ± 0.03	1.52 ± 0.03	2.5 ± 0.3	1.72 ± 0.01
3C 279	1.12 ± 0.08	4 ± 1	1.60 ± 0.03	1.49 ± 0.05	2.5 ± 0.3	1.67 ± 0.03
PKS 1329-049	1.22 ± 0.09	-	-	1.33 ± 0.09	-	-
PKS 1335-127	1.13 ± 0.04	3.5 ± 0.8	1.64 ± 0.06	1.42 ± 0.02	6.4 ± 0.7	2.1 ± 0.3
3C 309.1	1.41 ± 0.13	-	-	1.48 ± 0.07	-	-
PMN J1508-4953	0.6 ± 0.5	3.5 ± 0.7	1.5 ± 0.3	1.32 ± 0.07	-	-
PKS 1510-08	1.13 ± 0.06	3.3 ± 0.4	1.85 ± 0.09	1.31 ± 0.02	10.7 ± 6.5	1.39 ± 0.05
PKS 1622-29	1.08 ± 0.08	-	-	1.39 ± 0.03	7.1 ± 0.9	$1.91^{+0.4}_{-0.2}$
3C 345	1.33 ± 0.02	3.8 ± 0.8	1.80 ± 0.04	1.97 ± 0.22	1.2 ± 0.2	1.64 ± 0.04
Mrk 501	1.57 ± 0.10	-	-	2.09 ± 0.01	21 ± 14	2.46 ± 0.13

Table 4. *Cont.*

Object	Radio			X-rays		
Parameter ->	$\Gamma_{1,l}$	$E_b, 10^{-4}$ eV	$\Gamma_{1,h}$	$\Gamma_{3,l}$	$E_b,$ keV	$\Gamma_{3,h}$
S5 1803+784	1.17 ± 0.04	4.0 ± 0.8	1.51 ± 0.06	1.52 ± 0.04	-	-
3C 380	1.53 ± 0.09	3.7 ± 0.5	1.95 ± 0.15	1.67 ± 0.17	2.3 ± 0.9	2.1 ± 0.3
PKS 1830-21	1.74 ± 0.04	-	-	1.40 ± 0.03	5.3 ± 0.7	1.21 ± 0.05
8C 1849+670	1.26 ± 0.19	3.7 ± 1.6	1.53 ± 0.11	1.72 ± 0.10	6.3 ± 5.3	1.49 ± 0.1
B1921-293	1.32 ± 0.01	4.9 ± 0.3	1.72 ± 0.03	1.82 ± 0.10	-	-
4C +73.18	1.04 ± 0.55	2.2 ± 0.7	1.75 ± 0.08	1.76 ± 0.05	-	-
PKS 2005-489	1.13 ± 0.12	5.5 ± 0.9	1.9 ± 0.3	1.42 ± 0.02	7.1 ± 0.6	2.4 ± 0.3
PKS 2008-159	1.64 ± 0.14	-	-	1.40 ± 0.02	6.4 ± 0.9	$2.0^{+0.5}_{-0.3}$
QSO B2013+370	1.51 ± 0.20	-	-	1.16 ± 0.20	22 ± 18	2.2 ± 0.8
PKS 2052-47	1.39 ± 0.04	4.5 ± 1.5	1.41 ± 0.07	1.38 ± 0.02	11.0 ± 0.4	$2.2^{+0.5}_{-0.4}$
PKS 2145+06	1.35 ± 0.04	3.9 ± 0.4	1.97 ± 0.05	1.28 ± 0.06	2.0 ± 0.5	1.43 ± 0.01
[HB 89] 2142-758	0.7 ± 0.2	2.0 ± 0.2	1.46 ± 0.05	0.8 ± 0.3	1.8 ± 0.3	1.49 ± 0.11
4C +31.63	1.0 ± 0.4	3.0 ± 0.9	1.49 ± 0.154	1.62 ± 0.06	-	-
II Zw 171	1.27 ± 0.16	-	-	1.81 ± 0.05	-	-
PKS 2149-306	1.12 ± 0.09	-	-	1.22 ± 0.03	14 ± 5	1.61 ± 0.13
NGC 7213	-1.67 ± 0.15	-	-	1.69 ± 0.03	-	-
PKS 2227-088	0.97 ± 0.06	4.2 ± 0.9	1.61 ± 0.08	1.3 ± 1.5	<85	1.69 ± 0.32
[HB 89] 2230+114	1.39 ± 0.04	5.7 ± 0.8	2.0 ± 0.2	1.40 ± 0.02	7.1 ± 2.1	1.96 ± 0.11
3C 454.3	0.37 ± 0.05	3.0 ± 0.2	1.21 ± 0.03	1.92 ± 0.20	1.2 ± 0.3	1.62 ± 0.03
PKS 2331-240	0.9 ± 0.07	3.6 ± 0.3	1.57 ± 0.12	1.64 ± 0.04	7.6 ± 2.2	2.4 ± 0.3
4C +71.078	0.73 ± 0.05	3.2 ± 0.5	1.20 ± 0.15	1.33 ± 0.03	6.7 ± 1.8	1.73 ± 0.07
B2 2023+33	1.37 ± 0.15	7.4 ± 0.8	-0.2 ± 0.8	1.46 ± 0.17	-	-

4. AGN Sample: Interconnection of Radio and X-ray Spectra

As we can see from the Table 4 33 of 55 AGN have at least one of the photon indices in the model of Planck spectrum coincident with one of the photon indices of the X-ray spectral model. In total, 23 AGN of this subsample follow the pattern when the single Γ_1 or the upper $\Gamma_{1,h}$ photon index of the Planck spectrum coincides within the error levels with the single Γ_3 or lower $\Gamma_{3,l}$ photon index of the X-ray spectrum, i.e., in both radio and X-ray ranges we see the lower segments of the synchrotron and IC/SSC components.

Such a situation we observe for S5 0212+735, PKS0312-770, PKS 2331-240, PKS0537-441, PMN J0623-6436, 3C 309.1, 4C +31.63, 4C +32.14, 4C +50.11, 4C +73.18, PKS 1329-049, PKS 2005-489, PKS 2227-088, PKS 0405-12, PKS 0426-380, PKS 0528+134, PKS 0723-008, B2 0552+39A, S5 1039+81, S5 1803+784, B1921-293, B2 2023+33, PMN J1508-4953, and others. In total, 8 objects of this subsample follow the pattern when the single Γ_1 or the upper $\Gamma_{1,h}$ photon index of the Planck spectrum coincides within the error levels with the upper $\Gamma_{3,h}$ photon index of the X-ray spectrum, i.e., PKS 1143-696, PG1222+216, 3C 273, 3C 279, and others. Three objects have both upper and lower photon indices of the Planck spectrum coincident with the X-rays ones: PKS 0537-286, [HB 89] 2230+114 and 3C 380; such a situation can be interpreted as if we can see the $\Gamma_{1,l,h}$ and $\Gamma_{3,l,h}$ segments in the both Planck and Swift spectra. However, for some of them, namely 3C 273, [HB 89] 2142-758, etc., the values of the lower photon indices are below 1, what can be interpreted for the radio range as a sign of the self-absorption (see Figure 1 for the 3C273 case).

The subsample with no correspondence between the photon indices is also quite significant and consists of 22 blazars. There can be different explanations for each of these objects. For instance, as shown in [14], the single synchrotron zone model is inappropriate for 3C 454.3. Mrk 501 is the HSP blazar for which we see the lower part of synchrotron hump in the Planck spectrum and the upper one in X-rays. The IC/SSC spectrum of Mrk 501 is situated above the energies we consider here (see Figure 1). Similarly, to explain the 4C +49.22 spectral properties the two-zone SSC model is needed [15]. PKS 1830-21 is a gravitationally lensed blazar and its spectra thus can be distorted by the influence of a lensing object [16]. The classification type of PKS 0521-36 is not surely BL Lac, there are some signs of a Broad Line Radio Galaxy (BLRG) also [17], and thus the accretion disk/corona counterpart in its X-ray spectrum can be quite significant. S5 0716+714 is an ISP blazar for which the significant part of the Planck spectrum lies in the self-absorbed zone and in X-rays we see a dip-like intersection of synchrotron and SSC spectra at $E_b \approx 6$ keV [8].

5. Conclusions

In the simple one-zone model of multi-wavelength radiation of radio-loud AGN the observable two-hump spectra with νL_ν-maxima in low-energy ($\sim 10^{-1} - 10^4$ eV) and high-energy ($\geq 10^7$ eV) bands can be naturally explained by the non-thermal synchrotron and IC/SSC emission of relativistic leptons, accelerated at relativistic shocks, and during magnetic field reconnections in relativistic jets. Typical power-law spectra of accelerated leptonic cosmic-rays result in similar predicted slopes or photon indices of low-energy radio and high energy X-ray and γ-ray emission. Meantime, in some radio-loud AGN Comptonised thermal X-ray luminosity of accretion disk+corona complex can be comparable to the non-thermal X-ray jet luminosity. Joint analysis of radio and X-ray spectra of such AGN open a possibility to disentangle jet and disk+corona contributions and to clarify the radiative processes in AGN. We carry out a comparative analysis of Planck radio spectra and Swift /XRT+BAT X-ray spectra for the sample of 55 beamed AGN (Blazars and Seyferts 1) from the Swift BAT 105-Month Hard X-ray Survey. For 33 of 55 AGN we confirm predicted by one-zone model coincidences of the photon indices of radio and X-ray spectra. As expected, comparison of radio and X-ray data can help to disentangle jet and disk+corona contribution in case of LSP and some of ISP radio-loud AGN, in which transition from synchrotron to IC/SSC dominant contribution take place in sub-keV region.

Author Contributions: Conceptualization, E.F., B.H. and A.D.P.; methodology, E.F., B.H. and V.V.; software, E.F. and A.V.; investigation, writing—original draft preparation, E.F., B.H., A.D.P. and V.V.; writing—review and editing, A.V. All authors have read and agreed to the published version of the manuscript.

Funding: This research received no external funding.

Institutional Review Board Statement: Not applicable.

Informed Consent Statement: Not applicable.

Data Availability Statement: Sample of AGN is available from the authors.

Acknowledgments: We are grateful to the anonymous reviewer for very attentive and helpful comments and suggestions that helped us significantly improve the quality of the manuscript. We acknowledge reusing of figure 6 from article The Fermi blazar sequence of G. Ghisellini et al., MNRAS 2017, 469, 255–266.

Conflicts of Interest: The authors declare no conflict of interest.

Appendix A. Planck Spectral Models

The Planck spectra were fitted using the two models:

- Simple power-law *pow*;
- Broken power-law *bknpo*.

The results of our fitting are shown in the Table A1. In several exclusive cases when the both models mentioned above gave the fits with the $\chi/$d.o.f. > 2.0 we tried to apply the more complicated model, namely, the three-segment broken power-law one, with the three different photon indices and two breaks. These cases are marked by upper digits in the Table A1 and described in details below the Table. To estimate the statistical significance of our models, we suppose the simplest one of them, i.e., the single power-law as a null-hypothesis. Using the Fisher test (*ftest* in the XSPEC) we calculated the null-hypothesis probability for every object of the sample; these probabilities are shown in the last column of the Table A1.

Table A1. The model parameters for the Planck spectra.

Object	Power-Law		Broken Power-Law				
Parameter ->	Γ_1	$\chi^2/d.o.f.$	$\Gamma_{1,l}$	E_b, 10^{-4} eV	$\Gamma_{1,h}$	$\chi^2/d.o.f.$	P_{Null}
[HB 89] 2142-758	1.29 ± 0.05	103.4/5	0.7 ± 0.2	2.0 ± 0.2	1.46 ± 0.05	6.6/4	0.2 %
[HB 89] 2230+114	1.52 ± 0.05	81.4/5	1.39 ± 0.04	5.7 ± 0.8	2.2 ± 0.2	5.7/4	0.2 %
PMN J0623-6436	1.38 ± 0.08	26.1/4	$0.87^{+0.17}_{-0.23}$	2.1 ± 0.1	1.59 ± 0.08	4.9/3	3.7%
PMN J0525-2338	1.61 ± 0.12	10.8/2	0.6 ± 0.6	2.5 ± 0.3	2.5 ± 0.5	1.1/1	20.7%
PMN J1508-4953	0.9 ± 0.2	10.8/2	0.6 ± 0.5	3.5 ± 0.7	1.51 ± 0.46	1.3/1	22.6%
3C 273 [1]	1.35 ± 0.02	4520.6/6	0.96 ± 0.01	3.8 ± 0.2	1.76 ± 0.03	16.6/5	<0.01%
3C 279	1.47 ± 0.03	43.9/7	1.12 ± 0.08	4 ± 1	1.60 ± 0.03	7.9/6	0.2%
3C 309.1	1.41 ± 0.13	0.71/2	-	-	-	-	-
3C 345	1.61 ± 0.03	61.6/6	1.33 ± 0.06	3.8 ± 0.5	1.80 ± 0.04	3.1/4	0.3%
3C 380	1.7 ± 0.05	24.9/5	1.53 ± 0.09	3.7 ± 0.5	1.95 ± 0.15	6.6/4	3%
3C 454.3	0.95 ± 0.03	426.7/7	0.31 ± 0.05	3.0 ± 0.20	1.21 ± 0.03	6.2/5	<0.01%
4C +31.63	1.2 ± 0.2	20.0/5	1.0 ± 0.35	3.0 ± 0.9	1.49 ± 0.15	4.0/3	9%
4C +32.14	1.62 ± 0.14	2.9/3	-	-	-	-	-
4C +49.22	1.29 ± 0.05	-	-	-	-	-	-
4C +50.11	1.60 ± 0.05	-	-	-	-	-	-
4C +71.078	1.33 ± 0.05	347.3/5	0.73 ± 0.05	3.2 ± 0.5	1.92 ± 0.05	0.8/3	0.01%
4C +73.18	1.65 ± 0.07	9.4/5	1.04 ± 0.55	2.2 ± 0.7	1.75 ± 0.08	2.9/3	17%
8C 1849+670	1.26 ± 0.10	24.4/5	1.25 ± 0.19	3.7 ± 1.5	1.53 ± 0.11	3.6/3	5.6%
S5 0212+735	1.6 ± 0.6	7.0/3	1.57 ± 0.08	4 ± 0.6	2.0 ± 0.3	1.9/2	14.6%
S5 0716+714	1.18 ± 0.06	22.2/4	0.75 ± 0.28	3.4 ± 0.6	1.28 ± 0.07	4.3/3	3.9%
S5 1039+81	1.38 ± 0.25	1.3/2	-	-	-	-	-
S5 1803+784	1.3 ± 0.05	32.6/5	1.17 ± 0.04	4.0 ± 0.8	1.51 ± 0.06	1.8/3	1.3%
PKS 0312-770	1.45 ± 0.08	35.8/4	0.4 ± 0.3	2.0 ± 0.2	1.95 ± 0.15	2.7/3	0.9%
PKS 1127-14	1.16 ± 0.08	141.9/4	0.6 ± 0.4	2.9 ± 0.6	1.7 ± 0.08	0.9/2	0.6%
PKS 1143-696	1.3 ± 0.2	17.3/4	1.12 ± 0.35	3.3 ± 1.5	1.85 ± 0.27	6.0/3	10%
PKS 1329-049	1.22 ± 0.09	7.2/4	-	-	-	-	-
PKS 1335-127	1.35 ± 0.03	52.1/5	1.13 ± 0.04	3.5 ± 0.8	1.64 ± 0.06	4.5/3	2.5%
PKS 1510-08	1.38 ± 0.04	65.3/5	1.13 ± 0.06	3.9 ± 0.4	1.85 ± 0.09	7.4/4	0.5%
PKS 1622-29	1.08 ± 0.08	0.9/4	-	-	-	-	-
PKS 1830-21	1.74 ± 0.04	3.8/4	-	-	-	-	-
PKS 2005-489	1.31 ± 0.11	14.5/3	1.13 ± 0.12	5.5 ± 0.9	1.9 ± 0.3	4.7/2	17.8%
PKS 2008-159	1.64 ± 0.14	1.9/1	-	-	-	-	-
PKS 2052-47	1.48 ± 0.03	10.6/5	1.53 ± 0.04	4.5 ± 1.5	1.41 ± 0.07	8.6/4	38.9%
PKS 2145+06	1.57 ± 0.2	127.8/5	1.35 ± 0.04	3.9 ± 0.4	1.97 ± 0.05	4.4/3	0.6%
PKS 2149-306	1.52 ± 0.09	0.42/1	-	-	-	-	-

Table A1. *Cont.*

Object	Power-Law		Broken Power-Law				
Parameter ->	Γ_1	$\chi^2/d.o.f.$	$\Gamma_{1,l}$	E_b, 10^{-4} eV	$\Gamma_{1,h}$	$\chi^2/d.o.f.$	P_{Null}
PKS 2227-088	1.22 ± 0.04	15.7/4	0.97 ± 0.06	4.2 ± 0.9	1.61 ± 0.08	3.2/2	20%
PKS 2331-240	1.22 ± 0.04	28.0/3	0.9 ± 0.07	3.6 ± 0.3	1.57 ± 0.12	3.9/2	7.2%
PKS 0402-362 [2]	1.13 ± 0.02	316.3/7	0.65 ± 0.07	3.1 ± 0.5	1.37 ± 0.05	20.9/5	0.1%
PKS 0405-12	1.64 ± 0.17	1.3/1	-	-	-	-	-
PKS 0426-380	1.43 ± 0.06	15.9/6	0.7 ± 0.4	2.0 ± 0.6	1.52 ± 0.06	7.7/4	23%
PKS 0521-36	1.21 ± 0.02	38.1/5	1.11 ± 0.03	4.1 ± 0.7	1.32 ± 0.03	5.6/3	5.6%
PKS 0528+134	1.90 ± 0.15	4.5/3	-	-	-	-	-
PKS 0537-286	2.0 ± 0.4	0.5/1	-	-	-	-	-
PKS 0537-441	1.33 ± 0.02	44.7/7	1.10 ± 0.08	4.1 ± 0.8	1.43 ± 0.03	9.2/5	1.9%
PKS 0723-008	1.17 ± 0.04	91.7/5	0.85 ± 0.06	3.7 ± 0.6	1.48 ± 0.06	3.4/3	0.7%
B2 0552+39A	2.0 ± 0.1	7.1/3	2.5 ± 0.3	1.9 ± 0.5	1.78 ± 0.10	1.2/1	41.3%
B2 2023+33	1.28 ± 0.12	13.8/3	1.37 ± 0.15	7.4 ± 0.8	-0.2 ± 0.8	6.03/2	25%
B1921-293	1.43 ± 0.10	509.7/6	1.32 ± 0.01	4.9 ± 0.3	1.72 ± 0.03	7.9/4	0.02%
FBQS J1159+2914 [3]	1.17 ± 0.06	13.4/6	1.24 ± 0.19	unconstr.	1.15 ± 0.08	13.1/5	74%
II Zw 171	1.27 ± 0.16	0.03/0	-	-	-	-	-
Mrk 501	1.57 ± 0.10	2.6/2	-	-	-	-	-
Mrk 1501	1.34 ± 0.07	8.6/5	0.4 ± 0.8	1.6 ± 0.4	1.47 ± 0.09	4.2/4	11%
NGC 7213	-1.67 ± 0.15	0.59/0	-	-	-	-	-
PG 1222+216	1.17 ± 0.08	35.2/3	0.28 ± 0.22	2.4 ± 0.6	1.62 ± 0.12	1.3/1	19%
QSO B0309+411	1.0 ± 0.2	3.9/1	0.5 ± 0.6	6.0 ± 2.0	1.2 ± 0.3	0.23/0	15.6%
QSO B2013+370	1.51 ± 0.20	0.51/1	-	-	-	-	-

If the null-hypothesis probability is less than 50% we consider the alternative model (i.e., broken power-law one) as the best-fit one. Otherwise, $P_{null} > 50\%$ means that the single power-law fit is adequate and the broken power-law fit is excessively and statistically indistinguishable from the power-law one.

In case when the double broken power-law model was applied (AGN 3C 273, PKS 0402-362 and FBQS J1159+2914) we have calculated for it two "null-hypothesis" probabilities; one for the single power-law null hypothesis and the second one considering the broken power-law model as a null-hypothesis.

There are the comments 1, 2 and 3 to the Table:

1. *bkn2po* $\Gamma_{1,l} = 0.96 \pm 0.01$, $E_{b1} = 3.8*10^{-4}$ eV, $\Gamma_{1,i} = 1.76 \pm 0.02$, $E_{b2} = 10^{-3}$ eV and $\Gamma_{1,h} = 1.88 \pm 0.05$; $\chi^2/d.o.f. = 4.2/4$; with $P_{Null} = 3\%$ of the *bknpo* model relatively to *bkn2po* one and $P_{Null} < 10^{-4}\%$ relatively to the power-law one;

2. *bkn2po* $\Gamma_{1,l} = 0.7 \pm 0.1$, $E_{b1} = 2.5 \pm 0.3*10^{-4}$ eV, $\Gamma_{1,i} = 1.13 \pm 0.08$, $E_{b2} = 6.0 \pm 0.6*10^{-4}$ eV, $\Gamma_{1,h} = 1.45 \pm 0.05$ and $\chi^2/d.o.f. = 9.2/3$; with $P_{Null} = 15\%$ of the *bknpo* model relatively to *bkn2po* one and $P_{Null} = 0.1\%$ relatively to the power-law one;

3. *bkn2po* $\Gamma_{1,l} = 0.4 \pm 0.5$, $E_{b1} = 1.9 \pm 0.1*10^{-4}$ eV, $\Gamma_{1,i} = 3.4 \pm 0.8$, $E_{b2} = 2.3 \pm 0.2*10^{-4}$ eV, $\Gamma_{1,h} = 1.13 \pm 0.07$ and $\chi^2/d.o.f. = 7.9/4$; with $P_{Null} = 18\%$ of the *bknpo* model relatively to *bkn2po* one and $P_{Null} = 34.7\%$ relatively to the power-law one.

Appendix B. X-ray Spectral Models

The Swift/XRT+BAT spectra were fitted using the two models:

- Simple power-law with neutral hydrogen absorption *pow*tbabs*, with the fitting parameters: photon index Γ_3 and absorbing column density N_H;
- Broken power-law with neutral hydrogen absorption *bknpo*tbabs*, with the fitting parameters: two photon indices $\Gamma_{3,l}$ and $\Gamma_{3,h}$, break energy E_b and the neutral absorbing

column density N_H. The lower fitting values for N_H were set to the Galactic absorption values for particular object shown in the Tables 1–3. In the Table A2 we show the absorption excesses relatively to those values.

The results of our fitting are shown in the Table A2. To estimate the statistical significance of our models, we have performed the same as for the Planck spectra, namely we suppose that the simplest model of two, i.e., the single power-law is the null-hypothesis. Using the Fisher test we have calculated the null-hypothesis probability P_{Null} for every object of our sample; these probabilities are shown in the last column of the Table A2. We have determined the best-fit model in the same way as we did for the Planck spectra.

Table A2. The model parameters for the Swift spectra.

Object	Power-Law			Broken Power-Law					
Parameter ->	Γ_3	N_H *	$\chi^2/d.o.f.$	$\Gamma_{3,l}$	E_b **	$\Gamma_{3,h}$	N_H^*	$\chi^2/d.o.f.$	P_{Null}
[HB 89] 2142-758	1.44 ± 0.07	<100	74/56	0.8 ± 0.3	1.8 ± 0.3	1.49 ± 0.11	<100	68/54	10%
[HB 89] 2230+114	1.41 ± 0.03	7 ± 4	841/702	1.40 ± 0.02	7.1 ± 2.1	1.96 ± 0.11	7 ± 4	838/700	100%
PMN J0623-6436	1.63 ± 0.11	<4	85.7/63	-	-	-	-	-	-
3C 273	1.65 ± 0.01	3 ± 2	1279/829	1.52 ± 0.03	2.5 ± 0.3	1.72 ± 0.01	3 ± 2	1087/827	<0.01%
3C 279	1.59 ± 0.01	2 ± 1	1165/829	1.49 ± 0.05	2.5 ± 0.3	1.67 ± 0.03	2 ± 1	1087/827	<0.01%
3C 309.1	1.48 ± 0.07	<6	54.2/65	-	-	-	-	-	-
3C 345	1.64 ± 0.04	<2	258/270	1.97 ± 0.22	1.2 ± 0.2	1.64 ± 0.04	<2	251/268	2.5%
3C 380	1.89 ± 0.06	90 ± 10	711/691	1.67 ± 0.17	2.3 ± 0.9	2.1 ± 0.3	60 ± 10	702/689	2%
3C 454.3	1.63 ± 0.03	<4	281/270	1.92 ± 0.20	1.2 ± 0.3	1.62 ± 0.03	<4	272/268	2%
4C +31.63	1.62 ± 0.06	80 ± 60	118//110	-	-	-	-	-	-
4C +32.14	1.54 ± 0.03	<20	121/112	1.47 ± 0.07	25 ± 4	1.8 ± 0.3	<20	114/110	3.8%
4C +49.22	1.78 ± 0.02	<10	219/144	1.68 ± 0.05	5.4 ± 2.8	1.84 ± 0.04	<10	202/142	0.3%
4C +50.11	1.66 ± 0.07	<110	76.1/86	-	-	-	-	-	-
4C +71.078	1.49 ± 0.01	39 ± 4	957/498	1.33 ± 0.03	6.7 ± 1.8	1.73 ± 0.07	<42	602/496	<0.01%
4C +73.18	1.76 ± 0.05	<5	92.4/90	-	-	-	-	-	-
8C 1849+670	1.57 ± 0.03	<50	62.1/42	1.72 ± 0.10	6.3 ± 5.3	1.49 ± 0.11	<50	54.2/40	6.6%
S5 0212+735	1.41 ± 0.08	170 ± 100	85.5/73	-	-	-	-	-	-
PKS 0312-770	1.81 ± 0.03	<20	862/561	2.15 ± 0.08	2.7 ± 0.3	1.41 ± 0.09	<20	659/559	<0.01%
PKS 1127-14	1.59 ± 0.05	85 ± 16	140/111	1.49 ± 0.06	18.4 ± 9.9	2.7 ± 1.6	85 ± 16	113/109	<0.01%
PKS 1143-696	1.72 ± 0.06	<7	64.3/68	2.5 ± 1.4	unc.	1.79 ± 0.14	<7	62.8/66	41%
PKS 1329-049	1.38 ± 0.08	<55	34.3/32	1.33 ± 0.09	unc.	unc.	<40	32.0/30	35%
PKS 1335-127	1.47 ± 0.02	<2	759/619	1.42 ± 0.02	6.4 ± 0.7	2.1 ± 0.3	<2	646/617	<0.01%
PKS 1510-08	1.32 ± 0.02	<5	666/579	1.31 ± 0.02	10.7 ± 6.5	1.39 ± 0.05	<5	661/577	8.8%
PKS 1622-29	1.44 ± 0.02	<2	748/619	1.39 ± 0.03	7.1 ± 0.9	$1.91^{+0.4}_{-0.2}$	<2	659/617	<0.01%
PKS 1830-21	1.33 ± 0.02	<2	769.6/619	1.40 ± 0.03	5.3 ± 0.7	1.21 ± 0.05	<2	689/617	<0.01%
PKS 2005-489	1.54 ± 0.02	19 ± 4	993/619	1.42 ± 0.02	7.1 ± 0.6	2.4 ± 0.3	6 ± 1	654/617	<0.01%
PKS 2008-159	1.42 ± 0.02	<3	745/695	1.40 ± 0.02	6.4 ± 0.9	$2.0^{+0.5}_{-0.3}$	<3	730/693	<0.01%
PKS 2052-47	1.39 ± 0.02	25 ± 4	751/695	1.38 ± 0.02	11 ± 4	$2.2^{+0.5}_{-0.4}$	25 ± 4	737/693	0.2%
PKS 2145+06	1.39 ± 0.02	<6	744/695	1.28 ± 0.06	2.0 ± 0.5	1.43 ± 0.04	<6	730/693	<0.01%
PKS 2149-306	1.35 ± 0.01	9 ± 7	370/255	1.22 ± 0.03	13.6 ± 4.3	1.61 ± 0.13	9 ± 7	285/253	<0.01%
PKS 2227-088	1.46 ± 0.08	<14	43.1/37	1.33 ± 1.50	<85	1.69 ± 0.32	<14	40.2/35	29.6%
PKS 2331-240	1.73 ± 0.03	14 ± 5	384/293	1.64 ± 0.04	7.6 ± 2.2	2.4 ± 0.3	14 ± 5	298/291	<0.01%
PKS 0402-362	1.64 ± 0.04	34 ± 10	205/168	1.58 ± 0.04	9.5 ± 9.3	2.0 ± 0.5	24 ± 10	187/166	<0.01%

Table A2. *Cont.*

Object	Power-Law			Broken Power-Law					
Parameter ->	Γ_3	N_H *	$\chi^2/d.o.f.$	$\Gamma_{3,l}$	E_b **	$\Gamma_{3,h}$	N_H^*	$\chi^2/d.o.f.$	P_{Null}
PKS 0405-12	2.5 ± 0.1	<40	94.4/38	1.79 ± 0.15	21 ± 4	2.7 ± 0.3	<40	54.2/36	$<0.01\%$
PKS 0426-380	1.43 ± 0.05	<31	93/40	1.66 ± 0.07	4.1 ± 1.9	1.35 ± 0.07	<31	61.8/38	0.03%
PKS 0521-36	1.70 ± 0.02	<9	434/318	1.58 ± 0.03	4.8 ± 1.8	1.86 ± 0.06	<9	357/316	$<0.01\%$
PKS 0528+134	1.53 ± 0.06	430 ± 90	171/146	0.93 ± 0.28	$1.5^{+0.9}_{-0.2}$	1.49 ± 0.07	340 ± 40	168/144	28%
PKS 0537-286	1.29 ± 0.01	15 ± 14	124/86	1.20 ± 0.06	10.9 ± 5.2	1.36 ± 0.04	15 ± 14	107/84	0.2%
PKS 0537-441	1.66 ± 0.02	<5	306/292	2.0 ± 0.2	0.9 ± 0.2	1.68 ± 0.03	<3	297/290	1.3%
PKS 0723-008	1.62 ± 0.05	<3	79.6/90	-	-	-	-	-	-
B2 0552+39A	1.45 ± 0.11	26 ± 17	20.1/20	-	-	-	-	-	-
S5 0716+714	2.22 ± 0.15	20 ± 5	1731/848	2.61 ± 0.09	1.24 ± 0.05	2.17 ± 0.02	20 ± 5	1343/846	$<0.01\%$
S5 1039+81	1.55 ± 0.07	<12	50.5/29	1.25 ± 0.35	2.7 ± 1.8	1.65 ± 0.12	<12	44.4/27	17.6%
S5 1803+784	1.52 ± 0.04	<5	123/131	-	-	-	-	-	-
B1921-293	1.82 ± 0.10	<4	53.6/73	-	-	-	-	-	-
B2 2023+33	1.46 ± 0.17	<250	8.7/9	-	-	-	-	-	-
FBQS J1159+2914	1.54 ± 0.04	<13	218/206	1.53 ± 0.04	47^{+90}_{-19}	>1.6	<11	214/204	15.1%
II Zw 171	1.81 ± 0.05	<7	146/136	2.1 ± 1.0	unc.	1.82 ± 0.09	<7	144/134	40%
Mrk 501	2.10 ± 0.01	<1	652/457	2.04 ± 0.01	56 ± 25	2.7 ± 0.5	<1	642/455	7.2%
Mrk 1501	1.63 ± 0.03	<6	172/128	1.62 ± 0.04	74 ± 69	2.91 ± 0.16	<6	169/126	33%
NGC 7213	1.69 ± 0.03	14 ± 13	72.2/94	-	-	-	-	-	-
PG 1222+216	1.41 ± 0.03	<11	739/277	2.16 ± 0.17	1.1 ± 0.1	1.43 ± 0.03	<11	285/275	$<0.01\%$
PMN J0525-2338	1.5 ± 0.8	<95	7.2/5	-	-	-	-	-	-
PMN J1508-4953	1.32 ± 0.07	<300	30.3/23	>1.0	0.9 ± 0.2	1.30 ± 0.06	20 ± 10	19/21	0.7%
QSO B0309+411	1.79 ± 0.04	<60	158.9/128	2.7 ± 0.6	1.2 ± 0.1	1.80 ± 0.07	<60	151.4/126	4.8%
QSO B2013+370	1.57 ± 0.08	56 ± 25	75.3/54	1.6 ± 0.2	22 ± 18	2.2 ± 0.8	59 ± 25	65.3/52	2.5%

* 10^{19} cm^{-2}; ** keV.

Notes

[1] https://swift.gsfc.nasa.gov/results/bs105mon/ Accessed on 10 June 2020.

[2] https://simbad.u-strasbg.fr/simbad/ Accessed on 10 June 2020.

[3] https://www.isdc.unige.ch/heavens Accessed on 12 June 2020.

[4] https://www.swift.ac.uk/ Accessed on 12 June 2020.

[5] https://swift.gsfc.nasa.gov/results/bs105mon/ Accessed on 10 June 2020.

[6] https://heasarc.gsfc.nasa.gov/lheasoft/ Accessed on 15 July 2020.

References

1. Padovani, P.; Giommi, P. A sample-oriented catalogue of BL Lacertae objects. *MNRAS* **1995**, *277*, 1477–1490. [CrossRef]
2. Abdo, A.A.; Ackermann, M.; Agudo, I.; Ajello, M.; Aller, H.D.; Aller, M.F.; Angelakis, E.; Arkharov, A.A.; Axelsson, M.; Bach, U.; et al. The Spectral Energy Distribution of Fermi Bright Blazars. *ApJ* **2010**, *716*, 30–70. [CrossRef]
3. Giommi, P.; Polenta, G.; Lähteenmäki, A.; Thompson, D.J.; Capalbi, M.; Cutini, S.; Gasparrini, D.; González-Nuevo, J.; León-Tavares, J.; López-Caniego, M.; et al. Simultaneous Planck, Swift, and Fermi observations of X-ray and γ-ray selected blazars. *Astron. Astrophys.* **2012**, *541*, A160. [CrossRef]
4. Ghisellini, G.; Righi, C.; Costamante, L.; Tavecchio, F. The Fermi blazar sequence. *MNRAS* **2017**, *469*, 255–266. [CrossRef]
5. Pei, Z.; Fan, J.; Yang, J.; Huang, D.; Li, Z. The Estimation of Fundamental Physics Parameters for Fermi-LAT Blazars. *arXiv* **2021**, arXiv:2112.00530.
6. Grandi, P.; Palumbo, G.G.C. Jet and Accretion-Disk Emission Untangled in 3C 273. *Science* **2004**, *306*, 998–1002. [CrossRef] [PubMed]

7. Fedorova, E.; Hnatyk, B.I.; Zhdanov, V.I.; Del Popolo, A. X-ray Properties of 3C 111: Separation of Primary Nuclear Emission and Jet Continuum. *Universe* **2020**, *6*, 219. [CrossRef]
8. Zhu, S.F.; Brandt, W.N.; Luo, B.; Wu, J.; Xue, Y.Q.; Yang, G. The L_X-L_{uv}-L_{radio} relation and corona-disc-jet connection in optically selected radio-loud quasars. *MNRAS* **2020**, *496*, 245–268. [CrossRef]
9. Gao, H.; Lei, W.H.; Wu, X.F.; Zhang, B. Compton scattering of self-absorbed synchrotron emission. *MNRAS* **2013**, *435*, 2520–2531. [CrossRef]
10. Planck Collaboration; Lawrence, C.R. The Planck Early Release Compact Source Catalog. In Proceedings of the American Astronomical Society Meeting Abstracts #217, Seattle, WA, USA, 9–13 January 2011; Volume 217, p. 243.07. Available online: https://www.isdc.unige.ch/heavens/ (accessed on 15 July 2020).
11. La Mura, G.; Busetto, G.; Ciroi, S.; Rafanelli, P.; Berton, M.; Congiu, E.; Cracco, V.; Frezzato, M. Relativistic plasmas in AGN jets. From synchrotron radiation to γ-ray emission. *Eur. Phys. J. D* **2017**, *71*, 95, [CrossRef]
12. Urry, C.M. Multiwavelength properties of blazars. *Adv. Space Res.* **1998**, *21*, 89–100. [CrossRef]
13. Willingale, R.; Starling, R.L.C.; Beardmore, A.P.; Tanvir, N.R.; O'Brien, P.T. Calibration of X-ray absorption in our Galaxy. *MNRAS* **2013**, *431*, 394–404. [CrossRef]
14. Anjum, M.S.; Tammi, J. Nonlinear synchrotron self-compton modelling of blazars. In Proceedings of the 2015 Fourth International Conference on Aerospace Science and Engineering (ICASE), Islamabad, Pakistan, 2–4 September 2015; pp. 1–8. [CrossRef]
15. Cutini, S.; Ciprini, S.; Orienti, M.; Tramacere, A.; D'Ammando, F.; Verrecchia, F.; Polenta, G.; Carrasco, L.; D'Elia, V.; Giommi, P.; et al. Radio-gamma-ray connection and spectral evolution in 4C +49.22 (S4 1150+49): the Fermi, Swift and Planck view. *MNRAS* **2014**, *445*, 4316–4334. [CrossRef]
16. Nair, S.; Narasimha, D.; Rao, A.P. PKS 1830-211 as a Gravitationally Lensed System. *ApJ* **1993**, *407*, 46. [CrossRef]
17. D'Ammando, F.; Orienti, M.; Tavecchio, F.; Ghisellini, G.; Torresi, E.; Giroletti, M.; Raiteri, C.M.; Grandi, P.; Aller, M.; Aller, H.; et al. Unveiling the nature of the γ-ray emitting active galactic nucleus PKS 0521-36. *MNRAS* **2015**, *450*, 3975–3990. [CrossRef]

Article

Promise of Persistent Multi-Messenger Astronomy with the Blazar OJ 287

Mauri J. Valtonen [1,2,*], Lankeswar Dey [3,*], Achamveedu Gopakumar [3], Staszek Zola [4], S. Komossa [5], Tapio Pursimo [6], Jose L. Gomez [7], Rene Hudec [8,9], Helen Jermak [10] and Andrei V. Berdyugin [2]

[1] Finnish Centre for Astronomy with ESO, University of Turku, 20014 Turku, Finland
[2] Tuorla Observatory, Department of Physics and Astronomy, University of Turku, 20014 Turku, Finland; andber@utu.fi
[3] Department of Astronomy and Astrophysics, Tata Institute of Fundamental Research, Mumbai 400005, India; gopu@tifr.res.in
[4] Astronomical Observatory, Jagiellonian University, ul. Orla 171, 30-244 Krakow, Poland; staszek.zola@gmail.com
[5] Max-Planck-Institut für Radioastronomie, Auf dem Hügel 69, 53121 Bonn, Germany; astrokomossa@gmx.de
[6] Nordic Optical Telescope, Apartado 474, 38700 Santa Cruz de La Palma, Spain; tpursimo@not.iac.es
[7] Instituto de Astrofisica de Andalucia—CSIC, Glorieta de la Astronomia s/n, 18008 Granada, Spain; jlgomez@iaa.es
[8] Faculty of Electrical Engineering, Czech Technical University, Technicka 1902/2, 16627 Praha, Czech Republic; rene.hudec@gmail.com
[9] Engelhardt Astronomical Observatory, Kazan Federal University, 18 Kremlyovskaya Street, 420008 Kazan, Russia
[10] Astrophysics Research Institute, Liverpool John Moores University, Liverpool L3 5RF, UK; H.E.Jermak@ljmu.ac.uk
[*] Correspondence: mvaltonen2001@yahoo.com (M.J.V.); lankeswar.dey@tifr.res.in (L.D.)

Citation: Valtonen, M.J.; Dey, L.; Gopakumar, A.; Zola, S.; Komossa, S.; Pursimo, T.; Gomez, J.L.; Hudec, R.; Jermak, H.; Berdyugin, A.V. Promise of Persistent Multi-Messenger Astronomy with the Blazar OJ 287. *Galaxies* **2022**, *10*, 1. https://doi.org/10.3390/galaxies10010001

Academic Editor: Santanu Mondal

Received: 2 December 2021
Accepted: 19 December 2021
Published: 22 December 2021

Publisher's Note: MDPI stays neutral with regard to jurisdictional claims in published maps and institutional affiliations.

Abstract: Successful observations of the seven predicted bremsstrahlung flares from the unique bright blazar OJ 287 firmly point to the presence of a nanohertz gravitational wave (GW) emitting supermassive black hole (SMBH) binary central engine. We present arguments for the continued monitoring of the source in several electromagnetic windows to firmly establish various details of the SMBH binary central engine description for OJ 287. In this article, we explore what more can be known about this system, particularly with regard to accretion and outflows from its two accretion disks. We mainly concentrate on the expected impact of the secondary black hole on the disk of the primary on 3 December 2021 and the resulting electromagnetic signals in the following years. We also predict the times of exceptional fades, and outline their usefulness in the study of the host galaxy. A spectral survey has been carried out, and spectral lines from the secondary were searched for but were not found. The jet of the secondary has been studied and proposals to discover it in future VLBI observations are mentioned. In conclusion, the binary black hole model explains a large number of observations of different kinds in OJ 287. Carefully timed future observations will be able to provide further details of its central engine. Such multi-wavelength and multidisciplinary efforts will be required to pursue multi-messenger nanohertz GW astronomy with OJ 287 in the coming decades.

Keywords: BL lacertae objects; OJ 287; supermassive black holes; accretion discs; gravitational waves; jets

1. Introduction

Supermassive black hole (SMBH) binary systems are expected to be common in the Universe [1–7]. Inspiral gravitational waves from SMBH binaries with total mass $\gtrsim 10^9\,M_\odot$ and orbital periods between months and years should lie in the nanohertz (nHz)—microhertz range [8–10]. The routine detection of nHz gravitational waves (GWs) from SMBH binary systems is the primary goal of rapidly maturing Pulsar Timing Array (PTA) experiments [11–13]. The eventual detection of nHz GW sources will complement the vibrant field of GW astronomy, inaugurated by the routine detections of hectohertz GWs from stellar-mass black hole (BH) binaries by the LIGO-Virgo collaboration [14,15]. This collaboration also heralded the era of multi-messenger GW astronomy by the observations of hectohertz GWs from a binary neutron star merger (GW170817) and its electromagnetic (EM) counterparts (EM170817) [16,17]. The eventual detection of continuous nHz GWs from individual SMBHBs and observations of their electromagnetic (EM) counterparts can make profound contributions to the emerging field of multi-messenger GW astronomy [10,18]. EM observations in recent years have suggested the existence of many candidate binary SMBH systems with varying orbital periods (e.g., OJ 287 with 12 yr period [19]; 3C 66B with 1 yr period in the radio [20] but also see [21,22]; SDSS J1201+3003 identified from a tidal disruption event optical lightcurve [23]; PG1302-102 identified in the optical Catalina transient survey [24] see also [25]; and see, for example, Charisi et al. [26], who found 33 further quasars with statistically significant periodic variability). Some of these systems and candidates are in the nHz GW regime which should be detectable in GWs by the PTA consortium during the SKA era [27–29]. However, none of these systems had as detailed modeling and testable predictions as OJ 287, and no lightcurve has had the very long-term coverage that was obtained for OJ 287.

Persistent multi-wavelength electromagnetic observations will be crucial to pursue multi-messenger GW astronomy with OJ 287 in the coming decades. In what follows, we provide a brief introduction to the binary black hole (BBH) central engine description for OJ 287 and discuss the context and the need to pursue a variety of electromagnetic observations. These efforts should allow us to place tight observational constraints on the various aspects of the BBH central engine description for OJ 287 that will be crucial for pursuing multi-messenger GW astronomy with OJ 287. We now describe briefly our BBH description for OJ 287 and ongoing multi-wavelength observational campaigns on the source.

1.1. OJ 287 and Its BH Binary Central Engine Description

The bright blazar OJ 287 (RA: 08:54:48.87, Dec: +20:06:30.6), situated at a redshift of $z = 0.306$, was first discovered as a radio source during the course of the Ohio sky survey in the 1960s [30]. However, the optical light curve of OJ 287 goes all the way back to the year 1888 as it had been unintentionally photographed often in the past due to its proximity to the ecliptic plane [19,31]. Interestingly, the optical light curve of OJ 287, extending over 130 years, exhibits unique quasi-periodic high-brightness flares (outbursts) with a period of $\sim$12 years [19,32,33]. Further, the blazar shows a long-term variation in its apparent magnitude with a period of $\sim$60 years [32] (see Figure 1).

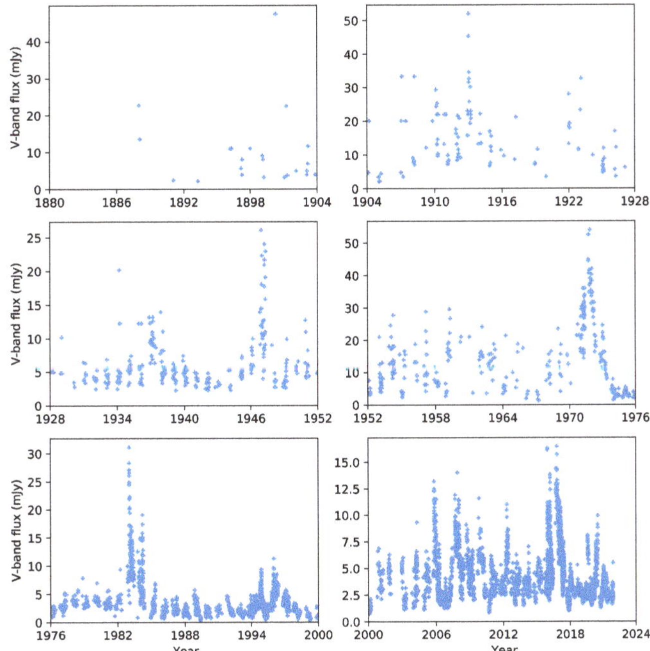

Figure 1. Optical lightcurve of OJ 287 between 1888 and 2021. The figure includes a large number of previously unpublished data points by two of the authors (R.H., S.Z.). All point have been transformed to the V-band, including data in Laine et al. [34].

In 1982, the binary nature of OJ 287 was recognized by Aimo Sillanpää, while constructing historical light curves for the quasars in the Tuorla–Metsähovi variability survey which had begun two years earlier [35]. It was obvious that the next major outburst was about to happen and the blazar monitoring community was alerted. This first OJ 287 campaign was fruitful [36,37]. A more detailed explanation of the binary BH (BBH) scenario was subsequently proposed to explain the quasi-periodic flares in OJ 287 by associating them with increased accretion flow to the primary BH. This was due to the accretion disk perturbations of the secondary BH during its pericenter passages in an orbit which lies in the disk plane [19]. Moreover, a notable prediction of the model was that OJ 287 should show an outburst again in the autumn of 1994. This was verified by the second campaign called OJ-94 [38].

In 1994, Lehto & Valtonen [39] recognized that the flares in OJ 287 were not exactly periodic, and that the systematics of the past flares are better understood if the flares are double-peaked and the two peaks are separated by ∼1–2 years. This led to the proposal of a new SMBHB central engine model (LV96 model hereafter) for OJ 287 where a supermassive secondary BH is orbiting the more massive primary BH in a relativistic eccentric orbit with a redshifted orbital period of ∼12 years and where the orbital plane is inclined with respect to the accretion disk of the primary at a large angle. In this model, the double-peaked flares occur due to the impacts of the secondary BH on the accretion disk of the primary BH twice during each orbit. The third campaign carried out by the OJ-94 group verified the flare on October 1995. It came within the narrow two-week time window of the prediction [40,41].

Each impact releases hot bubbles of gas from both sides of the disk which then expand and cool down until they become optically thin [39,42]. At this point, bremsstrahlung radiation from the entire volume becomes observable causing the optical brightness of the blazar to increase rapidly [43]. The resulting flare can last for a few weeks and should reach its peak within the course of a few days.

The LV96 BBH central engine model of OJ 287 with subsequent fine-tuning of the orbital and disk parameters has been very successful in explaining the past observed optical flares, including many which have been discovered in historical data since 1994. See Dey et al. [33] for a collection of flare light curves. Moreover, this model has accurately predicted the time of arrival of future flares expected during the years 2005, 2007, 2015, and 2019 which triggered new observing campaigns [32,33,44–46]. The 2005 flare came a week ahead of the orbital schedule calculated by Sundelius et al. [44] and corrected for the astrophysical time shift between the disk impact and the flare by LV96 [47,48]. The latest prediction of the 2019 *Eddington flare* was successfully observed with the *Spitzer space telescope* [34] which showed that the orbital times are now known with the accuracy of ± 4 h.

The model also predicted that the excess emission during the impact flares is unpolarised, and it has the bremsstrahlung spectrum at $\sim 3 \times 10^5$ K. At the time (in the spring of 1995 when the model was constructed) it was only known that during the rapid brightening of the 1983 flare the polarisation decreased with rising flux: the first point at ~ 16.5 mJy in the V-band had polarisation of 8.2% while the peak at ~ 33 mJy had polarisation of 0.8% and the four points measured around the peak had the average polarisation of 4% [36]. This is exactly what is expected if the excess emission is unpolarised. The same trend was seen again in 2007 and 2015 [49,50]. In 1984, 1994, 1995, 2005 and 2019 the polarisation measurements were not obtained for the primary peak, partly due to seasonal and weather restrictions.

The spectral slope over a wide range of frequencies was measured during the 2005 flare and the expected spectral index of $\alpha_\nu \sim -0.2$ was found. It is quite different from the usual spectral index of $\alpha_\nu \sim -1.3$ over the same spectral range [51,52]. The spectral index $\alpha_\nu \sim -0.2$ was confirmed during the 2015 and 2019 flares. The complete change in the nature of emission in the flare as compared with the usual out-of-flare emission excludes the possibility that the flares are somehow related to Doppler boosting variations in a turning jet [53,54].

As an example, the 2015 flare was observed by the Swift observatory [55]. The first observation prior to the flare on JD 2457354 found the flux 1.92 mJy in the UV channel M2. Almost simultaneously it was observed in several optical bands, including the V-band, where the flux was recorded as 5.67 mJy [56]. The highest M2 flux value in this set of Swift observations was recorded on JD 2457360.65: it was 5.55 mJy, while the nearly simultaneous V-flux (about 0.5 d earlier) was 10.0 mJy [57]. Our R-band magnitude 13.36 on JD 2457360.65 is in agreement with this V-flux. We take the first observations as the base level, and obtain the flare contributions in M2 and V-bands of 3.63 mJy and 4.33 mJy, respectively. The spectral index between these two wavebands becomes $\alpha_\nu \sim -0.2$, the bremsstrahlung value for the temperature in question.

The observation of the bremsstrahlung optical flare during early September 2007 confirmed the highly-relativistic nature of the binary BH in OJ 287. This is because the BBH model would have predicted the arrival of OJ 287's impact flare on Earth around early October 2007 if the effects of GW emission had been ignored while describing the relativistic trajectory of the secondary BH in the blazar [46,49]. Thereafter, the observation of the *GR centenary flare* during November-December 2015 was used to constrain the dimensionless Kerr parameter of OJ 287's primary BH [33,50]. The updated BBH parameters of the OJ 287 system, determined using post-Newtonian (PN) equations of motion, are given below: primary BH mass $m_1 = 18.35 \times 10^9 M_\odot$, secondary BH mass $m_2 = 150 \times 10^6 M_\odot$, primary BH Kerr parameter $\chi_1 = 0.38$, orbital eccentricity $e = 0.657$, and orbital period (redshifted) $p = 12.06$ years [33]. We now describe the various EM observational campaigns that allowed us to extract additional astrophysical details of the BBH central engine description for the blazar.

1.2. Multi-Wavelength Emission of OJ 287 and Previous and Ongoing Monitoring Campaigns

OJ 287 has been the target of intense multi-wavelength coverage and monitoring in recent decades. Early optical observations first revealed very bright high-states reaching up to ∼12 mag. Dedicated optical monitoring campaigns [58–60] and the addition of photographic plate data dating back to the late 19th century [31] then revealed that bright optical maxima recur semi-periodically. These observations led to first suggestions of the presence of a binary SMBH in OJ 287 [19] and its later confirmation in subsequent campaigns, as mentioned above [48,61]. An ongoing optical monitoring program of OJ 287 combines optical data taken at multiple ground-based observatories and provides coverage at sub-daily cadence except for epochs when OJ 287 is unobservable due to its proximity to the Sun [33,62]. Figure 1 shows the optical lightcurve of OJ 287 from late 19th century to the present epoch. Based on the latest optical high-state discovered in this program in late 2015 and interpreted as a binary impact flare [50], a new broad-band radio—X-ray monitoring program of OJ 287 was initiated and is described below. Most of the time, the optical emission of OJ 287 is dominated by synchrotron emission from the jet, as known from SED modeling [63] and high levels of polarization [64]. However, it shows a decrease in polarization, expected in the binary SMBH model, at epochs where an additional thermal bremsstrahlung contribution from the disk impacts contributes to the emission, for example, [49]. Optical polarimetric monitoring of OJ 287 was carried out during the epoch 2016–2017 following changes in polarization and led to the interpretation of the 2016–2017 optical high-state as a binary after-flare [61].

Based on the faintness of its optical Balmer lines from the broad-line region (BLR; [65,66]), OJ 287 is classified as a BL Lac object. Given its spectral-energy distribution (SED), it is of LBL type (low-frequency-peaked blazar; [67]). Optical emission lines from the narrow-line region (NLR) like [OIII]λ5007 are only faintly present [66].

As a blazar, OJ 287 is a bright radio emitter [30] with a structured, relativistic jet [68–70] that is highly polarized [71]. On average, the jet is pointing close to our line of sight [68,72], but it shows large swings in its position angle [73,74] that can be understood in the context of the binary SMBH model [75].

OJ 287 was first detected in the γ-ray regime with the CGRO/EGRET satellite observatory [76] and shows repeated flaring in the Fermi γ-ray band for example, [63,72,77] along with more quiescent epochs not detected by Fermi. The first very-high energy (VHE; $E > 100$ GeV) detection was obtained in 2017 with VERITAS [78]; an observation triggered by the measurement of an exceptionally bright X-ray outburst (Figure 2) discovered with Swift in the course of the dedicated long-term, multi-frequency monitoring program MOMO (Multiwavelength Observations and Modelling of OJ 287; [79,80]).

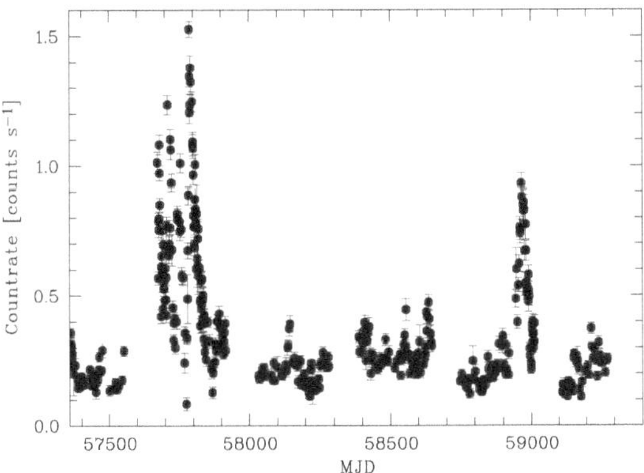

Figure 2. X-ray (0.3–10 keV) lightcurve of OJ 287 obtained in the course of the MOMO program with Swift between 2015 December and 1 March 2020. Two bright outbursts were discovered in 2016–2017 and 2020. Adopted from Komossa et al. [81].

OJ 287 is a bright and variable X-ray emitter observed with most major X-ray observatories and first detected with the Einstein satellite [82]. The X-ray spectrum of OJ 287, based on XMM-Newton observations between 2005 and 2020 (energy range: 0.3–10 keV), is characterized by a super-soft (low-energy) synchrotron component and a hard (high-energy) inverse Compton (IC) component [83]. The synchrotron component dominates the X-ray outburst states [80]. High-energy emission up to 70 keV was detected with NuSTAR with an unusually soft power-law spectrum of photon index $\Gamma_X = 2.4$ accompanying the 2020 outburst [80]. OJ 287 shows an X-ray jet extended by $20''$ [84].

Starting in late 2015, OJ 287 has been the target of the dedicated multi-wavelength monitoring program MOMO [79–81,83,85] (see the review by Komossa et al. [86]). MOMO is aimed at characterizing the blazar properties and facets of the binary SMBH in detail. The program employs multiple frequencies in the radio band between 2–40 GHz obtained with the Effelsberg observatory, UV–optical Swift data in 6 filters, and the Swift X-ray band (0.3–10 keV). The monitoring cadence is as short as 1–4 days. Additional deeper observations with multiple ground and space-based observatories are triggered in the case of exceptional outburst or low-states discovered with Swift and/or in the radio band. The community is alerted in form of *Astronomer's Telegrams* about particular states of OJ 287 (ATel #8411, #9629, #10043, #12086, #13658, #13702, #13785, and #14052). MOMO is the densest multi-wavelength monitoring program of OJ 287 involving X-rays, with >1000 datasets so far taken, providing spectra, broad-band SEDs, and lightcurves, during all activity states of OJ 287 [81,86].

Some of the main results of the program MOMO are as follows: Two bright outbursts were identified in the course of MOMO in 2016–2017 and 2020 (Figure 2); the brightest so far recorded in X-rays [79–81]). These outbursts were non-thermal in nature, and they are consistent [80] with after-flares predicted by the binary SMBH model, when disk impacts of the secondary lead to the triggering of new jet activity of the primary SMBH. OJ 287 has turned out to be one of the most spectrally variable blazars in the X-ray regime (Figure 3) with power-law spectral indices as steep as $\Gamma_X = 3$ [81,83]. Supersoft synchrotron emission dominates the outburst states. Another key features of the MOMO lightcurve is a pronounced, symmetric UV–optical deep fade in November–December 2017 (Figure 3) of still unknown nature [80,81]. Structure function analyses of Swift data between 2015–2020 has revealed characteristic break times of 4–39 days depending on activity state and waveband [81]. Discrete correlation analysis between UV, optical and X-ray bands during the same epoch shows lags and leads in X-rays w.r.t. the UV, attributed to an IC component, where external Comptonization is favored over synchrotron-self-Compton emission [81,86].

The MOMO program continues as OJ 287 nears its next predicted binary SMBH impact flare. Results obtained so far including >1000 broad-band SEDs will serve as a useful data base to identify new activity states of OJ 287 driven by blazar variability in comparison to binary-driven activity, and allow detailed modelling of the broad-band emission from the radio to the high-energy regime at each epoch. We now take a closer look at the present description of the primary BH accretion disk and propose a detailed scenario to describe the outcome of secondary BH impact with the disk.

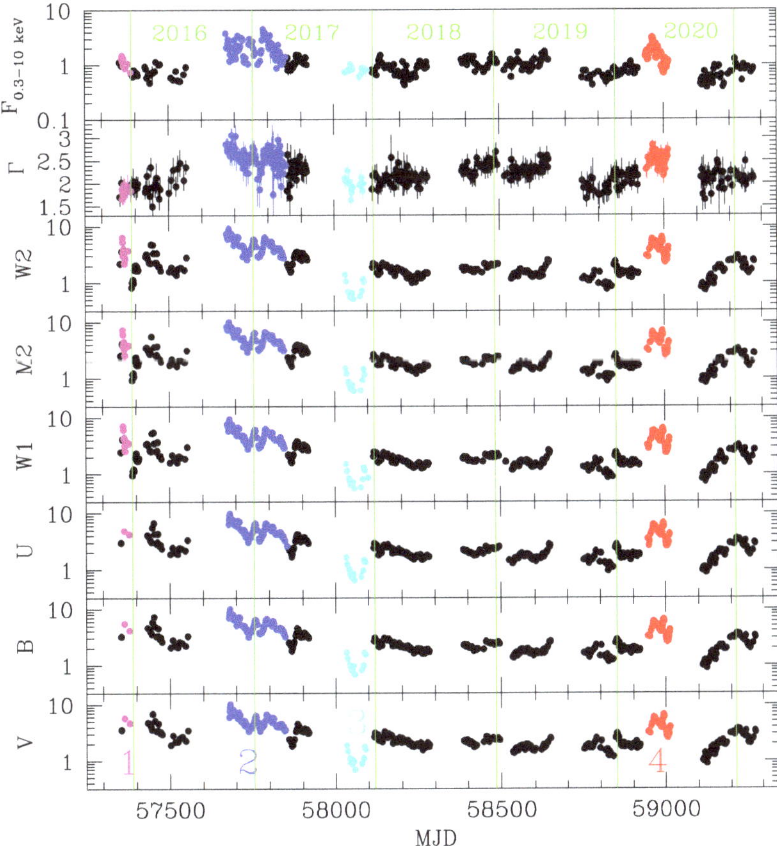

Figure 3. Multi-wavelength X-ray–optical lightcurve of OJ 287 taken with Swift since December 2015. The majority of data was obtained during the MOMO program. Panels from top to bottom: Observed X-ray flux in the band (0.3–10) keV in units of 10^{-11} erg/s/cm^2, X-ray powerlaw photon index Γ_x, Swift UVOT UV and optical fluxes (corrected for Galactic extinction) in units of 10^{-11} erg/s/cm^2. Every year, observations are interrupted by $\sim$3 months when OJ 287 is in close proximity to the Sun such that Swift cannot observe. Selected epochs are marked in color: (1) The epoch near the 2015 binary impact flare, (2) the bright 2016–2017 outburst, (3) the 2017 optical-UV deep fade, and (4) the 2020 April outburst. Epochs (2) and (4) were interpreted as possible binary after-flares. Epochs like the deep fade in 2017 (feature 3) and the new low-state in September 2020 can be used to obtain host galaxy imaging of OJ 287 when the blazar emission itself is least disturbing. Adopted from Komossa et al. [81].

2. Accretion Disk in OJ 287

Our binary model of OJ 287 is based on the interpretation of the huge optical flares as impacts of a secondary black hole on accretion disk of the primary. While the emission level of OJ 287 is between 1 and 2 mJy in the optical V-band most of the time, the emission rises to 13 mJy (in the 2015 flare) or similar values in the peaks that we interpret as impact flares, that is, they represent an order of magnitude increase in the optical flux. The increase occurs on the time scale of a day or so which is about a fraction of 0.0002 of the 12 yr (9 yr redshift corrected) binary orbital period. This implies that the emission region in the source responsible for these huge flux variations, cannot be greater than one light day in size. Most of the work on the binary black hole model has concentrated on the orbit of the secondary.

However, it is equally important that we also find out the basic properties of the accretion disk from the way the disk reacts to the impacts, and how these one-light-day emission regions are activated.

In the original binary black hole model of Lehto & Valtonen [39] the disk was taken as a standard disk with two parameters, viscosity α and accretion rate $\dot{m}$. Fixed values were given for these parameters. When more impacts were observed later, both in the future and in the past historical light curve, it became possible to solve the values of these parameters together with the orbit solution. Altogether it is a 9 parameter problem which has a unique solution as soon as the exact times of 10 flares are known. Mathematically it is equivalent to a ninth order equation which has nine unique roots. The method of solution is iterative, similar to the Newton-Raphson algorithm [46]. From the detailed study of the 2015 flare, the values of the two standard parameters were determined to be $\alpha = 0.26$ and $\dot{m} = 0.08$ in Eddington units.

In the model of Lehto & Valtonen [39], a spherical bubble of hot plasma is released from the disk by the impact. Ivanov et al. [42] later demonstrated that actually two bubbles are released, one on each side of the disk. Therefore each impact is detectable via a radiating bubble on the front side of the disk. The model further specified that the bubble expands isotropically and therefore retains its spherical shape. The simulation of Ivanov et al. [42] casts doubt on whether this is true at an early stage of the bubble, but this was not considered important, as the bubble becomes optically thin only at a later time, which results in the observed great increase in brightness.

Valtonen et al. [87] noted that the maximum brightness of the 2015 flare should have been 26 mJy in the standard model. The fact that the peak value was only half of this value is most easily explained if the bubble is compressed along the line of sight. Thus it appears necessary that the expansion of the bubble is not completely unconstrained, but that it is influenced possibly by the general magnetic fields in the disk. In that case, it would be interesting to know what happens to the flux of the bubble from the very beginning if its expansion is constrained all the way. From Ivanov et al. [42] one could conclude that initially the expansion is more jet-like, a flow perpendicular to the disk rather than a gentle release of bubbles. Ivanov et al. [42] called them fountains. In the following we consider what difference this makes to the impact flare light curve.

3. Anisotropically Expanding BH Impact Generated Bubbles

According to Lehto and Valtonen (LV in the following, [39]), the compressed plasma inside the accretion disk, following the impact, bursts out of the disk with the speed $v_{perp} = (6/7)v_{sec}$ (LV Equation (7)) and the time to come out of the disk is $\tau_{dyn} = h/v_{perp}$ (LV Equation (8)), where v_{sec} is the speed of the secondary black hole at the impact and h is the semi-height of the disk. The flux from this bubble follows $S \sim \epsilon^{ff} V_{bubble}$ (LV Equation (14)), if the bubble is fully transparent. The quantities ϵ^{ff} and V_{bubble} are the free-free emissivity and the volume of the bubble, respectively. However, we see only the fraction $1/\tau$ of the bubble, where τ is its optical depth. Here $\tau = d\epsilon^{ff}T^{-1}$, if d is the geometrical depth and T the temperature of the bubble. Therefore the flux goes as

$$S_{seen} \sim \epsilon^{ff} V_{bubble}/(d\epsilon^{ff}T^{-1}) . \tag{1}$$

We can substitute $T \sim V_{bubble}^{-1/3}$ in adiabatic expansion which means that

$$S_{seen} \sim V_{bubble}^{2/3}/d . \tag{2}$$

We can put $V_{bubble} \sim r^2 d$, where r is the bubble radius ($d \sim r$ in spherical geometry). The adiabatic expansion speed $v \sim T^{1/2} \sim V_{bubble}^{-1/6}$ and

$$d \sim vt \sim V_{bubble}^{-1/6} t . \tag{3}$$

Now, we have to specify the geometry of the bubble: For a tube of fixed radius seen end-on, $V_{bubble} \sim dr^2$, $r = constant$, or $S_{seen} \sim d^{-1/3}$, and since $d \sim d^{-1/6}t$, it follows that $d \sim t^{6/7}$. Thus

$$S_{seen} \sim t^{-2/7}. \tag{4}$$

Flux drops with time roughly as $S_{seen} \sim t^{-1/3}$.

For a tube of fixed depth seen end-on, $V_{bubble} \sim dr^2$, $d = constant$ or $S_{seen} \sim r^{4/3}$, and since $r \sim r^{-1/3}t$, it follows that $r \sim t^{4/3}$. Thus

$$S_{seen} \sim t^{16/9}. \tag{5}$$

Flux increases with time nearly as $S_{seen} \sim t^2$ (cf. Figure 2 in [87])

For standard spherical bubbles of radius r the corresponding relation is [43]

$$S_{seen} \sim V_{bubble}^{2/3}/r \sim r \sim t^{2/3}. \tag{6}$$

Table 1 lists the disk impacts between 1971 and 2031. The first two columns come from [33], the numbers from the third column onwards are derived in the next section.

Table 1. Times of disk impacts and bremsstrahlung flares as well as the radii of the H_α bubbles, impact speed on the disk, the delay of the H_α flare and the time of the H_α flare

Impact Time	Time of Flare	$R_{H\alpha}$ (AU)	v_{rel}/c	Delay (yr)	$H\alpha$ Flare
1971.11	1971.13	2860	0.34	0.72	1971.83
1972.71	1972.93	3930	0.20	1.66	1974.37
1982.83	1982.96	3160	0.30	0.81	1983.74
1984.07	1984.12	3250	0.28	0.91	1984.98
1994.48	1994.59	3610	0.23	1.31	1995.79
1995.82	1995.84	2940	0.33	0.76	1996.58
2005.11	2005.74	4360	0.16	2.31	2007.42
2007.68	2007.69	2740	0.36	0.65	2008.33
2013.48	2015.87	5000	0.12	3.52	2016.80
2019.56	2019.57	2740	0.36	0.65	2021.21
2021.92	2022.55	4360	0.16	2.31	2024.23
2031.40	2031.41	2900	0.33	0.74	2032.14

In the spring of 2005, there was a major flare, 6.6 mJy at 2005.12, one of the biggest deviations from the standard binary black hole model during years 1996–2010 [44,45,88]. Figure 4 shows the observations (monthly averages) from 2005, prior to the main flare in October of that year. The disk impact occurred at 2005.11 (9 February 2005). With a time translation such that 3 December 2021 corresponds to 9 February 2005, we generate a predicted light curve for the 2021/22 season. For the values appropriate for 2005 and 2021 impacts, which take place at the same radial distance from the center, $\tau_{dyn} \sim 20$ days. The flaring related directly to the disk impact should start around 23 December 2021, and the flares should be recognizable as bremsstrahlung flares in the optical—UV region, with no counterpart in radio nor X-rays. The on-going and planned observational campaigns should be able to provide excellent constraints on our detailed modeling. So far, the period of early December 2021 was covered with Swift in the ongoing MOMO project and with ground-based optical observations. The optical–UV low-state detected in 2–14 December implies a small shift in the onset of the flare (work in preparation).

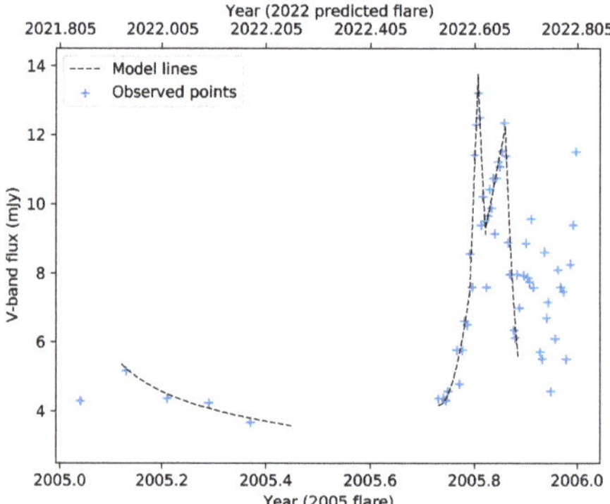

Figure 4. The optical light curve of OJ 287 during 2005 in V-band mJy units. Plot shows the monthly averaged data points for the early part of 2005 while the latter part, with faster variability, has 3 day averaged data points. The dashed curve is the theoretical line based on present work and Valtonen et al. [87]. Additionally, the theoretical curve provides the expected light-curve of OJ 287 during 2021–2022 and we mark its epochs on the upper x-axis while employing the time shift of the SMBH binary central engine description Dey et al. [33]. This is motivated by the fact that the secondary BH impact distances from the center should be roughly the same during 2005 and 2021. The first big flare during 2022 is not expected to be visible from any Earth-based facility while the secondary peak may become barely visible by early September 2022.

We now move on to explore other observational implications of the expanding impact bubble. This requires an extension of LV96 [39], where the brightness evolution of the bubbles was terminated at the epoch when the bremsstrahlung flares emerge. However, at a later stage when the bubbles have cooled enough, they may become visible again as hydrogen line flares. In what follows, we study the timing of these proposed $H\alpha$ flares, and connect them with one epoch when the hydrogen lines were seen to be exceptionally bright, by a factor of ten above the normal.

4. Possible Spectroscopic Implications of the Evolving BH Impact Generated Bubbles

After the bremsstrahlung flare the radiating bubble cools until it falls below the jet continuum level and is not observed any more [87]. However, at a later time, the bubble has cooled enough to become an HII region. This may appear as a sudden increase of the strength of the Balmer lines for a limited period of time. At the end of 1984, Sitko & Junkkarinen [65] observed such an increased line flux, approximately a factor of ten higher than normal [66]. To our knowledge, this was a unique occasion when such an $H\alpha$ flare has been seen. Therefore it is tempting to associate this epoch with the dying moments of one the bubbles that had generated a bremsstrahlung flare at an earlier time.

In order to estimate when the bubbles left over from disk impact may reach the $H\alpha$ stage, we calculate the required increase in the bubble radius until its temperature has dropped to 10^4 K. Using Table 3 of Dey et al. [33] and Table 1 of Valtonen et al. [87], and assuming an adiabatically expanding bubble, we obtain the radii of the bubbles listed in column 3 of Table 1. In Valtonen et al. [87] the expansion speed of the bubble is given as $\sim v_{rel}/4$ where v_{rel} is the speed at which the secondary impacts the accretion disk. In fact, the expansion speed value $v_{rel}/4.6$ is used in the following since it gives the correct timing for the 1984 $H\alpha$ flare, 1984.98 [65]. This value is well within the uncertainty of the expansion speed, extracted from the numerical simulations by Ivanov et al. [42]. These values are given in column 4 of Table 1, while column 5 gives the time of the expansion up to this stage. In column 6 we give the estimated time of the beginning of the $H\alpha$ flare.

The epochs of observation in the line monitoring program of Nilsson et al. [66] and in the program by some of the present authors at the *Nordic Optical Telescope* (T.P.) as well at the *Rozhen Telescope* of Bulgarian Academy of Sciences (S.Z.) are given in Table 2.

Table 2. The epochs of the line monitoring program.

Time	Telescope	Strong Lines?	Secondary Line
2005.26	VLT	no	(8059)
2005.89	VLT	no	(8092)
2006.25	VLT	no	(8059)
2006.93	VLT	no	(8637)
2007.26	VLT	no	(9405)
2008.01	VLT	no	[7418]
2008.27	VLT	no	[7787]
2010.12	NOT	no	6438
2010.81	NOT	no	6494
2010.88	NOT	no	6499
2011.01	NOT	no	6506
2011.05	NOT	no	6508
2011.11	NOT	no	6512
2011.31	NOT	no	6521
2011.35	NOT	no	6522
2011.75	NOT	no	6538
2011.88	NOT	no	6542
2012.14	NOT	no	6549
2012.99	NOT	no	6564
2013.18	NOT	no	6566
2015.94	Rozhen	no	(8273)
2016.26	Rozhen	no	(8300)
2021.82	NOT	no	6708

The expected flares in the time span and close to these observations were at 2007.42, 2008.33, 2016.80 and 2021.21. In the first two cases, the observational epochs were too early by 0.16 and 0.05 yrs, respectively, in order to catch the beginning of the flares. The last two were too early by 0.54 yr and too late by 0.61 yr. Therefore, if the $H\alpha$ flares started as expected, we would not have seen them, even if the flares lasted several months.

In order to calculate the line flux, we proceed as follows. The emissivity of the Balmer $H\beta$ line may be taken as

$$\epsilon_{H\beta} = 8.3 \times 10^{-26} \, n^2 \, \text{erg cm}^{-3} \, \text{s}^{-1}, \tag{7}$$

where n is the number density of Hydrogen atoms. We calculate the line flux using the density for the 1984 line flare, $n = 3.2 \times 10^8$ and the bubble radius 3250 AU from Table 1. The result is

$$S_{H\beta} = 2.2 \times 10^{-15} \, \text{erg s}^{-1}. \tag{8}$$

Sitko & Junkkarinen [65] find $S_{H\beta} \sim 1.2 \times 10^{-15}$ erg s^{-1}, but state the value as uncertain. The more certain $S_{H\alpha} = 24 \times 10^{-15}$ erg s^{-1}, divided by some number of order of 10 [65] takes us close to the above estimate. The radius and the number density are different in 2024 than in 1984, but the different factors practically cancel each other. Therefore we expect the 1984 $H\alpha$ line flare to repeat itself in 2024.

5. Explaining Observed Gamma Ray Flares

Here, we will distinguish the classical gamma-ray regime, as observed by the Fermi satellite, with many gamma flares detected (typically in the $E \sim 100$ MeV regime); and the VHE emission ($E > 100$ GeV) that this Section mostly focuses on. At VHE, OJ 287 was only detected once, by VERITAS. The VERITAS observation was triggered by the Swift detection of an exceptional X-ray high-state in the course of the MOMO program [79,80]. Otherwise, OJ 287 is monitored in gamma-rays with Fermi, with many large and small flares.

Classical gamma ray flares have been reported many times in OJ 287 [77,78]. For example, the 2009 classical flare coincides with the last of the predicted tidal flares following the 2007 bremsstrahlung flare [45] while the 2017 VHE flare came soon after the last of the predicted tidal flares following the 2015 bremsstrahlung flare [89]. We can predict the next VHE gamma ray flare following the same principles.

In order to study the orbital phase of the 2017 VHE flare in our model, we have calculated the response of the OJ 287 accretion disk to the 2013 secondary crossing, using a disk of 146,880 particles. The particles leaving the disk are counted, and the time (*time*1) and distance (r) from the center are noted down. Assuming a constant transit speed to the jet (taken to be at $r = 3000$ AU), the jet reaction time (*time*2) is calculated. Figure 5 corresponds to the transit speed of $c/6$ in the observers frame which equals to $0.22c$ in the OJ 287 frame. This transit speed value fits best with the two observed light curve peaks at 2016.2 and 2016.8. The present calculation differs from Pihajoki et al. [89] in that the surface density of the disk follows Lehto & Valtonen [39] rather than a uniform density which was used in Pihajoki et al. [89]. Pihajoki et al. [89] also did not take the transit time into account, so that it would have to be done collectively for all particles for a comparison with observations. Note that the second peak becomes stronger when a realistic surface density model is used, in agreement with observations.

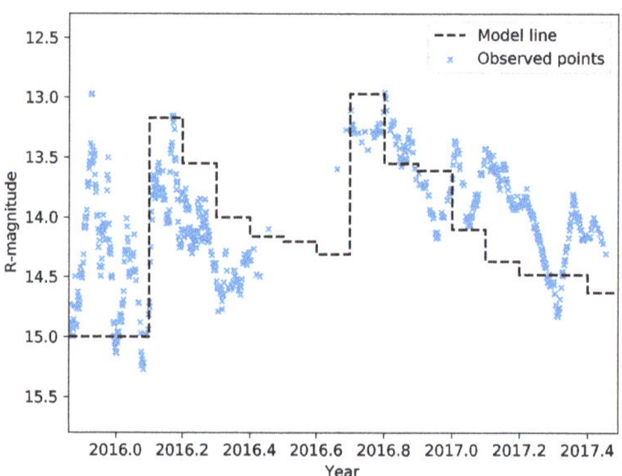

Figure 5. Simulation of the response of the accretion disk to the crossing of the secondary black hole through the disk before the 2015 bremsstrahlung flare. Particles which escape the disk are counted, and they are placed on the time axis assuming that the particle contributes to the light curve after traveling the distance of ~15,000 AU with a constant speed of $0.22c$ where c is the speed of light. The exact value of travel distance depends on the radial distance from the center where the particle is released from the disk. The particle count in each time box (*time*2) is turned into a magnitude, assuming that the relation between the particle number and the total brightness in each box is linear. The first flare in observations at the end of 2015 is different (impact flare) and is not modeled by this process.

The tidal flares of the light curve of OJ 287 following the 2005 and 2022 bremsstrahlung flares were calculated by Pihajoki et al. [89]. No time transformation was applied so that the plots in Figure 4 of Pihajoki et al. [89] are for *time*1, as defined above. Using the same transit speed of particles to the jet as above we should add 1.0 yr to the time axis. For the 2005 impact the two main flares in the predicted light curve are at 2007.4 and 2008.6. Even though these are placed at the summer period of few or no observations (OJ 287 is close the sun at the beginning of August), a strong brightening of the source was observed in the following autumn (both 2007 and 2008) which agrees with the tidal flare explanation. The interpretation of the light curve is made complicated by the fact that the bremsstrahlung flare of 2007.7 falls in the same time interval as the tidal flare [48].

As to the 2022 case, the predicted light curve has maxima at 2023.5 and 2024.2 [89]. Note that the nature of the simulation by Pihajoki et al. [89] is such that the tidal flares appear to get weaker with time. This is an artifact of the simulation technique which does not replace the escaped particles by new ones in the disk. Because of practically the same impact distance in 2005 and 2022, the tidal responses should be rather similar. We therefore expect that VHE gamma rays would again be detectable in the spring of 2024. The prediction cannot be more precise until we see the corresponding flare starting in the optical region. Further, we expect to see such flare again in the X-ray regime, as in 2017 when the very bright X-ray flare detected with Swift triggered the VHE observation. Given the ongoing monitoring program with Swift with very dense coverage and >1000 SEDs already obtained for example, [80,81], very detailed timing and comparison with the 2017 flare will then be possible. This ongoing project, or any X-ray monitoring with future X-ray missions, can then also be used to trigger new VHE observations.

In addition, OJ 287 observations can reveal interesting astrophysical information when it enters low luminosity state and this is what we focus on in the next section.

6. Optical Multicolor Photometry at Minimum Light

From time to time, the brightness of OJ 287 falls to a very low light level. It is hard to predict when these times will happen again, even though it is thought that the low light level may be associated with the secondary moving in front of the primary [90]. The absolute minimum light observed so far took place in the spring of 1989. In Figure 6 we re-plot the observations [90], and fit smooth curves through them. When the brightness of the AGN falls towards the brightness of the host galaxy, the color of OJ 287 is expected to change gradually from the typical jet color to the color of the host galaxy as the fade proceeds. Takalo et al. [90] conclude that at the brightness level V = 17.4, the host galaxy contributes at least half of the brightness of the source. Using the usual conversion V − K = 3.2, K-correction and evolution correction of 0.6 and the distance modulus m − M = 41, this implies the absolute K-band magnitude of $M_K = -26.8 \pm 0.3$ [91].

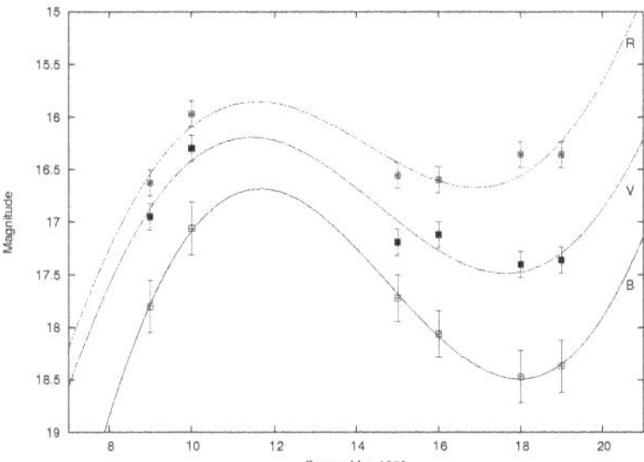

Figure 6. The observations of the 1989 fade of OJ 287 with the *Nordic Optical Telescope* and the *Jacobus Kapteyn Telescope* at La Palma, with 2σ errorbars. Here we plot the magnitudes in three filters B, V and R: the I band data were too few to show a significant variation. The lines are fifth order polynomial fits through the data points. We note that the dip on 18 May was deepest in the B band and less deep going to longer wavelengths. There is a possible time delay in the sense that the dip starts from longer wavelengths and happens later at shorter wavelengths. At 37 GHz the minimum took place on 3 May 1989.

Using surface photometry in the K-band with the *Nordic Optical Telescope* Nilsson et al. [92] find $K = -14.65 \pm 0.6$, leading to $M_K = -25.85 \pm 0.6$ with the K-correction -0.5, which is somewhat, but not significantly, fainter than the value measured by the color method. However, this may not be the total brightness of the host. It is well known that galaxies of this brightness category have extensive haloes which contribute as much as one magnitude to the host. For example, the absolute K-magnitude of NGC 4889 is $M_K = -25.4$ [93] (using m − M = 34.7 and V − K = 3.2) measured out to roughly the same 30 kpc linear distance as the surface photometry of OJ 287 in Nilsson et al. [92] while its total magnitude out to 100 kpc distance is about $M_K = -26.8$ [94]. Typically the outer envelope contributes about 0.8 magnitudes to the total brightness [95–97]. Actually, if taken to the redshift of $z = 0.3$, NGC 4889 would probably appear very much like the OJ 287 host. Even the estimated mass of its central black hole is practically the same as the primary BH mass in OJ 287 [94].

We also note that the color photometry method is more likely to capture the faint halo light than the surface photometry method since the latter is seriously influenced by the high background light level generated by the fringes of the strong point source in the center [98]. The signal-to-noise ratio is better at longer wavelengths than in the traditional optical range which may be the reason why it is difficult to get a host magnitude measurement in the shorter wavelength bands below the infrared K-band [92]. For example, the *Hubble Space Telescope* image gives the range from $M_K = -24.7$ to $M_K = -26.3$, converted from the I-band to the K-band in the usual way [98]. The magnitude of the i-band image taken at *Gran Telescopio Canarias* corresponds to $M_K = -25.2 \pm 0.3$ [92].

It is interesting to see where this magnitude by multicolor method places OJ 287 in the black hole mass—host galaxy mass diagram. Considering that OJ 287 is at a much higher redshift than galaxies normally displayed in this diagram [99] and that the correlation is redshift dependent [100], we find that OJ 287 follows almost exactly the well established correlation (Figure 7).

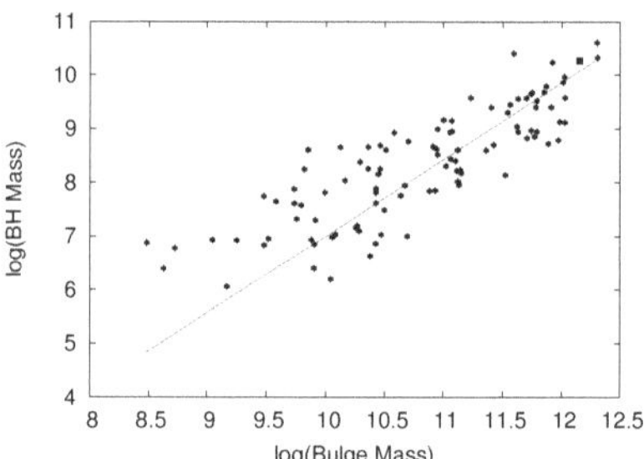

Figure 7. The galaxy sample of Saglia et al. [99] with three additional galaxies in the black hole mass vs. galaxy stellar mass diagram (stars), while the position of OJ 287 is given as a solid square. The dashed line represents the correlation found by Portinari et al. [100] for galaxies in the upper-right hand part of the graph around redshift 0.5.

When do we expect to be able to repeat the host galaxy brightness measurement by the photometric method? It depends on the reason behind the fade. At the time, in 1989, the secondary was near apocentre of its orbit, and the major axis of the orbit was close to perpendicular to the disk. This configuration minimizes the tidal perturbation of the disk by secondary. The next time a similar configuration arises is around 2032 (see Figure 8).

Takalo et al. [90] suggests that during the fade the secondary moves between us and the primary, and this causes a temporary misalignment of the jet, leading to decreased Doppler boosting that shows up as a fade. This is a sensible proposition as we may find as follows: Suppose the secondary passes within 100 AU of the jet. Then the jet is gravitationally deflected by ∼4°. Using the parameters from Hovatta et al. [101], and adding 4° to the viewing angle, we find that the jet becomes fainter by about 3 magnitudes. That is, if we have the normal V-band magnitude ∼15, in the fade it may drop to ∼18 which is actually what happened during the 1989 fade. We cannot perform a more detailed calculation since we do not know the inclination of the orbit of the secondary. This proposition could be studied further, as the jet swinging back and forth should appear also in radio maps, with the proper delay for each frequency, as well as in optical polarisation with a small delay. A comparison of optical and radio light curves at the fades should also establish the relative positions of the radio and optical cores in the jet. The 1989 fade suggests that they lie close to each other.

We may ask when this configuration repeats itself. Figure 8 and its caption lists such times. Since the optical emission region is in the jet a little way from the center, the times given by the model are about half-a-year earlier than the response in the optical emission. The 1998 and 2008 fades were discussed in Pietilä et al. [102] and Valtonen et al. [88] while the 2020.05 alignment, with the same response time as in 1989, would place the fade in the summer period of 2020 when we had no observations (Figure 8).

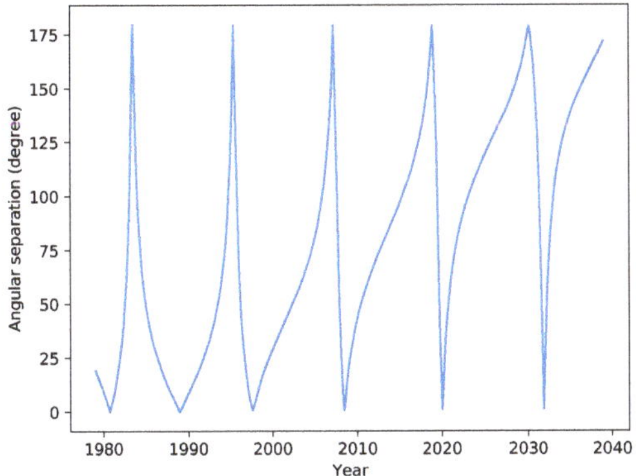

Figure 8. Temporal evolution of the angular separation between the primary jet in OJ 287 and the line joining the primary and secondary BHs. The separation is calculated assuming that the jet precession follows the disk model of Dey et al. [75]. Minimum separation angle occurs at 1980.73, 1988.95, 1997.61, 2008.45, 2020.05, and 2031.85.

Obviously, a fade would be a good time to repeat the surface photometry study of the host, too. In addition, epochs of very faint intrinsic emission from this blazar (jet) can be used to trigger host galaxy imaging because the central point source and its point spread function (PSF) is least affecting the host measurements at such epochs. For instance, in November 2017 and September 2020 such optical–UV low-states were identified in Swift monitoring (Figure 3) and new ones will be detected in the course of the ongoing MOMO program as well as in ground-based optical monitoring campaigns.

7. The Radio Jet

OJ 287 has a prominent radio jet which is known for its large swings in the position angle in the sky [73]. These swings are well understood if they result from the variations in the orientation of the inner accretion disk, and if these variations are translated into the direction of the jet [74,75]. Actually, these large swings were predicted by Valtonen et al. [88] before they were observed. The high resolution data at 86 GHz is compared with a theory line in Figure 9 (see Dey et al. [75] for details). A 3.4 yr delay is found between the change of the disk axis and the direction of the jet. This is because the point at which the jet position angle is measured is typically 200 microarcsec from the radio core, and it takes some time to transit the information of the disk orientation change to this distance in the jet. The actual jet speed in OJ 287 is discussed for example, in Valtonen et al. [103]. In this case we find the relativistic $\Gamma \sim 7$. This is somewhat lower than the values measured by other methods. For example, Savolainen et al. [104] give $\Gamma \sim 9.3$, Jorstad et al. [68] find $\Gamma \sim 16.5$, and Hovatta et al. [101] find $\Gamma \sim 15.4$. Anyway, our $\Gamma \sim 7$ is close enough, and gives further support for the model. If the disk model was not accurate, or if the principles of jet-disk connection were not valid, there is no reason why the value coming out of this model would be anywhere near the values obtained via quite different methods.

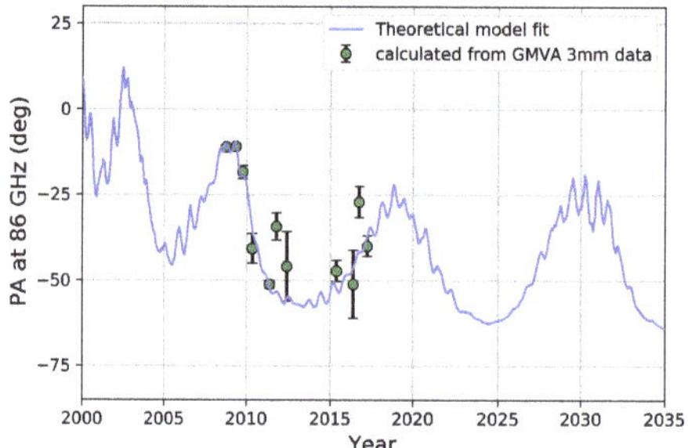

Figure 9. The position angle of the radio jet in 86 GHz observations (points) compared with the model line. The best-fit parameters are taken from Table 1 of Dey et al. [75] and we have used the disk model to calculate the theoretical model line. The viewing angle varies with time but is typically similar to the value found in Hovatta et al. [101]. The delay time between the change of the jet direction at its origin and the appearance of the change in 86 GHz observations is 3.4 yr.

The region of the jet which we observe depends on the frequency because the resolution of the global VLBI maps is proportional to the observing frequency. Unfortunately, this technique cannot be extended as far as the optical frequencies. However, some information about the jet may be obtained by monitoring of optical polarization. Valtonen et al. [105] demonstrate that the optical polarization angle is a good proxy for the position angle of the radio jet. A big position angle jump occurred in 1995 in the optical polarization while a similar jump took place in 2009 in 15 GHz radio maps. Figure 10 shows how the optical polarization is modeled by our swinging jet model. The time delay is one quarter of the 86 GHz delay, so the distance of the region emitting in optical should be closer to the center by the same amount.

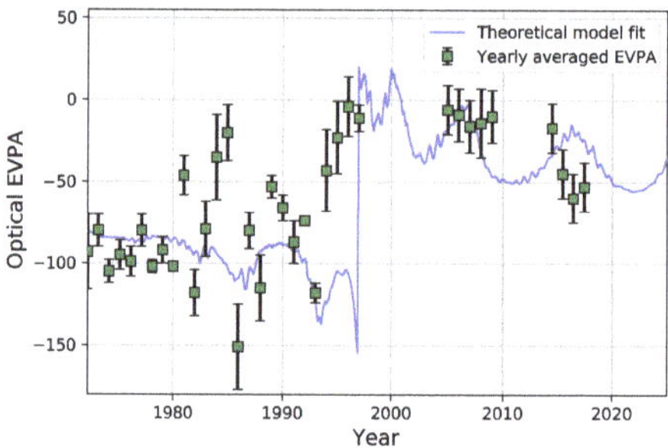

Figure 10. The annually averaged optical position angles of OJ 287 (points) together with the same jet model as in the radio jet observations. Note that the position angle has large fluctuations after the time of large optical flares, 1983/84 and 1994/95. The model parameters are the same as for the radio flares, except that time delay is 0.92 years, that is, the optical emission region is closer to the center than the observed radio jet region at 86 GHz. This region cannot be resolved in optical observations.

In the high resolution map from April/May 2014 [70], the 86 GHz component C2 at 90 microarcsec from the center has the position angle $-8 \pm 45°$. The latter component more or less coincides with the point in the jet where our model places the optical emission region (Figure 11). The innermost component C1b at 40 microarcsec lies at the position angle $13 \pm 10°$, interestingly close to the theoretically calculated position angle the secondary jet at the position angle $14.6 \pm 4°$ of Dey et al. [75] (see Figure 11). Below, we explore other possibilities of searching for observational evidence of the secondary black hole.

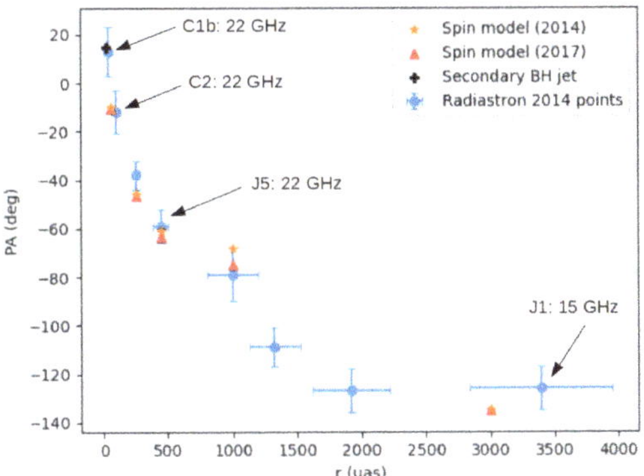

Figure 11. The distance and position angle PA of radio components in the Radioastron VLBI map by Gomez et al. [70] with error bars. The 22 GHz data are used when available. The component J1 comes from 15 GHz observations. Superposed are the theoretical positions of the corresponding components at two epochs (stars and triangles) in the spin model [74,75]. The position of the expected secondary jet is also marked (cross). Its position angle is independent of epoch.

8. On the Presence of Secondary BH Accretion Disk

The mass ratio of the two black holes in the OJ 287 binary is 122. Thus for the same $\dot{m}$ and α, the primary disk should be 122 times brighter. However, if the mass accretion rate is 122 times greater in the secondary than in the primary, both disks could be equally bright. This would put the brightness of the secondary over the Eddington limit, which may be possible at least in some accretion disk models [106]. It is well known that tidal perturbations increase the accretion flow considerably, see, for example, Sillanpää et al. [19].

We may use Sundelius et al. [45] and Pihajoki et al. [89] to estimate the accretion rate to the secondary. In these simulations one particle represents $\sim$100 solar mass of gas and the gas is accreted at the average rate of $\sim$5 particles per year, that is, the mass accretion rate is about 5×10^2 solar mass per year. This is approximately the mass contained within the Bondi-Hoyle accretion radius for the secondary, and much of it is likely to remain bound to the secondary when it leaves the disk. At what rate this matter is accreted to the black hole itself is another problem: it may well be that only some fraction of the gaseous envelope around the secondary black hole finds its way to the center before it suffers another collision with the disk. On the other hand, the steady state accretion rate of the primary is much lower, $\sim$0.6 solar mass per year [87]. Thus the higher accretion rate to the secondary may compensate for its lower mass, in which case the secondary may not be all that much fainter than the primary.

However, the secondary is truncated by tides which may reduce its optical emission below the brightness of the primary, since the optical emission is associated primarily with larger radii in standard disk models. The accretion by the secondary is also not steady, and its brightness may vary considerably depending on the epoch of observation. Not seeing it at one epoch does not mean that it could not be seen at all.

At present, there are several ways by which the secondary could show up. In spectral observations there could be extra lines that match the redshift of the secondary at that particular instant in time. Because of the high orbital speed, these lines would be found quite far apart from the primary redshift in the spectrum. For example in November 2021 the secondary should have had redshift 0.38, in contrast to 0.307 in the primary. However, the relative redshift changes constantly, and moreover, the secondary spends a great deal of time behind the accretion disk and cannot be seen. After the 3 December crossing to the far side of the disk we should not see it for another ten years, and even after that only briefly. This is because precession now lines up the secondary orbit such that the pericenter is towards us, and the secondary spends only about 1/10 of its orbital time in the pericenter part of the orbit.

Table 2, column 4, gives the expected wavelengths (in angstroms) of the secondary $H\beta$ or $H\alpha$ lines (the latter in square brackets). During the periods when the secondary should be behind the disk, we give the line for the opposite viewing direction, since we cannot be absolutely sure that we have identified the viewing direction correctly. Our interpretation in this respects relies heavily on Takalo et al. [90]. These values are in curved brackets, and refer to $H\alpha$. In Figure 12, we show the theoretically calculated redshift of the secondary BH.

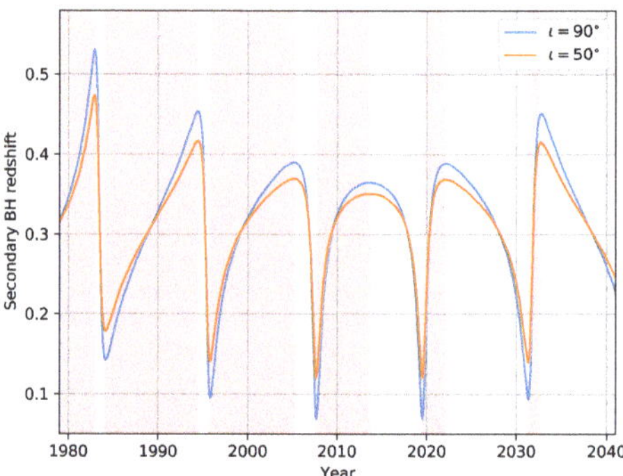

Figure 12. The redshift of the secondary black hole as a function of time. Two inclination values are shown: 90° (edge-on) and 50°. The model does not specify the inclination. The shaded regions are the time intervals when the secondary is in front of the accretion disk and thus visible to us. The identification of the "front side" is based on taking the secondary to be in front of us during the 1989 deep fade [90].

Among the VLT data, there is only one epoch where the expected secondary line would be within the observed wavelength range, and no line is seen. However, in this case we would need to have had the viewing direction wrong. Thus there is nothing in the VLT spectra that excludes the possibility of secondary lines in our current model. In the NOT and Rozhen spectra we are in the right wavelength region for $H\beta$, and the NOT spectra are taken at the time when we should see the secondary lines (see Figure 13). We do not see any of them. In fact, only a single primary line is detected weakly. Thus this non-detection does not give strong observational limits.

The second way to find signs of the secondary is to look for the jet of the secondary. It may arise as soon as a new accretion disk has formed after a disk impact. Since the new accretion disk is formed from the material pulled out of the primary accretion disk, its axis should share the orientation of the axis of the mean disk of the primary. This is in contrast to the primary jet, which wobbles considerably around the axis of the mean disk under the influence of the companion [75]. A search for radio emission in a constant direction at the position angle $\sim15°$ may reveal an indication of such a jet. The length of the jet should scale with the mass of the black hole, so that the secondary jet may be a factor ~122 shorter than the primary jet [107,108]. In the primary jet the outermost component J1 at 15 GHz is found at 3400 microarcsec from the center; thus the secondary jet may be expected in the scale of ~30 microarcsec. This is the scale which has just been reached at VLBI [70] and will be studied in coming years.

Additionally, some of the flares may arise in the secondary jet rather than in the primary jet. They could be recognized for their short variability time scale. However, since the secondary jet is expected to represent only a fraction of the flux of the primary jet, the fast variations would appear at a low amplitude, consistent with Valtaoja et al. [109] who found 1–10% peak-to-peak amplitude variations during the 1983/84 flaring season, with a 15.7 min period. Pihajoki et al. [89] find that variations down to 3.5 d may be associated with the primary while the variability of the secondary could be faster by at least a factor of 122, depending on the secondary spin. For a maximally spinning secondary the 15.7 min period may arise [110]. It is the period of the innermost stable orbit about the secondary black hole in the jet reference frame.

Since the secondary disk is likely to overflow its Roche lobe, we may also find interesting X-ray emission from the ring in the primary disk where the overflow stream terminates, akin to X-ray binaries. It is also worth searching for the direct accretion disk signature of the primary in the SED including a strong disk contribution in X-rays persistently detected in AGN with SMBHs of that mass.

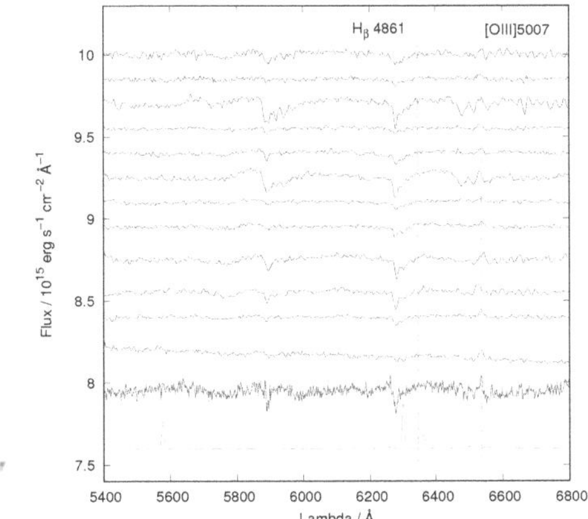

Figure 13. Optical spectra of OJ 287 taken at the *Nordic Optical Telescope*. The continuum has been fitted with a low-order polynomial and subtracted and the spectra have been shifted in vertical direction for clarity. The positions of λ4861 H$_\beta$ and λ50006 [OIII] at the redshift of OJ 287 are indicated by vertical lines. The red curve below shows the positions of prominent atmospheric emission lines. Apparent deep absorption structures in the spectrum are telluric lines. The epochs are (from top to bottom): 12 February 2010, 23 October 2010, 18 November 2010, 4 January 2011, 17 January 2011, 1 October 2011, 11 February 2011, 23 April 2011, 7 May 2011, 16 November 2011, 19 February 2012, 27 December 2012, and 6 March 2013.

9. Are There More OJ 287-like Systems?

The remaining lifetime of the OJ 287 black hole binary before a merger is 10^4 yr. Comparing this with the merger time scale of 10^7 yr, there is only a chance of $\sim 10^{-3}$ to catch the system at this late stage of evolution. Is this a problem? It depends on the parent population of OJ 287-like systems. As an order of magnitude estimate, it has been deduced that there are something like 10^4 binary black hole systems in the sky in the OJ 287 brightness category [111]. If this is the parent population, some ten of them would be found in the final stages of merger [112].

How could we find other OJ 287-like systems? We define an OJ 287-like binary black hole system such that it shows repeated bremsstrahlung flares due to the impacts on the accretion disk. The main requirement is that the mass ratio of the two black holes is rather large, so that the smaller black hole does not disrupt the accretion disk entirely in the process [51]. This is not very restrictive as large galaxies act as attractors of mergers with smaller galaxies, and together with the black hole mass-galaxy mass correlation this produces typically unequal pairs of black holes.

OJ 287 is a blazar with a jet pointing nearly toward us which increases its brightness many-fold, by about 3 magnitudes by our estimate above. The jets of other similar systems would likely point in some other direction. Thus it is unlikely that other OJ 287-like systems are found in lists of brightest radio sources, such as the Ohio Survey which gives the initial to the name of OJ 287.

However, if such a source is discovered through the repeated bremsstrahlung flares, it will take time before enough flares are seen to solve the orbit. In the case of OJ 287 we are lucky that the source lies in the ecliptic zone which has been frequently photographed over 130 yrs. No such luck can be expected for the other OJ 287-like systems. It is also rather likely that these other systems contain smaller binary components by mass than OJ 287, with weaker bremsstrahlung flares. While other candidate binary black hole systems have been identified in recent years (review by [113]), none of them were explained by involving optical bremsstrahlung flares, and none of them so far has the dense polarimetry monitoring that OJ 287 has seen. Thus we may find more OJ 287-like systems, but not before full-sky monitoring programs have been operational for some time. We now briefly list GW aspects of our description for OJ 287.

10. Pulsar Timing Array Response to nHz GWs from OJ 287's SMBH Binary Engine

The above listed observational campaigns in many electromagnetic observational windows clearly point to the presence of a nHz GW emitting SMBH binary in OJ 287. Therefore, it is important to develop prescriptions to model the expected pulsar timing array (PTA) response to nHz inspiral GWs from systems like OJ 287 [10]. The fact that the orbital eccentricity and primary BH spin can substantially influence the BH binary orbital evolution, as demonstrated in [33,114], prompted us to employ an improved version of GW phasing approach to model temporally evolving GW polarization states from such systems [115–117]. This is required as the PTA relevant GW-induced pre-fit pulsar timing residual, namely PTA response to inspiral GWs, induced by an isolated GW source at a given epoch t may be written as [118]

$$R(t) = \int_0^t \left[h(t' - \tau_P) - h(t') \right] dt',\tag{9}$$

where τ_P is the geometric time delay in the solar system barycenter (SSB) frame of the underlying millisecond pulsar while t and t' are the SSB relevant coordinate time variables. Further, $h(t)$ provides the effective GW strain induced by an isolated source at an epoch t and depends on the intrinsic parameters of the GW source like the total mass, spin values, orbital eccentricity and angles that specify its sky location, as detailed in [116]. In other words, PTA response of a pulsar to inspiral GWs from OJ 287 like system at a time t depends not only on the strength of the inspiral GWs at the SSB at t but also on the GW amplitude at location of the pulsar at time $t - \tau_P$. This is another reason why we need to employ PN approximation to track the orbital evolution and the resulting GW polarization states while modeling PTA response to inspiral GWs from SMBH binaries [116].

In an ongoing effort, we are modeling temporally evolving quadrupolar order $h(t)$ from comparable mass BH binaries inspiraling along PN-accurate eccentric orbits under the influence of dominant order spin-orbit interactions due to the primary BH spin. This is being done by adapting and extending the GW phasing efforts of Damour et al. [115], Susobhanan et al. [116]. It may be recalled that this approach usually employs Keplerian-type parametric solution to model the conservative dynamics of SMBH binaries in eccentric orbits. Thereafter, the effects of GW emission are incorporated by adapting the method of variation of constants, as detailed in [115]. In practice, we invoke the arguments of energy and angular momentum balance that lead to differential equations for orbital frequency and eccentricity with the help of orbital averaged PN-accurate far-zone energy and angular momentum fluxes [117].

In Figure 14, we plot the expected PTA response of PSR J0751 + 1807 to inpsiral nHz GWs associated with our SMBH binary central engine description for OJ 287 as detailed in [33]. The plot is based on Equation (9) where we have employed quadrupolar order $h(t)$ expression for spinning BH binaries in eccentric orbits from Königsdörffer & Gopakumar [119] while the orbital parameters are as detailed in Dey et al. [33]. The sharp features depend on the orbital eccentricity and pulsar distance and provide unique signatures of the expected PTA response to nHz GWs from such SMBH binaries. Unfortunately, the expected timing residuals are substantially below what we can achieve with the current timing campaigns on PSR J0751 + 1807. However, it should be possible to find and time milli-second pulsars with such exquisite timing precision during the SKA era [29]. Therefore, it should be possible to monitor persistent inspiral GWs from SMBH binaries like OJ 287 during the SKA era. This development should naturally allow us to pursue persistent multi-messenger GW astronomy with crucial inputs from the ongoing, planned and proposed multi-wavelength electromagnetic observations of OJ 287. This is because such observational campaigns should allow us to constrain various astrophysical parameters of the binary BH central engine description for the unique blazar OJ 287.

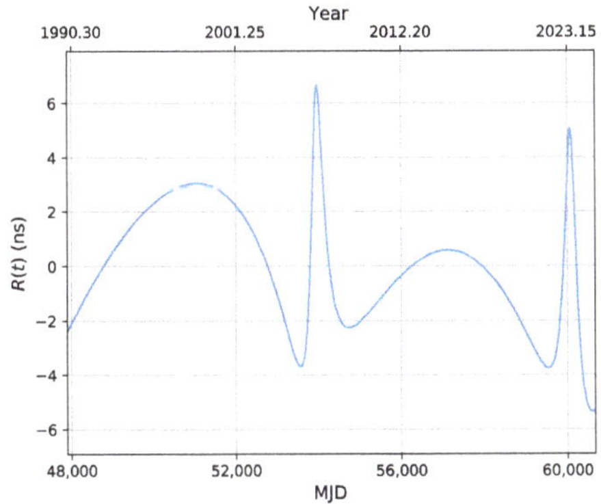

Figure 14. Timing residuals induced by inspiral GWs from OJ 287's SMBH binary on a typical PTA pulsar PSR J0751 + 1807. We let the pulsar distance to be 1.4 Kpc [120] while SMBH binary parameters as listed in Section 1.1 with its luminosity distance to be 1.6 Gpc. Timing of milli-second pulsars with SKA are expected to achieve similar timing residuals.

11. Discussion

This article emphasizes the importance of continuing monitoring of OJ 287 at all wavelengths of the electromagnetic spectrum as well as in gravitational waves. Even though the binary orbit has now been solved [33] and confirmed [34], there are many aspects of the system which still need verification and from which we can learn interesting astrophysics.

(1) The confirmation of the 2022 optical flare: This flare does not appear in any of the simple schemes that have been proposed over the years as alternatives to the LV96 binary black hole model. Difficult as it is to observe from the ground, we should just be able to observe the tail end of this flare. Polarisation observations play an important role here since the tail end of the flare should very highly polarised [36,50,59].

(2) The detection of the signals from the 3 December 2021 disk impact: This is energetically a huge event which has at previous times been studied only by occasional photometric measurements. This time we expect a full coverage from optical polarisation to X-rays which are the expected key indicators of the impact. So far, in ongoing MOMO/Swift and ground-based monitoring observations, OJ 287 was found in a low-state in early December 2021 (Komossa et al., and Zola et al., in prep.). The timing of the impact also depends on the detailed model, which will be improved after the 2021/22 observing season.

(3) The $H\alpha$ flares: This has been caught only once, accidentally by Sitko & Junkkarinen [65]. Later spectroscopic surveys, some of which are reported in this paper for the first time, have not been equally lucky. Now we give a rather precise prediction for the $H\alpha$ flare which follows the 3 December 2021, disk impact.

(4) The secondary black hole: The secondary black hole has never been directly detected. One of the problems is the short separation between the two black holes in the sky, typically 10 microarcseconds, depending on its orbital phase in an eccentric orbit. It will be several years before this resolution is achieved in radio VLBI. However, an attempt was made to detect the secondary by its spectral lines, but so far we only report a null result. It is a difficult task due to the high background level of the jet component. New attempts

should be made during periods of low light level. Here we have made a prediction when such occasions may arise.

(5) Radio jet of the secondary: Even though the immediate surroundings of the secondary are hard to resolve, the current resolution in radio VLBI should be just enough to see the jet. A strong radio component was recently reported [70] at the expected position of the secondary jet [75]. It remains to be seen if this component behaves as expected. One problem is that the current direction of the primary jet is not very far from the suspected secondary jet. However, if the primary jet behaves according to our models [75], the primary jet will move away from the secondary jet, and the identification becomes more clear. Thus the continuing radio VLBI monitoring of OJ 287 is essential.

(6) Radio jet of the primary: The radio jet of OJ 287 has been mapped in radio VLBI since the 1980s. The origin of the jet and its connection to the central black hole and its accretion disk are still open problems. In the LV96 binary black hole model it is possible to calculate the direction of the spin axis of the primary as well as the normal to the innermost disk. They follow each other more or less [75], with the disk normal showing a little more short time scale wobble than the spin axis. If we make the reasonable assumption that the initial jet direction is determined either by the spin axis (spin model) or the disk normal (disk model), we obtain a unique model for the jet which wobbles predictably. When projected onto the sky, we obtain position angles at any point in time, and since the resolution at higher frequencies is better than at lower frequencies in the global VLBI, we get snapshots of the jet at different distances from the center. After solving for the unknown viewing angles, we get a full picture of the bulk flow in the jet.

(7) Optical polarisation angle: It is usually assumed that the optical emission arises in the jet closer to the black hole than the radio emission. We have applied the same model for the optical polarisation that works for the radio position angle, and find that the region of optical emission is about four times closer to the central black hole than the inner 86 GHz emission knot. It is interesting also that the optical polarisation gives the direction of the innermost jet. This is not surprising since the jet has been found to have helical magnetic field geometry [70]. When compressed along the jet axis in shock fronts, the field lies practically perpendicular to the jet axis and the polarisation is lined up with the jet. It is important to continue regular polarisation monitoring programs to learn more about the inner jet.

(8) Optical fades: The times of low light level in the jet are useful also for the detection of the host galaxy. The absolute magnitude of the host has been measured by surface photometry out to $\sim$30 kpc from the galaxy center, beyond which it is difficult to measure any light above the background generated by the bright point source, the AGN itself. Its brightness measured out to this distance is among the highest ever detected, and therefore it is likely that the total brightness of OJ 287 host is also among the highest. The brightness may also be measured by multicolor photometry at low light, because the color of the host is expected to be very different from the color of the jet. Thus the combined color of the source becomes more and more like galaxy colors when the source fades. This method suggests a brightness increase by one magnitude if the outer envelope of the host galaxy is included.

(9) High-energy emission and broad-band SEDs: Very dense, simultaneous X-ray–UV–optical monitoring has allowed us to identify the latest binary after-flares in 2017 and 2020, based on the brightest X-ray flares so far discovered from OJ 287 and other arguments [80]. Ongoing monitoring will reveal new outstanding X-ray–UV outbursts. Spectroscopy, SED modelling, cross-correlation analyses and structure function analyses [81,86] will allow us to probe in detail the binary and outburst physics across the electromagnetic spectrum, and will allow to search for an accretion disk contribution of the secondary at select epochs. Ongoing high-energy monitoring could also serve as trigger criterion for new VHE observations.

(10) Very high energy gamma rays: VHE gamma rays have been detected in OJ287 only once. This occurred after the two big tidal flares which followed the 2015 impact flare. Even though we do not know the mechanism which produced this event, it may useful

to monitor OJ 287 again after the tidal flares which follow the 2022 impact flare. More frequent monitoring would be also useful to see if the two phenomena are really connected.

(11) Fast variability: The shortest time scale of periodic or quasi-periodic variability that may be associated with the primary jet is 3.5 days. This has been seen in many occasions. The time scale of variability should be $\sim$122 times shorter in the secondary jet, due to the smaller mass of the secondary by this factor, or it could be even as low as $\sim$15.7 min if the secondary spins much faster than the primary. Thus the study of fast variability in radio, optical and X-rays may reveal details of the secondary jet which are otherwise hard to recover.

(12) Pulsar Timing Array: It will be important to develop accurate and efficient algorithms to model the expected PTA signals from SMBH binaries like OJ 287. Further, data analysis approaches that can efficiently look for such nHz GW sources in the PTA datasets of SKA era will need to be developed [10]. Additionally, it will be desirable to search for milli-second pulsars close to the Sky location of OJ 287 as they are expected to be more sensitive to inspiral GWs from OJ 287. Investigations should be pursued to figure out the possibility of tracking the trajectory of secondary BH during the ngEHT era.

(13) It has been argued that OJ 287 may produce a few muon neutrinos with energies more than 100 TeV in instruments like IceCube-Gen2 during ten years of Fermi flaring epochs, influenced by the modeling of the 2017 flare of TXS 0506 + 056 [121,122]. The neutrino studies will be an important part of multi-messenger astronomy with OJ 287 in the coming decades.

Author Contributions: M.J.V. and L.D. have been responsible for the main design of this article as well as for theoretical developments. L.D. and A.G. were in charge of the gravitational wave part, S.Z. of optical photometry, S.K. of the MOMO project and X-ray parts, T.P. and S.Z. of optical spectroscopy, R.H. of historical magnitude measurements, H.J. and A.V.B. of optical polarimetry and J.L.G. of radio mapping. All authors have read and agreed to the published version of the manuscript.

Funding: L.D. and A.G. acknowledge the support of the Department of Atomic Energy, Government of India, under project identification # RTI 4002. A.G. is grateful for the financial support and hospitality of the Pauli Center for Theoretical Studies and the University of Zurich. S.Z. acknowledge the grant NCN 2018/29/B/ST9/01793. R.H. acknowledges support by GACR grant 13-33324S and by the MSMT Mobility project 8J18AT036.

Institutional Review Board Statement: Not applicable.

Informed Consent Statement: Not applicable.

Data Availability Statement: Not applicable.

Acknowledgments: We thank Kari Nilsson for a valuable contribution in data reduction and producing one of the figures. We also appreciate the help that Vilppu Piirola has provided and continues to provide in optical polarisation observations of OJ 287. We also thank Gene Byrd and Shirin Haque for reading the manuscript and for valuable inputs.

Conflicts of Interest: The authors declare no conflict of interest.

References

1. Begelman, M.C.; Blandford, R.D.; Rees, M.J. Massive black hole binaries in active galactic nuclei. *Nature* **1980**, *287*, 307. [CrossRef]
2. Valtaoja, L.; Valtonen, M.J.; Byrd, G.G. Binary Pairs of Supermassive Black Holes: Formation in Merging Galaxies. *Astrophys. J.* **1989**, *343*, 47. [CrossRef]
3. Mikkola, S.; Valtonen, M.J. Evolution of binaries in the field of light particles and the problem of two black holes. *Mon. Not. R. Astron. Soc.* **1992**, *259*, 115. [CrossRef]
4. Valtonen, M.J. Triple black hole systems formed in mergers of galaxies. *Mon. Not. R. Astron. Soc.* **1996**, *278*, 186. [CrossRef]
5. Quinlan, G.D. The dynamical evolution of massive black hole binaries I. Hardening in a fixed stellar background. *New Astron.* **1996**, *1*, 35. [CrossRef]
6. Milosavljevic, M.; Merritt, D. Formation of Galactic Nuclei. *Astrophys. J.* **2001**, *563*, 34. [CrossRef]
7. Volonteri, M.; Haardt, F.; Madau, P. The Assembly and Merging History of Supermassive Black Holes in Hierarchical Models of Galaxy Formation. *Astrophys. J.* **2003**, *582*, 559. [CrossRef]

8. Sesana, A.; Haardt, F.; Madau, P.; Volonteri, M. Low-Frequency Gravitational Radiation from Coalescing Massive Black Hole Binaries in Hierarchical Cosmologies. *Astrophys. J.* **2004**, *611*, 623. [CrossRef]

9. Burke-Spolaor, S.; Blecha, L.; Bogdanovic, T.; Comerford, J.M.; Lazio, J.; Liu, X.; Maccarone, T.J.; Pesce, D.; Shen, Y.; Taylor, G. Supermassive Black Hole Pairs and Binaries. *Sci. Next Gener. Very Large Array* **2018**, *517*, 677.

10. Burke-Spolaor, S.; Taylor, S.R.; Charisi, M.; Dolch, T.; Hazboun, J.S.; Holgado, A.M.; Kelley, L.Z.; Lazio, T.J.W.; Madison, D.R.; McMann, N.; et al. The astrophysics of nanohertz gravitational waves. *Astron. Astrophys. Rev.* **2019**, *27*, 5. [CrossRef]

11. Arzoumanian, Z.; Baker, P.T.; Blumer, H.; Becsy, B.; Brazier, A.; Brook, P.R.; Burke-Spolaor, S.; Chatterjee, S.; Chen, S.; Cordes, J.M.; et al. The NANOGrav 12.5 yr Data Set: Search for an Isotropic Stochastic Gravitational-wave Background. *Astrophys. J.* **2020**, *905*, L34. [CrossRef]

12. Goncharov, B.; Shannon, R.M.; Reardon, D.J.; Hobbs, G.; Zic, A.; Bailes, M.; Curylo, M.; Dai, S.; Kerr, M.; Lower, M.E.; et al. On the Evidence for a Common-spectrum Process in the Search for the Nanohertz Gravitational-wave Background with the Parkes Pulsar Timing Array. *Astrophys. J.* **2021**, *917*, L19. [CrossRef]

13. Chen, S.; Caballero, R.N.; Guo, Y.J.; Chalumeau, A.; Liu, K.; Shaifullah, G.; Lee, K.J.; Babak, S.; Desvignes, G.; Parthasarathy, A.; et al. Common-red-signal analysis with 24-yr high-precision timing of the European Pulsar Timing Array: Inferences in the stochastic gravitational-wave background search. *Mon. Not. R. Astron. Soc.* **2021**, *508*, 4970. [CrossRef]

14. Abbott, R.; Abbott, T.D.; Abraham, S.; Acernese, F.; Ackley, K.; Adams, A.; Adams, C.; Adhikari, R.X.; Adya, V.B.; Affelt, C.; et al. GWTC-2: Compact Binary Coalescences Observed by LIGO and Virgo during the First Half of the Third Observing Run. *Phys. Rev. X* **2021**, *11*, 021053. [CrossRef]

15. Abbott, R.; Abbott, T.D.; Acernese, F.; Ackley, K.; Adams, C.; Adhikari, N.; Adhikari, R.X.; Adya, V.B.; Affeldt, C.; Agarwal, D.; et al. GWTC-3: Compact Binary Coalescences Observed by LIGO and Virgo during the Second Part of the Third Observing Run. *arXiv* **2021**, arXiv:2111.03606.

16. Abbott, B.P.; Abbott, R.; Abbott, T.D.; Acernese, F.; Ackley, K.; Adams, C.; Adams, T.; Addesso, P.; Adhikari, R.X.; Adya, V.B.; et al. GW170817: Observation of Gravitational Waves from a Binary Neutron Star Inspiral. *Phys. Rev. Lett.* **2017**, *119*, 161101. [CrossRef] [PubMed]

17. Abbott, B.P.; Abbott, R.; Abbott, T.D.; Acernese, F.; Ackley, K.; Adams, C.; Adams, T.; Addesso, P.; Adhikari, R.X.; Adya, V.B.; et al. Multi-messenger Observations of a Binary Neutron Star Merger. *Astrophys. J.* **2017**, *848*, L12.

18. Xin, C.; Mingarelli, C.M.F.; Hazboun, J.S. Multimessenger Pulsar Timing Array Constraints on Supermassive Black Hole Binaries Traced by Periodic Light Curves. *Astrophys. J.* **2021**, *915*, 97. [CrossRef]

19. Sillanpää, A.; Haarala, S.; Valtonen, M.J.; Sundelius, B.; Byrd, G.G. OJ 287: Binary Pair of Supermassive Black Holes. *Astrophys. J.* **1988**, *325*, 628. [CrossRef]

20. Sudou, H.; Iguchi, S.; Murata, Y.; Taniguchi, Y. Orbital Motion in the Radio Galaxy 3C 66B: Evidence for a Supermassive Black Hole Binary. *Science* **2003**, *300*, 1263. [CrossRef]

21. Jenet, F.A.; Lommen, A.; Larson, S.L.; Wen, L. Constraining the Properties of Supermassive Black Hole Systems Using Pulsar Timing: Application to 3C 66B. *Astrophys. J.* **2004**, *606*, 799. [CrossRef]

22. Iguchi, S.; Okuda, T.; Sudou, H. A Very Close Binary Black Hole in a Giant Elliptical Galaxy 3C 66B and its Black Hole Merger. *Astrophys. J.* **2010**, *724*, L166. [CrossRef]

23. Liu, F.K.; Li, S.; Komossa, S. A Milliparsec Supermassive Black Hole Binary Candidate in the Galaxy SDSS J120136.02+300305.5. *Astrophys. J.* **2014**, *786*, 103. [CrossRef]

24. Graham, M.J.; Djorgovski, S.G.; Stern, D.; Glikman, E.; Drake, A.J.; Mahabal, A.A.; Donalek, C.; Larson, S.; Christensen, E. A possible close supermassive black-hole binary in a quasar with optical periodicity. *Nature* **2015**, *518*, 74. [CrossRef]

25. Zhu, X.-J.; Thrane, E. Toward the Unambiguous Identification of Supermassive Binary Black Holes through Bayesian Inference. *Astrophys. J.* **2020**, *900*, 117. [CrossRef]

26. Charisi, M.; Bartos, I.; Haiman, Z.; Price-Whelan, A.M.;Graham, M.J.; Bellm, E.C.; Laher, R.R.; Marka, S. A population of short-period variable quasars from PTF as supermassive black hole binary candidates. *Mon. Not. R. Astron. Soc.* **2016**, *463*, 2145. [CrossRef]

27. Zhu, X.-J.; Cui, W.; Thrane, E. The minimum and maximum gravitational-wave background from supermassive binary black holes. *Mon. Not. R. Astron. Soc.* **2019**, *482*, 2588. [CrossRef]

28. Dey, L.; Gopakumar, A.; Valtonen, M.; Zola, S.; Susobhanan, A.; Hudec, R.; Pihajoki, P.; Pursimo, T.; Berdyugin, A.; Piirola, V.; et al. The Unique Blazar OJ 287 and Its Massive Binary Black Hole Central Engine. *Universe* **2019**, *5*, 108. [CrossRef]

29. Feng, Y.; Li, D.; Zheng, Z.; Tsai, C.-W. Supermassive binary black hole evolution can be traced by a small SKA pulsar timing array. *Phys. Rev. D* **2020**, *102*, 023014. [CrossRef]

30. Dickel J.R.; Yang, K.S.; McVittie, G.C.; Swenson, G.W., Jr. A survey of the sky at 610.5 MHz. II. The region between declinations +15 and +22 degrees. *Astron. J.* **1967**, *72*, 757. [CrossRef]

31. Hudec R.; Basta, M.; Pihajoki, P.; Valtonen, M. The historical 1900 and 1913 outbursts of the binary blazar candidate OJ287. *Astron. Astrophys.* **2013**, *559*, 20. [CrossRef]

32. Valtonen, M.J.; Lehto, H.J.; Sillanpää, A.; Nilsson, K.; Mikkola, S.; Hudec, R.; Basta, M.; Teräsranta, H.; Haque, S.; Rampadarath, H. Predicting the Next Outbursts of OJ 287 in 2006–2010. *Astrophys. J.* **2006**, *646*, 36–48. [CrossRef]

33. Dey, L.; Valtonen, M.J.; Gopakumar, A.; Zola, S.; Hudec, R.; Pihajoki, P.; Ciprini, S.; Matsumoto, K.; Sadakane, K.; Kidger, M.; et al. Authenticating the Presence of a Relativistic Massive Black Hole Binary in OJ 287 Using Its General Relativity Centenary Flare: Improved Orbital Parameters. *Astrophys. J.* **2018**, *866*, 11. [CrossRef]

34. Laine, S.; Dey, L.; Valtonen, M.; Gopakumar, A.; Zola, S.; Komossa, S.; Kidger, M.; Pihajoki, P.; Gomez, J.L.; Caton, D.; et al. Spitzer Observations of the Predicted Eddington Flare from Blazar OJ 287. *Astrophys. J. Lett.* **2020**, *894*, L1. [CrossRef]
35. Kidger, M. *Cosmological Enigmas: Pulsars, Quasars, and Other Deep-Space Questions*; The Johns Hopkins University Press: Baltimore, MD, USA, 2007.
36. Smith, P.S.; Balonek, T.J.; Heckert, P.A.; Elston, R.; Schmidt, G.D. UBVRI field comparison stars for selected active quasars and BL Lacertae objects. *Astron. J.* **1985**, *90*, 1184–1187. [CrossRef]
37. Sillanpää, A.; Teerikorpi, P.; Haarala, S.; Korhonen, T.; Efimov, I.S.; Shakhovskoi, N.M. Similar structures in the outbursts of OJ 287 in 1972 and 1983. *Astron. Astrophys.* **1985**, *147*, 67–70.
38. Sillanpää, A.; Takalo, L.O.; Pursimo, T.; Lehto, H.J.; Nilsson, K.; Teerikorpi, P.; Heinämäki, P.; Kidger, M.; de Diego, J.A.; Gonzalez-Perez, J.N.; et al. Confirmation of the 12-year optical outburst cycle in blazar OJ 287. *Astron. Astrophys.* **1996**, *305*, L17.
39. Lehto, H.J.; Valtonen, M.J. OJ 287 Outburst Structure and a Binary Black Hole Model. *Astrophys. J.* **1996**, *460*, 207. [CrossRef]
40. Valtonen, M.J. The OJ 287 binary model and the expected outburst in November 1995. In Proceedings of the Workshop on Two Years of Intensive Monitoring of OJ 287 and 3C 66A 1996, Oxford, UK, 11–14 September 1995.
41. Sillanpää, A.; Takalo, L.O.; Pursimo, T.; Nilsson, K.; Heinämäki, P.; Katajainen, S.; Pietilä, H.; Hanski, M.; Rekola, R.; Kidger, M.; et al. Double-peak structure in the cyclic optical outbursts of blazar OJ 287. *Astron. Astrophys.* **1996**, *315*, L13–L16.
42. Ivanov, P.B.; Igumenshchev, I.V.; Novikov, I.D. Hydrodynamics of Black Hole-Accretion Disk Collision. *Astrophys. J.* **1998**, *507*, 131–144. [CrossRef]
43. Pihajoki, P. Black hole accretion disc impacts. *Mon. Not. R. Astron. Soc.* **2016**, *457*, 1145–1161. [CrossRef]
44. Sundelius, B.; Wahde, M.; Lehto, H.J.; Valtonen, M.J. Long-time Brightness Variations of 0J287 in the Binary Black Hole Model. *ASP-CS* **1996**, *110*, 99.
45. Sundelius, B.; Wahde, M.; Lehto, H.J.; Valtonen, M.J. A Numerical Simulation of the Brightness Variations of OJ 287. *Astrophys. J.* **1997**, *484*, 180–185. [CrossRef]
46. Valtonen, M.J. New Orbit Solutions for the Precessing Binary Black Hole Model of OJ 287. *Astrophys. J.* **2007**, *659*, 1074–1081. [CrossRef]
47. Valtonen, M.J.; Nilsson, K.; Sillanpää, A.; Takalo, L.O.; Lehto, H.J.; Keel, W.C.; Haque, S.; Cornwall, D.; Mattingly, A. The 2005 November Outburst in OJ 287 and the Binary Black Hole Model. *Astrophys. J.* **2006**, *643*, L9–L12 [CrossRef]
48. Valtonen, M.J. Sillanpää 2005–2010 Multiwavelength Campaing of OJ287. *AcPol* **2011**, *51*, 76.
49. Valtonen, M.J.; Lehto, H.J.; Nilsson, K.; Heidt, J.; Takalo, L.O.; Sillanpää, A.; Villforth, C.; Kidger, M.; Poyner, G.; Pursimo, T.; et al. A massive binary black-hole system in OJ287 and a test of general relativity. *Nature* **2008**, *452*, 851–853. [CrossRef] [PubMed]
50. Valtonen, M.J.; Zola, S.; Ciprini, S.; Gopakumar, A.; Matsumoto, K.; Sadakane, K.; Kidger, M.; Gazeas, K.; Nilsson, K.; Berdyugin, A.; et al. (OJ287-15/16 Collaboration). Primary Black Hole Spin in OJ 287 as Determined by the General Relativity Centenary Flare. *Astrophys. J.* **2016**, *819*, L37. [CrossRef]
51. Valtonen, M.J.; Ciprini, S.; Lehto, H.J. On the masses of OJ287 black holes. *Mon. Not. R. Astron. Soc.* **2012**, *427*, 77–83. [CrossRef]
52. Kidger, M.; Zola, S.; Valtonen, M.; Lähteenmäki, A.; Järvelä, E.; Tornikoski, M.; Tammi, J.; Liakos, A.; Poyner, G. Far-infrared photometry of OJ 287 with the Herschel Space Observatory. *Astron. Astrophys.* **2018**, *610*, A74. [CrossRef]
53. Villata, M.; Raiteri, C.M.; Sillanpää, A.; Takalo, L.O. A beaming model for the OJ 287 periodic optical outbursts. *Mon. Not. R. Astron. Soc.* **1998**, *293*, L13–L16. [CrossRef]
54. Rieger, F.M. On the Geometrical Origin of Periodicity in Blazar-type Sources. *Astrophys. J.* **2004**, *615*, L5–L8. [CrossRef]
55. Ciprini, S.; Perri, M.; Verreccha, F.; Valtonen, M. Fermi-LAT detection of hard spectrum and enhanced gamma-ray emission from the BL Lac object PKS 1717+177. *Astronomer's Telegr.* **2015**, *8401*, 1.
56. MacPherson, E.; Ister, J.C.; Urry, M.; Coppi, P.; Bailyn, C.; Dincer, T. SMARTS Enhanced Optical & Infrared Activity in Blazar OJ 287. *Astronomer's Telegr.* **2015**, *8392*, 1.
57. Shappee, B.J.; Stanek, K.Z.; Holoien, T.W.-S.; Brown, J.S.; Kochanek, C.S.; Godoy-Rivera, D.; Basu, U.; Prieto, J.L.; Bersier, D.; Dong, S.; et al. Strong Optical Flare from Blazar OJ 287 Detected by ASAS-SN. *Astronomer's Telegr.* **2015**, *8372*, 1.
58. Pursimo T.; Takalo, L.O.; Sillanpää, A.; Kidger, M.; Lehto, H.J.; Heidt, J.; Charles, P.A.; Aller, H.; Aller, M.; Beckmann, V.; et al. Intensive monitoring of OJ 287. *Astron. Astrophys. Suppl. Ser.* **2000**, *146*, 141–155. [CrossRef]
59. Villforth C.; Nilsson, K.; Heidt, J.; Takalo, L.O.; Pursimo, T.; Berdyugin, A.; Lindfors, E.; Pasanen, M.; Winiarski, M.; Drozdz, M.; et al. Variability and stability in blazar jets on time-scales of years: Optical polarization monitoring of OJ 287 in 2005–2009. *Mon. Not. R. Astron. Soc.* **2010**, *402*, 2087–2111. [CrossRef]
60. Pursimo T.; Takalo, L.O.; Sillanpää, A.; Kidger, M.; Lehto, H.J.; Heidt, J.; Charles, P.A.; Aller, H.; Aller, M.; Beckmann, V.; et al. VizieR Online Data Catalog: Intensive monitoring of OJ 287 (Pursimo+, 2000). *VizieR Online Data Catalog* **2021**, *J/A+AS*, 141–146.
61. Valtonen, M.J.; Zola, S.; Jermak, H.; Ciprini, S.; Hudec, R.; Dey, L.; Gopakumar, A.; Reichart, D.; Caton, D.; Gazeas, K.; et al. Polarization and Spectral Energy Distribution in OJ 287 during the 2016/17 Outbursts. *Galaxies* **2017**, *5*, 83. [CrossRef]
62. Zola, S.; Valtonen, M.J.; Bhatta, G.; Goyal, A.; Debski, B.; Baran, A.; Krzesinski, J.; Siwak, M.; Ciprini, S.; Gopakumar, A.; et al. A Search for QPOs in the Blazar OJ287: Preliminary Results from the 2015/2016 Observing Campaign. *Galaxies* **2016**, *4*, 41. [CrossRef]
63. Abdo A.A.; Ackermann, M.; Ajello, M.; Atwood, W.B.; Axelsson, M.; Baldini, L.; Ballet, J.; Barbiellini, G.; Bastieri, D.; Baughman, B.M.; et al. Bright Active Galactic Nuclei Source List from the First Three Months of the Fermi Large Area Telescope All-Sky Survey. *Astrophys. J.* **2009**, *700*, 597–622. [CrossRef]

64. D'Arcangelo, F.D.; Marscher, A.P.; Jorstad, S.; Smith, P.S.; Larionov, V.M.; Hagen-Thorn, V.A.; Williams, G.G.; Gear, W.K.; Clemens, D.P.; Sarcia, D.; et al. Synchronous Optical and Radio Polarization Variability in the Blazar OJ287. *Astrophys. J.* **2009**, *697*, 985–995. [CrossRef]

65. Sitko, M.L.; Junkkarinen, V.T. Continuum and line fluxes of OJ 287 at minimum light. *Publ. Astron. Soc. Pac.* **1985**, *97*, 1158–1162. [CrossRef]

66. Nilsson, K.; Takalo, L.O.; Lehto, H.J.; Sillanpää, A. H-alpha moni-toring of OJ 287 in 2005–2008. *Astron. Astrophys.* **2010**, *516*, A60 [CrossRef]

67. Padovani, P.; Giommi, P. The Connection between X-Ray– and Radio-selected BL Lacertae Objects. *Astrophys. J.* **1995**, *444*, 567. [CrossRef]

68. Jorstad, S.G.; Marscher, A.P.; Lister, M.L.; Stirling, A.M.; Cawthorne, T.V.; Gear, W.K.; Gomez, J.L.; Stevens, J.A.; Smith, P.S.; Forster, J.R.; et al. Polarimetric Observations of 15 Active Galactic Nuclei at High Frequencies: Jet Kinematics from Bimonthly Monitoring with the Very Long Baseline Array. *Astron. J.* **2005**, *130*, 1418–1465. [CrossRef]

69. Lee, J.W.; Lee, S.S.; Algaba, J.-C.; Hodgson, J.; Kim, J.-Y.; Park, J.; Kino, M.; Kim, D.-W.; Kang, S.; Yoo, S.; et al. Interferometric Monitoring of Gamma-Ray Bright AGNs: OJ 287. *Astrophys. J.* **2020**, *902*, 104. [CrossRef]

70. Gomez, J.L.; Traianou, E.; Krichbaum, T.P.; Lobanov, A.; Fuentes, A.; Lico, R.; Zhao, G.-Y.; Bruni, G.; Kovalev, Y.Y.; Lähteenmäki, A.; et al. Probing the innermost regions of AGN jets and their magnetic fields with RadioAstron. V. Space and ground millimeter-VLBI imaging of OJ 287. *arXiv* **2021**, arXiv:2111.11200.

71. Goddi C.; Marti-Vidal, I.; Messias, H.; Bower, G.C.; Broderick, A.E.; Dexter, J.; Marrone, D.P.; Moscibrodzka, M.; Nagai, H.; Algaba, J.C.; et al. Polarimetric Properties of Event Horizon Telescope Targets from ALMA. *Astrophys. J. Lett.* **2021**, *910*, L14. [CrossRef]

72. Hodgson, J.A.; Krichbaum, T.P.; Marscher, A.P.; Jorstad, S.G.; Rani, B.; Marti-Vidal, I.; Bach, U.; Sanchez, S.; Bremer, M.; Lindqvist, M.; et al. Location of γ-ray emission and magnetic field strengths in OJ 287. *Astron. Astrophys.* **2017**, *597*, A80. [CrossRef]

73. Agudo, I.; Marscher, A.P.; Jorstad, S.G.; Gomez, J.L.; Perucho, M.; Piner, P.G.; Rioja, M.; Dodson, R. Erratic Jet Wobbling in the BL Lacertae Object OJ287 Revealed by Sixteen Years of 7 mm VLBA Observations. *Astrophys. J.* **2011**, *747*, 63. [CrossRef]

74. Valtonen, M.J.; Pihajoki, P. A helical jet model for OJ287. *Astron. Astrophys.* **2013**, *557*, A28. [CrossRef]

75. Dey, L.; Valtonen, M.J.; Gopakumar, A.; Lico, R.; Gomez, J.; Susobhanan, A.; Komossa, S.; Pihajoki, P. Explaining temporal variations in the jet PA of the blazar OJ 287 using its BBH central engine model. *Mon. Not. R. Astron. Soc.* **2021**, *503*, 4400–4412. [CrossRef]

76. Shrader, C.R.; Hartman, R.C.; Webb, J.R. Probable detection of high-energy gamma-ray emission from OJ 287 during a major optical flare. *Astron. Astrophys. Suppl. Ser.* **1996**, *120*, 599–602.

77. Agudo, I.; Jorstad, S.G.; Marscher, A.P.; Larionov, V.M.; Gomez, J.L.; Lähteenmäki, A.; Gurwell, M.; Smith, P.S.; Wiesemeyer, H.; Thum, C.; et al. Location of γ-ray Flare Emission in the Jet of the BL Lacertae Object OJ287 More than 14 pc from the Central Engine. *Astrophys. J.* **2011**, *726*, L13. [CrossRef]

78. O'Brien, S.; For the VERITAS Collaboration. for the VERITAS Collaboration VERITAS detection of VHE emission from the optically bright quasar OJ 287. *arXiv* **2017**, arXiv:1708.02160.

79. Komossa, S.; Grupe, D.; Schartel, N.; Gallo, L.; Gomez, J.L.; Kollatschny, W.; Kriss, G.; Leighly, K.; Longinotti, A.L.; Parker, M.; et al. The Extremes of AGN Variability. *IAUS New Front. Black Hole Astrophys.* **2017**, *324*, 168–171. [CrossRef]

80. Komossa, S.; Grupe, D.; Parker, M.L.; Valtonen, M.J.; Gómez, J.L.; Gopakumar, A.; Dey, L. The 2020 April-June super-outburst of OJ 287 and its long-term multiwavelength light curve with Swift: Binary supermassive black hole and jet activity. *Mon. Not. R. Astron. Soc.* **2020**, *498*, L35. [CrossRef]

81. Komossa, S.; Grupe, D.; Gallo, L.C.; Gonzalez, A.; Yao, S.; Hollett, A.R.; Parker, M.L.; Ciprini, S. MOMO IV: The complete Swift X-ray and UV/optical light curve and characteristic variability of the blazar OJ 287 during the last two decades. *Astrophys. J.* **2021**, *923*, 51. [CrossRef]

82. Madejski, G.M.; Schwartz, D.A. Studies of BL Lacertae Objects with the Einstein Observatory: The Soft X-Ray Spectra of OJ 287 and PKS 0735+178. *Astrophys. J.* **1988**, *330*, 776. [CrossRef]

83. Komossa, S.; Grupe, D.; Parker, M.L.; Gómez, J.L.; Valtonen, M.J.; Nowak, M.A.; Jorstad, S.G.; Haggard, D.; Chandra, S.; Ciprini, S.; et al. X-ray spectral components of the blazar and binary black hole candidate OJ 287 (2005-2020). *Mon. Not. R. Astron. Soc.* **2021**, *504*, 5575. [CrossRef]

84. Marscher, A.P.; Jorstad, S.G. The Megaparsec-scale X-ray Jet of The BL Lac Object OJ287. *Astrophys. J.* **2011**, *729*, 26. [CrossRef]

85. Komossa, S.; Ciprini, S.; Dey, L.; Gallo, L.C.; Gomez, J.L.; Gonzalez, A.; Grupe, D.; Kraus, A.; Laine, S.; Parker, M.L.; et al. Supermassive Binary Black Holes and the Case of OJ 287. *Publ. Astron. Obs. Belgrade* **2021**, *100*, 29–42. Available online: https://publications.aob.rs/100/pdf/029-042.pdf (accessed on 15 November 2021).

86. Komossa, S.; Grupe, D.; Kraus, A.; Gallo, L.C.; Gonzalez, A.G.; Parker, M.L.; Valtonen, M.J.; Hollett, A.R.; Bach, U.; Gomez, J.L.; et al. Project MOMO: Multiwavelength Observations and Modeling of OJ 287. *Universe* **2021**, *7*, 261. [CrossRef]

87. Valtonen, M.J.; Zola, S.; Pihajoki, P.; Enestam, S.; Lehto, H.J.; Dey, L.; Gopakumar, A.; Drozdz, M.; Ogloza, W.; Zejmo, M.; et al. Accretion Disk Parameters Determined from the Great 2015 Flare of OJ 287. *Astrophys. J.* **2019**, *882*, 88. [CrossRef]

88. Valtonen, M.J.; Lehto, H.J.; Takalo, L.O.; Sillanpää, A. Testing the 1995 Binary Black Hole Model of OJ287. *Astrophys. J.* **2011**, *729*, 33. [CrossRef]

89. Pihajoki, P.; Valtonen, M.; Ciprini, S. Short time-scale periodicity in OJ 287. *Mon. Not. R. Astron. Soc.* **2013**, *434*, 3122–3129. [CrossRef]

90. Takalo, L.O.; Kidger, M.; de Diego, J.A.; Sillanpää, A.; Piirola, V.; Teräsranta, H. A sudden fade of OJ 287. *Astron. Astrophys. Suppl. Ser.* **1990**, *83*, 459.
91. Valtonen, M.J.; Zola, S.; Ciprini, S.; Kidger, T.; Pursimo, T.; Gopakumar, A.; Matsumoto, K.; Sadakane, K.; Caton, D.B.; Nilsson, K.; et al. Host galaxy magnitude of OJ287 from its colours at minimum light. *Mon. Not. R. Astron. Soc.* 2021, Submitted.
92. Nilsson, K.; Kotilainen, J.; Valtonen, M.; Gomez, J.L.; Castro-Tirado, A.J.; Gopakumar, A.; Jeong, S.; Kidger, M.; Komossa, S.; Mathur, S.; et al. The Host Galaxy of OJ 287 Revealed by Optical and Near-infrared Imaging. *Astrophys. J.* **2020**, *904*, 102. [CrossRef]
93. Sandage, A. Absolute Magnitudes of E and so Galaxies in the Virgo and Coma Clusters as a Function of U-B Color. *Astrophys. J.* **1972**, *176*, 21. [CrossRef]
94. Graham, A.; Scott, N. The M_{BH}-$L_{spheroid}$ Relation at High and Low Masses, the Quadratic Growth of Black Holes, and Intermediate-mass Black Hole Candidates. *Astrophys. J.* **2013**, *764*, 151. [CrossRef]
95. De Vaucouleurs, G.; De Vaucouleurs, A. Photometry of Intergalactic Matter in the Coma Cluster. *Astrophys. Lett.* **1970**, *5*, 219.
96. Läsker, R.; Ferrarese, L.; van de Ven, G. Supermassive Black Holes and Their Host Galaxies. II. The Correlation with Near-infrared Luminosity Revisited. *Astrophys. J.* **2014**, *780*, 70. [CrossRef]
97. Huang, S.; Ho, L.C.; Peng, C.Y.; Li, Z.-Y.; Barth, A.J. The Carnegie-Irvine Galaxy Survey. III. The Three-component Structure of Nearby Elliptical Galaxies. *Astrophys. J.* **2013**, *766*, 47. [CrossRef]
98. Yanny, B.; Jannuzi, B.T.; Impey, C. Hubble Space Telescope Imaging of the BL Lacertae Object OJ 287. *Astrophys. J.* **1997**, *484*, L113–L116 [CrossRef]
99. Saglia, R.P.; Opitsch, M.; Erwin, P.; Thomas, J.; Beifiori, A.; Fabricius, M.; Mazzalay, X.; Nowak, N.; Rusli, S.P.; Bender, R. The SINFONI Black Hole Survey: The Black Hole Fundamental Plane Revisited and the Paths of (Co)evolution of Supermassive Black Holes and Bulges. *Astrophys. J.* **2016**, *818*, 47. [CrossRef]
100. Portinari, L.; Kotilainen, J.; Falomo, R.; Decarli, R. On the cosmological evolution of the black hole-host galaxy relation in quasars. *Mon. Not. R. Astron. Soc.* **2012**, *420*, 732–744. [CrossRef]
101. Hovatta, T.; Valtaoja, E.; Tornikoski, M.; Lähteenmäki, A. Doppler factors, Lorentz factors and viewing angles for quasars, BL Lacertae objects and radio galaxies. *Astron. Astrophys.* **2009**, *494*, 527–537. [CrossRef]
102. Pietilä, H.; Takalo, L.O.; Tosti, G.; Benitez, E.; Chiattelli, B.; Corradi, R.L.M.; Cox, G.; de Diego, J.A.; de Francesco, G.; Dultzin-Hacyan, D.; et al. OJ 287 and the predicted fade of 1998. *Astron. Astrophys.* **1999**, *345*, 760–768.
103. Valtonen, M.J.; Gopakumar, A.; Mikkola, S.; Wiik, K.; Lehto, H.J. Black hole binary OJ287 as a testing platform for general relativity. *arXiv* 2012, arXiv:1208.4524.
104. Savolainen, T.; Homan, D.C.; Hovatta, T.; Kadler, M.; Kovalev, Y.Y.; Lister, M.L.; Ros, E.; Zensus, J.A. Relativistic beaming and gamma-ray brightness of blazars. *Astron. Astrophys.* **2010**, *512*, A24. [CrossRef]
105. Valtonen, M.J.; Wiik, K. Optical polarization angle and VLBI jet direction in the binary black hole model of OJ287. *Mon. Not. R. Astron. Soc.* **2012**, *421*, 1861–1867. [CrossRef]
106. Heinzeller, D.; Duschl, W.J. On the Eddington limit in accretion discs. *Mon. Not. R. Astron. Soc.* **2007**, *374*, 1146–1154. [CrossRef]
107. Ghosh, P.; Abramowicz, M.A. Electromagnetic extraction of rotational energy from disc-fed black holes: The strength of the Blandford-Znajek process. *Mon. Not. R. Astron. Soc.* **1997**, *292*, 887–895. [CrossRef]
108. Chen, Y.-Y.; Zhang, X.; Xiong, D.; Yu, X. Black Hole Mass, Jet Power, and Accretion in AGNs. *Astron. J.* **2015**, *150*, 8. [CrossRef]
109. Valtaoja, E.; Lehto, H.; Teerikorpi, P.; Korhonen, T.; Valtonen, M.; Teräsranta, H.; Salonen, E.; Urpo, S.; Tiuri, M.; Piirola, V.; et al. A 15.7-min periodicity in OJ287. *Nature* **1985**, *314*, 148–149. [CrossRef]
110. Pihajoki, P.; Valtonen, M.; Zola, S.; Liakos, A.; Drozdz, M.; Winiarski, M.; Ogloza, W.; Koziel-Wierzbowska, D.; Provencal, J.; Nilsson, K.; et al. Precursor Flares in OJ 287. *Astrophys. J.* **2013**, *764*, 5. [CrossRef]
111. Valtonen, M.J.; Mikkola, S.; Merritt, D.; Gopakumar, A.; Lehto, H.J.; Hyvönen, T.; Rampadarath, H.; Saunders, R.; Basta, M.; Hudec, R. Measuring the Spin of the Primary Black Hole in OJ287. *Astrophys. J.* **2010**, *709*, 725. [CrossRef]
112. Krolik, J.H.; Volonteri, M.; Dubois, Y.; Devriendt, J. Population Estimates for Electromagnetically Distinguishable Supermassive Binary Black Holes. *Astrophys. J.* **2019**, *879*, 110. [CrossRef]
113. Komossa, S.; Zensus J.A. Compact object mergers: Observations of supermassive binary black holes and stellar tidal disruption events. *IAUS Star Clust. Black Holes Galaxies Across Cosm. Time* **2016**, *312*, 13–25. [CrossRef]
114. Klein, A.; Boetzel, Y.; Gopakumar, A.; Jetzer, P.; de Vittori, L. Fourier domain gravitational waveforms for precessing eccentric binaries. *Phys. Rev. D* **2018**, *98*, 104043. [CrossRef]
115. Damour, T.; Gopakumar, A.; Iyer, B.R. Phasing of gravitational waves from inspiralling eccentric binaries. *Phys. Rev. D* **2004**, *70*, 064028. [CrossRef]
116. Susobhanan, A.; Gopakumar, A.; Hobbs, G.; Taylor, S.R. Pulsar timing array signals induced by black hole binaries in relativistic eccentric orbits. *Phys. Rev. D* **2020**, *101*, 043022. [CrossRef]
117. Dey, L.; Susobhanan, A.; Gopakumar, A.; Valtonen, M. Pulsar Timing Array Signals Due to Inspiralling Spinning Massive Black Hole Binaries in Relativistic Eccentric Orbits. In preparation.
118. Anholm, M.; Ballmer, S.; Creighton, J.D.E.; Price, L.R.; Siemens, X. Optimal strategies for gravitational wave stochastic background searches in pulsar timing data. *Phys. Rev. D* **2009**, *79*, 084030. [CrossRef]
119. Königsdörffer, C.; Gopakumar, A. Post-Newtonian accurate parametric solution to the dynamics of spinning compact binaries in eccentric orbits: The leading order spin-orbit interaction. *Phys. Rev. D* **2005**, *71*, 024039. [CrossRef]

120. Perera, B.B.P.; DeCesar, M.E.; Demorest, P.B.; Kerr, M.; Lentati, L.; Nice, D.C.; Oslowski, S.; Ransom, S.M.; Keith, M.J.; Arzoumanian, Z.; et al. The International Pulsar Timing Array: Second data release. *Mon. Not. R. Astron. Soc.* **2019**, *490*, 4666–4687. [CrossRef]
121. Oikonomou, F.; Murase, K.; Padovani, P.; Resconi, E.; Mészáros, P. High-energy neutrino flux from individual blazar flares. *Mon. Not. R. Astron. Soc.* **2019**, *489*, 4347. [CrossRef]
122. IceCube Collaboration; Aartsen, M.G.; Ackermann, M.; Adams, J.; Aguilar, J.A.; Ahlers, M.; Ahrens, M.; Al Samarai, I.; Altmann, D.; Andeen, K.; et al. Multimessenger observations of a flaring blazar coincident with high-energy neutrino IceCube-170922A. *Science* **2018**, *361*, eaat1378.

 *galaxies*

Article

γ-ray Flux and Spectral Variability of Blazar Ton 599 during Its 2021 Flare

Bhoomika Rajput * and Ashwani Pandey

Indian Institute of Astrophysics, Block II Koramangala, Bangalore 560034, India; ashwanitapan@gmail.com
* Correspondence: bhoomikarjpt2@gmail.com

Abstract: Blazars are known to emit exceptionally variable non-thermal emission over the wide range (from radio to γ-rays) of electromagnetic spectrum. We present here the results of our γ-ray flux and spectral variability study of the blazar Ton 599, which has been recently observed in the γ-ray flaring state. Using 0.1–300 GeV γ-ray data from the *Fermi* Gamma-ray Space Telescope (hereinafter *Fermi*), we generated one-day binned light curve of Ton 599 for a period of about one-year from MJD 59,093 to MJD 59,457. During this one year period, the maximum γ-ray flux detected was $2.24 \pm 0.25 \times 10^{-6}$ ph cm^{-2} s^{-1} at MJD 59,399.50. We identified three different flux states, namely, epoch A (quiescent), epoch B (pre-flare) and epoch C (main-flare). For each epoch, we calculated the γ-ray flux variability amplitude (F_{var}) and found that the source showed largest flux variations in epoch C with $F_{var} \sim 35\%$. We modelled the γ-ray spectra for each epoch and found that the Log-parabola model adequately describes the γ-ray spectra for all the three epochs. We estimated the size of the γ-ray emitting region as 1.03×10^{16} cm and determined that the origin of γ-ray radiation, during the main-flare, could be outside of the broad line region.

Keywords: galaxies; active; galaxies–quasars; individual (Ton 599)

Citation: Rajput, B.; Pandey, A. γ-ray Flux and Spectral Variability of Blazar Ton 599 during Its 2021 Flare. *Galaxies* **2021**, *9*, 118. https://doi.org/10.3390/galaxies9040118

Academic Editor: Santanu Mondal

Received: 15 November 2021
Accepted: 10 December 2021
Published: 13 December 2021

Publisher's Note: MDPI stays neutral with regard to jurisdictional claims in published maps and institutional affiliations.

1. Introduction

Blazars are the jetted subclass of the Active Galactic Nuclei (AGNs) that are understood as AGNs with very small viewing angles to the line of sight [1,2]. Blazars emit broadband electromagnetic radiation that ranges from radio to extremely high γ-ray energies. The jets of blazars are highly Doppler boosted, resulting in flux variations over the entire accessible electromagnetic wavebands. Flat spectrum radio quasars (FSRQs) and BL Lacertae (BL Lacs) objects are the two subclass of blazars. The difference between these subclasses is determined by the equivalent width (EW) of the emission lines in their optical spectra, with FSRQs having EW > 5 Å and BL Lacs having EW < 5 Å [3]. A more physical criterion for distinguishing between FSRQs and BL Lacs was proposed by [4] which is based on the ratio of broad line region (BLR) luminosity (L_{BLR}) to Eddington Luminosity (L_{Edd}). FSRQs have the value $L_{BLR}/L_{Edd} > 5 \times 10^{-4}$, whereas BL Lacs have the value $L_{BLR}/L_{Edd} < 5 \times 10^{-4}$.

The broad-band spectral energy distributions (SEDs) of blazars comprise of two humps; the low energy hump and the high energy hump. The low energy hump peaks at optical/UV/X-ray region and the high energy hump peaks at MeV/GeV/TeV region [5–7]. The genesis of the low energy hump is well understood through synchrotron emission mechanism of the relativistic electrons, while the high energy hump originates through inverse Compton (IC) emission process [6]. Based on the position of the synchrotron peak frequency (ν_{syn}), blazars are further classified as low synchrotron peaked (LSP; $\nu_{syn} < 10^{14}$ Hz), intermediate synchrotron peaked (ISP; 10^{14} Hz $< \nu_{syn} < 10^{15}$ Hz) and high synchrotron peaked (HSP; $\nu_{syn} > 10^{15}$ Hz) blazars.

Due to the Doppler boosting in blazars' jets, the observed emission S_{obs} relative to the emission in the co-moving frame S_{int} is defined as e.g., [8]

$$S_{obs} = S_{int}\delta^q \tag{1}$$

where $q = 3 + \alpha$ for a moving compact source and $q = 2 + \alpha$ for a stationary jet [9]. α is the spectral index which is defined as $f_\nu \propto \nu^{-\alpha}$. δ is the Doppler boosting factor which is described as $\delta = 1/\Gamma(1 - \beta\cos\theta)$, where Γ is bulk Lorentz factor ($\Gamma = 1/(1 - \beta^2)^{1/2}$), β is the speed of jet in units of the speed of light and θ is the viewing angle between observer's line of sight and jet's axis. The observed time is also shortened by the effect of Doppler boosting by a factor δ^{-1}. These two consequences of the Doppler boosting increase the chances of observed variations in blazars over a wide range of wavelengths and make them the brightest objects in the extragalactic sky.

The study of flux variability is a valuable tool for determining the size of the emission zone in blazars. Blazars exhibit variability over a wide range of timescales, from few minutes to several years, across the full wavelength range i.e., from radio to γ-rays [10–14]. The variability in blazars can be explained by the shock-in-jet model [15]. In this scenario, the inhomogeneities in the jet flow produce relativistic shocks. These shocks travel through the jet plasma at relativistic speed and accelerate the particles. The other scenario is the magnetic reconnection, which is responsible for the rapid variations in blazar jets and has been investigated extensively in recent years [16–19]. Within the reconnection region, magnetic reconnection possibly enables small-scale ultra-relativistic flow, also called jet-in-jet scenario [16,19,20]. The ultra-relativistic motion of small plasmoids causes additional Doppler boosting and results in very short and bright flares.

The launch of *Fermi* Telescope in 2008 gave us an unprecedented opportunity to explore the γ-ray regime in blazars. Using *Fermi* data at γ-ray energies, flux doubling time scales have been detected as short as few minutes in both FSRQs and BL Lacs [14,21–23]. Causality considerations in these circumstances point to highly compact γ-ray emission zones in blazars' jets. The detection of γ-ray photons with energy greater than 10 GeV in FSRQs leads to the conclusion that the γ-ray emission region should be located at outside the cavity formed by the broad emission line (BLR) [24]. Otherwise, pair production on UV photons emitted by (BLR) clouds should severely attenuate the γ-ray photons. Despite several studies, the physics of γ-ray variability in blazars is still a captivating subject of research.

Ton 599 (4FGL J1159.5 + 2914; [25]) is an FSRQ, located at redshift z = 0.725 [26] with R.A. = $11^h59^m31.8^s$ and Dec. = $+29°14'43.8''$. Ton 599 is a strongly polarized and a highly optically violent variable quasar [27]. It was first detected in high energy γ-rays by the Energetic Gamma Ray Experiment Telescope (EGRET) in the second EGRET catalogue [28] and later by the *Fermi* Large Area Telescope (*Fermi*-LAT) [29]. The correlation study between radio and γ-ray bands was carried out by [30] for this source to put the constraints on γ-ray emission region in parsec-scale jets. In 2017, for the first time, Ton 599 went through a protracted flaring condition spanning the full electromagnetic spectrum. During this flare, the maximum γ-ray flux observed was 1.26×10^{-6} ph cm^{-2} s^{-1} [31]. The detailed study of γ-ray flux variability during this flare was carried out by [31,32]. A multiwavelength study and the broad-band SED modelling of this flare was conducted by [33] using a two-component leptonic emission model. An EC mechanism with a dusty torus (DT) photon field producing seed photons was identified to be responsible for the GeV emission in this study. Ton 599 recently displayed a bright flare from July to September 2021, allowing us to investigate its γ-ray emission process and its consequences during this source's flaring condition.

In this paper, we describe the γ-ray analysis performed on the source Ton 599 utilising data collected over a one-year period from September 2020 to August 2021, with the goal of constraining the γ-ray emission region in blazars. The data used in this study is presented in Section 2. Section 3 describes the γ-ray light curve, followed by results in Section 4 and discussion in Section 5. Section 6 presents a summary of the results.

2. γ-ray Observations and Data Reduction

In this work, we used the γ-ray observations of Ton 599 taken by the *Fermi*-LAT for a period of $\sim$1 year, from 2020 September to 2021 August (MJD: 59,093–59,457). *Fermi*-LAT is a pair-conversion γ-ray telescope that can detect γ rays with energies ranging from 20 MeV to more than 1 TeV. It has a large field of view of about 2.4 sr and scans the full sky in every 3 h, covering 20% of the sky at any time [34]. We used the package ScienceTools v1.2.23 with the instrument response function P8R3_SOURCE_V3[1] for our analysis. We used the latest LAT Pass 8 data in the energy range 100 MeV to 300 GeV, where the photon-like events are classified as 'evclass = 128, evtype = 3'. The region of interest (ROI) is specified as a circle with a radius of $10°$ and is centred on the γ-ray position of the source. We used a maximum zenith angle of $90°$ to remove γ-ray contamination from the earth's limb. The latest isotropic model 'iso_P8R3_SOURCE_V3_v1' and the Galactic diffuse emission model 'gll_iem_v07' were used to analyze the data. The recommended condition '(DATA QUAL > 0)&&(LAT CONFIG= = 1)' was then used to construct the required good time intervals. Over the time period of interest, an unbinned likelihood analysis is performed to generate the 1-day binned γ-ray light curve of Ton 599. In the light curve, the source was considered to be detected if the test statistics (TS) > 9, which corresponds to a 3σ detection [35]. Our final one-day binned γ-ray light curve contains 256 confirmed measurements of Ton 599.

3. γ-ray Light Curve

The one-day binned γ-ray light curve of Ton 599 from 2020 September to 2021 August (MJD: 59,093–59,457) is shown in Figure 1. For the entire period of the light curve, we estimated the average γ-ray flux which was found to be $0.43 \pm 0.36 \times 10^{-6}$ ph cm^{-2} s^{-1}. Based on the average γ-ray flux, we visually identified three different flux states, namely, quiescent state, pre-flaring state, and the main flaring state, which were labelled as epoch A, epoch B, and epoch C, respectively. The duration of these epochs are marked by vertical lines in Figure 1 and the details of these epochs are given in Table 1. We classified epoch A as a quiescent epoch since the flux was lower than the average flux for the whole time span. During epochs B and C, the flux increased 2–3 times than the average flux, so we classified these epochs as flaring epochs. Over the whole time span of the light curve, the value of the flux was maximum during the flaring epoch C. The value of the maximum flux was determined to be $2.24 \pm 0.25 \times 10^{-6}$ ph cm^{-2} s^{-1} at MJD 59,399.50.

Table 1. The epochs considered for the γ-ray light curve study are listed in the table below. Here, N, $\bar{x}$, and F_{var} denote the total number of data points, the average γ-ray flux, and the fractional variability amplitude, respectively, for the epoch. The γ-ray flux is in the units of 10^{-6} ph cm^{-2} s^{-1}.

| Epochs | MJD | | Calendar Date (dd-mm-yyyy) | | Peak Flux | N | $\bar{x}$ | F_{var} | Remarks |
	Start	End	Start	End					
Epoch A	59,144	59,226	22-10-2020	12-01-2021	0.37 ± 0.16	42	0.15 ± 0.01	0.36 ± 0.12	Quiescent
Epoch B	59,315	59,360	11-04-2021	24-05-2021	1.16 ± 0.18	44	0.71 ± 0.02	0.22 ± 0.04	Flaring
Epoch C	59,374	59,411	09-06-2021	16-07-2021	2.24 ± 0.25	37	1.02 ± 0.03	0.35 ± 0.03	Flaring

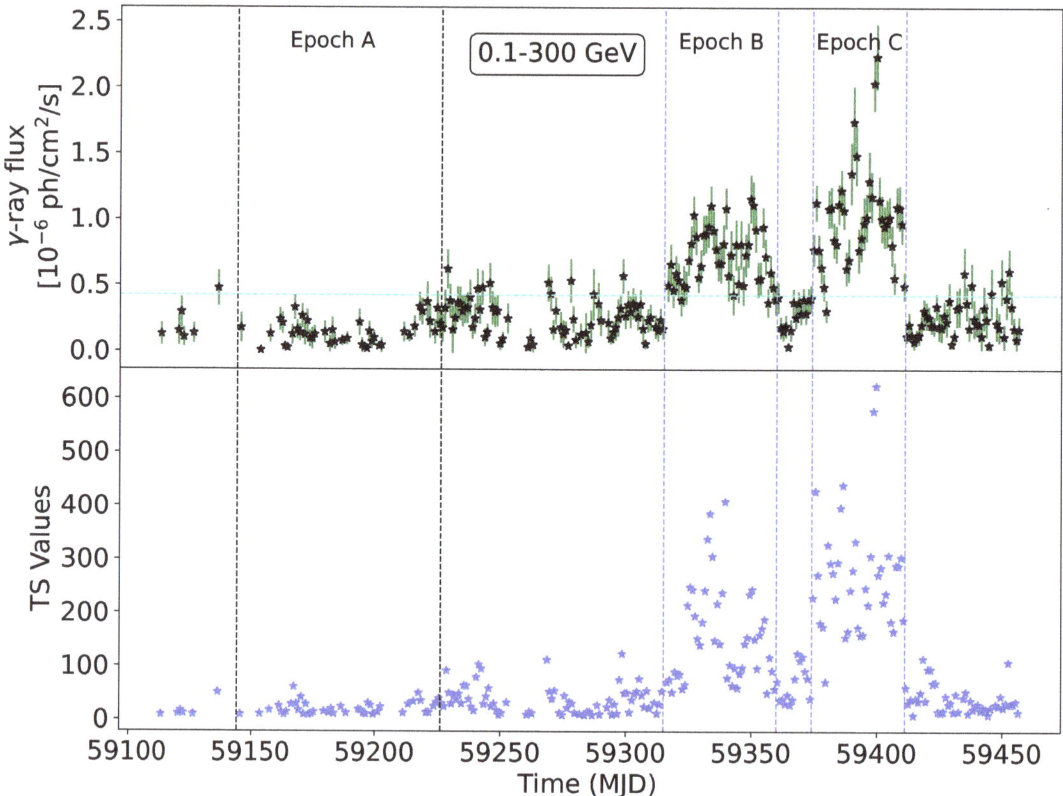

Figure 1. One-day binned γ-ray light curve of Ton 599 (**upper panel**) with the TS values corresponding to each flux value (**lower panel**). The vertical black lines refer to the quiescent epoch (Epoch A), whereas vertical blue lines refer to the γ-ray flaring epochs (Epoch B and Epoch C). In the upper panel the horizontal cyan line represents the average flux from September 2020 to August 2021.

4. Results

4.1. γ-ray Flux Variability

We used the fractional variability amplitude (F_{var}) to characterize the flux variability properties of Ton 599. The F_{var} is commonly used to quantify the intrinsic variations in blazar light curves and is defined as, e.g., [12,36,37]

$$F_{var} = \sqrt{\frac{S^2 - \overline{\sigma_{err}^2}}{\bar{x}^2}} \tag{2}$$

In the above equation S^2 is the sample variance and $\overline{\sigma_{err}^2}$ is mean square error. These are defined as

$$S^2 = \frac{1}{N-1} \sum_{i=1}^{N} (x_i - \bar{x})^2 \tag{3}$$

and

$$\overline{\sigma_{err}^2} = \frac{1}{N} \sum_{i=1}^{N} \sigma_{err,i}^2 \tag{4}$$

Here N is the total number of data points in the light curve and $\bar{x}$ is the arithmetic mean of the light curve.

The uncertainty in F_{var} is given by the following equation

$$err(F_{var}) = \sqrt{\left(\sqrt{\frac{1}{2N}}\frac{\overline{\sigma_{err}^2}}{\bar{x}^2 F_{var}}\right)^2 + \left(\sqrt{\frac{\overline{\sigma_{err}^2}}{N}}\frac{1}{\bar{x}}\right)^2} \tag{5}$$

For each epoch, we calculated the value of F_{var} separately by considering the average flux ($\bar{x}$) and the data points (N) within the epoch only. The values of N, $\bar{x}$, and F_{var} for each epoch are listed in Table 1. The errors calculated using Equation (5) are given in Table 1. For epoch A, the error in the value of F_{var} is relatively large which also indicates that it is a quiescent state. We considered the source to be variable in epochs if $F_{var} > 3 \times err(F_{var})$. According to our criteria for variations, the source was not variable in epoch A (quiescent state), while significant γ-ray flux variations were noticed during epochs B and C with F_{var} values of 22% and 35%, respectively.

4.2. γ-ray Spectral Fitting

We modeled the γ-ray spectrum of Ton 599 for each epoch to investigate the inherent distribution of electrons that causes the γ-ray emission during the epoch. We used the power-law (PL) and the log parabola (LP) models to fit the γ-ray spectra of Ton 599 using the maximum likelihood analysis. The PL model is defined as follows [38]:

$$\frac{dN(E)}{dE} = N_\circ \left(\frac{E}{E_\circ}\right)^{\Gamma} \tag{6}$$

where $N_\circ$ is the prefactor (normalization of the energy spectrum), $E_\circ$ is the pivot energy ($\sim$523.51 MeV) given in the 4FGL catalog, and Γ is the photon index, while, the LP model is given by

$$\frac{dN(E)}{dE} = N_\circ \left(\frac{E}{E_b}\right)^{-\alpha - \beta ln\left(\frac{E}{E_b}\right)} \tag{7}$$

In this equation $N_\circ$ is the normalization, E_b is the pivot energy (same as in PL), α is photon index at E_b and β is the curvature index that defines the curvature of the spectrum around the peak.

For each epoch, the model fitted γ-ray spectra of Ton 599 is shown in Figure 2, and the values of best fitted model parameters are given in Table 2. In Figure 2, the uncertainties are large at higher energies because of relatively low photon counts at these energies.

Table 2. Details of the PL and LP model fits for the three epochs. Here, γ-ray flux is in the units of 10^{-6} ph cm^{-2} s^{-1}.

Epoch	PL			LP				TS_{curve}
	Γ	Flux	$-$Log L	α	β	Flux	$-$Log L	
A	-2.381 ± 0.004	0.135 ± 0.002	146,206.160	2.045 ± 0.085	1.075 ± 0.473	0.111 ± 0.012	146,191.326	29.67
B	-2.142 ± 0.001	0.800 ± 0.002	92,649.820	2.051 ± 0.021	0.973 ± 0.117	0.735 ± 0.020	92,634.266	31.108
C	2.044 ± 0.002	1.084 ± 0.005	84,267.700	1.923 ± 0.002	0.958 ± 0.011	0.980 ± 0.003	84,245.258	44.884

To determine whether the γ-ray spectrum has a curvature or not, and which model (LP or PL) best describes the γ-ray spectrum of Ton 599, we calculated TS_{curve} (curvature of the test statistics) [38], which is defined as:

$$TS_{curve} = 2(logL_{LP} - logL_{PL}) \tag{8}$$

In the above equation, L represents the likelihood function. The value of TS_{curve} for each epoch is given in Table 2. We employed the $TS_{curve} > 16$ threshold (i.e., 4σ level; [35]) to determine the presence of statistically significant curvature in the γ-ray spectrum. We found that the LP model best describes the γ-ray spectra of Ton 599 for all three epochs.

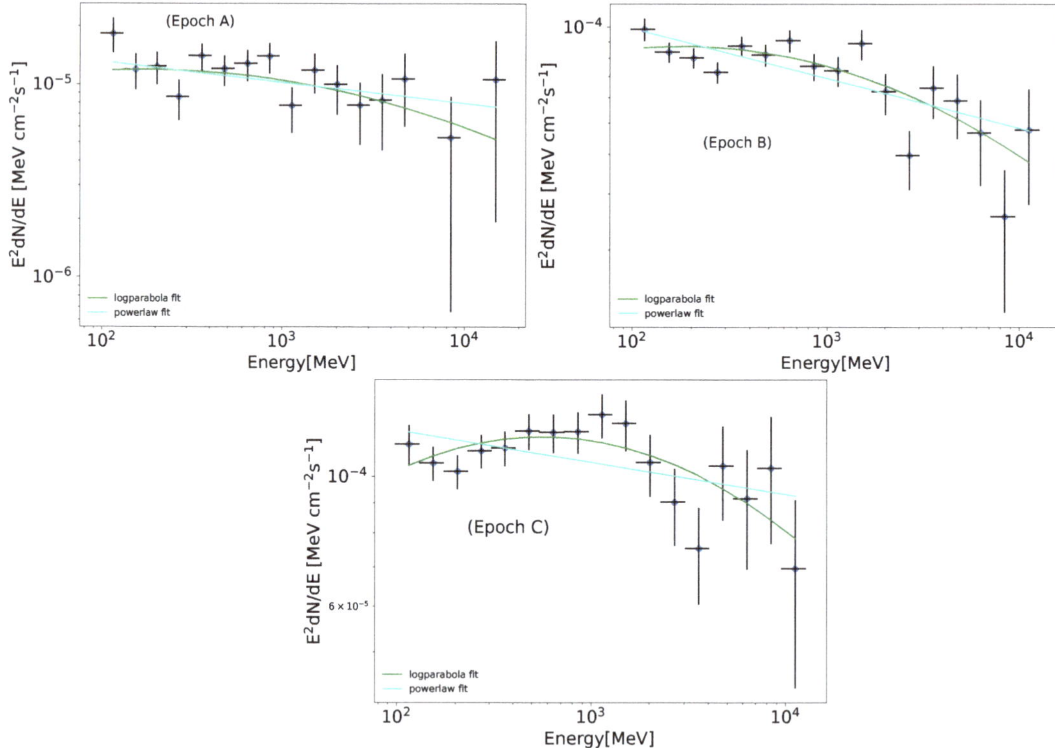

Figure 2. γ-ray spectrum for the epochs (**A–C**). The name of the epoch is mentioned in each plot.

4.3. Location of the γ-ray Emission Region

To determine the size of the γ-ray emitting region, the flux doubling time scale is usually estimated, e.g., [31,39]. We calculated the flux doubling time scale for the one-day binned γ-ray light curve during the main-flaring epoch (epoch C) of Ton 599 as follows:

$$F(t_2) = F(t_1) \times 2^{\Delta t / \tau_d} \tag{9}$$

In the above expression, $F(t_1)$ and $F(t_2)$ are the flux values at times t_1 and t_2 respectively, $\Delta t = t_2 - t_1$ and τ_d represents the flux doubling time.

We found a flux doubling timescale of $\sim$13.2 h during the epoch C for the blazar Ton 599. Using the flux doubling time scale, we estimated the size of the γ-ray emitting region for Ton 599 using the following expression:

$$r \leq c\tau_d \delta / (1 + z) \tag{10}$$

A gamma-ray Doppler factor of delta = 12.5 was calculated for Ton 599 by [40] using the multiwavelength data and the known radio Doppler factors with the assumption that the observed boosted emission is from the SSC model. This value is consistent with the lower limits (10.75 and 13.45) of delta obtained for the two bright flares of Ton 599 by [32]. Using $\delta = 12.5$, the size of the γ-ray emission region is estimated to be 1.03×10^{16} cm.

According to [41], the location of the γ-ray emission region with respect to the central super massive black hole (SMBH) can be approximated by using the relation R = r/ϕ, where r is the size of the γ-ray emitting blob and ϕ is the jet opening angle.

An intrinsic opening angle of 0.58° ($\sim$0.01 radian) was determined for Ton 599 by [42]. They estimated the jet opening angles for Fermi detected AGNs using 15.4 GHz Very Long

Baseline Array (VLBA) observations following two different methods: (a) by analyzing transverse jet profiles in the image plane and (b) by model fitting the data in the (u, v) plane. Using these values of r and ϕ, the location of the γ-ray emitting region for Ton 599 was found to be at 1.03×10^{18} cm from the SMBH.

5. Discussion

In this section we give our interpretation of the findings of above-mentioned analyses, as well as a discussion of them.

5.1. γ-ray Flux Variability

In the γ-ray band, blazars exhibit remarkable flux variability. For all the three epochs of the source Ton 599, we estimated the flux variability amplitude using one-day binned light curves. We found that during the main-flaring epoch (epoch C), the source showed largest variations with $F_{var} = 0.35 \pm 0.03$. During the pre flaring epoch (epoch B), the source was also variable with $F_{var} = 0.22 \pm 0.04$ and during the quiescent epoch (epoch A), the source was not significantly variable ($F_{var} = 0.36 \pm 0.12$) within the error bar. During the main-flare (epoch C), the source showed a maximum flux of $2.24 \pm 0.25 \times 10^{-6}$ ph cm^{-2} s^{-1} at MJD 59,399.50, which is larger than the maximum flux observed during the 2017 flare [31]. We found that the photon index value is 1.99 ± 0.10 during the main-flaring epoch (epoch C), when flux is highest, harder than the 4FGL value of 2.19 ± 0.01.

The observed γ-ray variability in the blazars could be attributed to both intrinsic and extrinsic factors. The distribution of relativistic electrons responsible for the emission is one of the intrinsic effects. These relativistic electrons, which can be accelerated to a Lorentz factor of up to 10^6, are responsible for non-thermal emission from blazar jets via synchrotron or inverse Compton emission processes. The distribution of these relativistic electrons and seed photon responsible for the inverse Compton emission process (synchrotron self Compton (SSC); [15,43] and external Compton (EC); [44,45]) are the intrinsic factors that cause short-term variability in jets. Extrinsic factors, such as the moving blob's high bulk Lorentz factor (Γ) $\sim$ 50, are in addition to intrinsic factors. Though the large bulk Lorentz factor is a favourable explanation in the case of BL Lacs, because the seed photons come from inside the jets and produce γ-rays through inverse Compton. On the other hand, the seed photons in FSRQ, emanate from outside the jets and produce γ-rays. However, pair-production through dense broad line region (BLR) can attenuate γ-ray emission. As a result, in the case of FSRQs, the large bulk Lorentz factor that causes γ-ray variability is not a plausible explanation [46]. The γ-ray flux variability in blazars jets can be explained by the magnetic reconnection (jet-in-jet) scenario, in which the mini jets generated in the jets can have a large bulk Lorentz factor and they can produce γ-ray flux variability on short time scale [16,19].

5.2. γ-ray Spectra

For each epoch, we modelled the γ-ray spectra of Ton 599 using the power law and the log parabola models. We used the TS_{curve} value to determine the best fitted model. Based on the TS_{curve} value, we found that the LP model best describes the γ-ray spectra of Ton 599 during all the three epochs indicating that the GeV γ-ray spectrum of Ton 599 is curved.

FSRQs usually have a curved GeV γ-ray spectrum, e.g., [47–49]. Several theories have been proposed, in the literature, to explain the curved γ-ray spectra of FSRQs. The attenuation of γ-ray photons by pair-production (inside the BLR) could explain the curved γ-ray spectrum [50]. A viable scenario for defining the curve in the γ-ray spectrum is the Klein-Nishina effect on the inverse Compton scattering of BLR photons through relativistic electrons present in the jet [51]. However, several investigations conducted to locate the γ-ray emission region discovered that the observed γ-ray spectra in FSRQs are not caused by IC scattering of BLR photons and that the γ-ray emission site is located outside the

BLR [19,52]. The curved γ-ray spectra of FSRQs could also be explained by the curved energy distribution of the electrons emitting the radiation [53,54].

5.3. Location of the γ-ray Emitting Region

We calculated the minimum γ-ray flux doubling timescale of 13.2 h during epoch C for Ton 599. Using this minimum doubling timescale, we estimated the size of the γ-ray emitting region as 1.03×10^{16} cm. We have also identified the location of the γ-ray emitting blob to be at a distance of 1.03×10^{18} cm from the SMBH. The size of the BLR for TON 599 was estimated as $\sim(2.11–2.45) \times 10^{17}$ cm by [32]. So, the location of the γ-ray emitting blob, found in this work, is outside of the BLR. During the 2017 γ-ray flare of this source, using two component leptonic model, Ref. [33] found that the seed photons for the GeV emission produced by the dusty torus (DT) and the GeV emitting region is located outside of the BLR. Our results are consistent with the findings of [33].

6. Summary

We investigated the γ-ray flux and spectral variability of the blazar Ton 599 during MJD 59,093 to MJD 59,457. For our study, we have chosen three epochs of different flux states: quiescent, pre-flaring, and main-flaring. The outcomes of the analysis of these epochs are summarised below.

- We estimated the flux variability amplitude for the specified epochs. The largest γ-ray flux variations were found for the main-flaring epoch (C) with $F_{var} = 0.35 \pm 0.03$. The source also showed variations in epoch B with $F_{var} = 0.22 \pm 0.04$. However, no significant flux variation was observed in the quiescent epoch.
- The γ-ray spectra were well fit by the LP model for all the three epochs.
- We estimated the size of the γ-ray emitting region as 1.03×10^{16} cm and the location of the γ-ray emitting blob could be outside of the BLR.

Author Contributions: Conceptualization, B.R.; Data curation, B.R. and A.P.; Formal analysis, B.R.; Investigation, B.R.; Project administration, A.P.; Resources, B.R.; Visualization, B.R.; Writing—original draft, B.R.; Writing—review & editing, A.P. All authors have read and agreed to the published version of the manuscript.

Funding: This research received no external funding.

Data Availability Statement: The data used in this work is publicly available from the Fermi-LAT data archive https://fermi.gsfc.nasa.gov/ssc/data/access/, accessed on 1 September 2021.

Acknowledgments: We thank the anonymous referees for their constructive comments, which helped us to improve the manuscript. This publication made use of data from the *Fermi* Gamma-ray Space Telescope, which accessed from the Fermi Science Support Center. We also acknowledge the use of High Performance Computing Facility (Nova cluster) at Indian Institute of Astrophysics.

Conflicts of Interest: The authors declare no conflict of interest.

Note

1. https://fermi.gsfc.nasa.gov/ssc/data/analysis/documentation/, accessed on 1 September 2021.

References

1. Urry, C.M.; Padovani, P. Unified Schemes for Radio-Loud Active Galactic Nuclei. *Publ. Astron. Soc. Pac.* **1995**, *107*, 803. [CrossRef]
2. Padovani, P.; Alexander, D.M.; Assef, R.J.; De Marco, B.; Giommi, P.; Hickox, R.C.; Richards, G.T.; Smolčić, V.; Hatziminaoglou, E.; Mainieri, V.; et al. Active galactic nuclei: What's in a name? *Astrophys. Space Phys. Rev.* **2017**, *25*, 2. [CrossRef]
3. Stocke, J.T.; Morris, S.L.; Gioia, I.M.; Maccacaro, T.; Schild, R.; Wolter, A.; Fleming, T.A.; Henry, J.P. The Einstein Observatory Extended Medium-Sensitivity Survey. II. The Optical Identifications. *Astrophys. J. Suppl.* **1991**, *76*, 813. [CrossRef]
4. Ghisellini, G.; Tavecchio, F.; Foschini, L.; Ghirland, A.G. The transition between BL Lac objects and flat spectrum radio quasars. *Mon. Not. R. Astron. Soc.* **2011**, *414*, 2674–2689. [CrossRef]
5. Fossati, G.; Maraschi, L.; Celotti, A.; Comastri, A.; Ghisellini, G. A unifying view of the spectral energy distributions of blazars. *Mon. Not. R. Astron. Soc.* **1998**, *299*, 433–448. [CrossRef]

6. Abdo, A.A.; Ackermann, M.; Agudo, I.; Ajello, M.; Aller, H.D.; Aller, M.F.; Angelakis, E.; Arkharov, A.A.; Axelsson, M.; Bach, U.; et al. The Spectral Energy Distribution of Fermi Bright Blazars. *Astrophys. J.* **2010**, *716*, 30–70. [CrossRef]

7. Mao, P.; Urry, C.M.; Massaro, F.; Paggi, A.; Cauteruccio, J.; Künzel, S.R. A Comprehensive Statistical Description of Radio-through-Gamma-Ray Spectral Energy Distributions of All Known Blazars. *Astrophys. J. Suppl.* **2016**, *224*, 26. [CrossRef]

8. Lin, C.; Fan, J.H.; Xiao, H.B. The intrinsic γ-ray emissions of Fermi blazars. *Res. Astron. Astrophys.* **2017**, *17*, 066. [CrossRef]

9. Lind, K.R.; Blandford, R.D. Semidynamical models of radio jets: Relativistic beaming and source counts. *Astrophys. J.* **1985**, *295*, 358–367. [CrossRef]

10. Gaur, H.; Gupta, A.C.; Strigachev, A.; Bachev, R.; Semkov, E.; Wiita, P.J.; Peneva, S.; Boeva, S.; Slavcheva-Mihova, L.; Mihov, B.; et al. Optical flux and spectral variability of blazars. *Mon. Not. R. Astron. Soc.* **2012**, *425*, 3002–3023. [CrossRef]

11. Rani, B.; Krichbaum, T.P.; Fuhrmann, L.; Böttcher, M.; Lott, B.; Aller, H.D.; Aller, M.F.; Angelakis, E.; Bach, U.; Bastieri, D.; et al. Radio to gamma-ray variability study of blazar S5 0716+714. *Astron. Astrophys.* **2013**, *552*, A11. [CrossRef]

12. Pandey, A.; Gupta, A.C.; Wiita, P.J. X-ray Intraday Variability of Five TeV Blazars with NuSTAR. *Astrophys. J.* **2017**, *841*, 123. [CrossRef]

13. Prince, R.; Majumdar, P.; Gupta, N. Long-term Study of the Light Curve of PKS 1510-089 in GeV Energies. *Astrophys. J.* **2017**, *844*, 62. [CrossRef]

14. Shukla, A.; Mannheim, K.; Patel, S.R.; Roy, J.; Chitnis, V.R.; Dorner, D.; Rao, A.R.; Anupama, G.C.; Wendel, C. Short-timescale γ-ray Variability in CTA 102. *Astrophys. J. Lett.* **2018**, *854*, L26. [CrossRef]

15. Marscher, A.P.; Gear, W.K. Models for high-frequency radio outbursts in extragalactic sources, with application to the early 1983 millimeter-to-infrared flare of 3C 273. *Astrophys. J.* **1985**, *298*, 114–127. [CrossRef]

16. Giannios, D. Reconnection-driven plasmoids in blazars: Fast flares on a slow envelope. *Mon. Not. R. Astron. Soc.* **2013**, *431*, 355–363. [CrossRef]

17. Sironi, L.; Petropoulou, M.; Giannios, D. Relativistic jets shine through shocks or magnetic reconnection? *Mon. Not. R. Astron. Soc.* **2015**, *450*, 183–191. [CrossRef]

18. Werner, G.R.; Uzdensky, D.A.; Cerutti, B.; Nalewajko, K.; Begelman, M.C. The Extent of Power-law Energy Spectra in Collisionless Relativistic Magnetic Reconnection in Pair Plasmas. *Astrophys. J. Lett.* **2016**, *816*, L8. [CrossRef]

19. Shukla, A.; Mannheim, K. Gamma-ray flares from relativistic magnetic reconnection in the jet of the quasar 3C 279. *Nat. Commun.* **2020**, *11*, 4176. [CrossRef]

20. Giannios, D.; Uzdensky, D.A.; Begelman, M.C. Fast TeV variability from misaligned minijets in the jet of M87. *Mon. Not. R. Astron. Soc.* **2010**, *402*, 1649–1656. [CrossRef]

21. Aharonian, F.; Akhperjanian, A.G.; Bazer-Bachi, A.R.; Behera, B.; Beilicke, M.; Benbow, W.; Berge, D.; Bernlöhr, K.; Boisson, C.; Bolz, O.; et al. An Exceptional Very High Energy Gamma-Ray Flare of PKS 2155-304. *Astrophys. J. Lett.* **2007**, *664*, L71–L74. [CrossRef]

22. Arlen, T.; Aune, T.; Beilicke, M.; Benbow, W.; Bouvier, A.; Buckley, J.H.; Bugaev, V.; Cesarini, A.; Ciupik, L.; Connolly, M.P.; et al. Rapid TeV Gamma-Ray Flaring of BL Lacertae. *Astrophys. J.* **2013**, *762*, 92. [CrossRef]

23. Ackermann, M.; Anantua, R.; Asano, K.; Baldini, L.; Barbiellini, G.; Bastieri, D.; Becerra Gonzalez, J.; Bellazzini, R.; Bissaldi, E.; Blandford, R.D.; et al. Minute-timescale > 100 MeV γ-ray Variability during the Giant Outburst of Quasar 3C 279 Observed by Fermi-LAT in 2015 June. *Astrophys. J. Lett.* **2016**, *824*, L20. [CrossRef]

24. Liu, H.T.; Bai, J.M. Absorption of 10-200 GeV Gamma Rays by Radiation from Broad-Line Regions in Blazars. *Astrophys. J.* **2006**, *653*, 1089–1097. [CrossRef]

25. Abdollahi, S.; Acero, F.; Ackermann, M.; Ajello, M.; Atwood, W.B.; Axelsson, M.; Baldini, L.; Ballet, J.; Barbiellini, G.; Bastieri, D.; et al. Fermi Large Area Telescope Fourth Source Catalog. *Astrophys. J. Suppl.* **2020**, *247*, 33. [CrossRef]

26. Hewett, P.C.; Wild, V. Improved redshifts for SDSS quasar spectra. *Mon. Not. R. Astron. Soc.* **2010**, *405*, 2302–2316. [CrossRef]

27. Fan, J.H.; Tao, J.; Qian, B.C.; Gupta, A.C.; Liu, Y.; Yuan, Y.H.; Yang, J.H.; Wang, H.G.; Huang, Y. Optical Photometrical Observations and Variability for Quasar 4C 29.45. *Publ. Astron. Soc. Jpn.* **2006**, *58*, 797–808. [CrossRef]

28. Thompson, D.J.; Bertsch, D.L.; Dingus, B.L.; Esposito, J.A.; Etienne, A.; Fichtel, C.E.; Friedlander, D.P.; Hartman, R.C.; Hunter, S.D.; Kendig, D.J.; et al. The Second EGRET Catalog of High-Energy Gamma-Ray Sources. *Astrophys. J. Suppl.* **1995**, *101*, 259. [CrossRef]

29. Abdo, A.A.; Ackermann, M.; Ajello, M.; Allafort, A.; Antolini, E.; Atwood, W.B.; Axelsson, M.; Baldini, L.; Ballet, J.; Barbiellini, G.; et al. The First Catalog of Active Galactic Nuclei Detected by the Fermi Large Area Telescope. *Astrophys. J.* **2010**, *715*, 429–457. [CrossRef]

30. Ramakrishnan, V.; León-Tavares, J.; Rastorgueva-Foi, E.A.; Wiik, K.; Jorstad, S.G.; Marscher, A.P.; Tornikoski, M.; Agudo, I.; Lähteenmäki, A.; Valtaoja, E.; et al. The connection between the parsec-scale radio jet and γ-ray flares in the blazar 1156+295. *Mon. Not. R. Astron. Soc.* **2014**, *445*, 1636–1646. [CrossRef]

31. Prince, R. Multi-frequency Variability Study of Ton 599 during the High Activity of 2017. *Astrophys. J.* **2019**, *871*, 101. [CrossRef]

32. Patel, S.R.; Chitnis, V.R.; Shukla, A.; Rao, A.R.; Nagare, B.J. Temporal Variability and Estimation of Jet Parameters for Ton 599. *Astrophys. J.* **2018**, *866*, 102. [CrossRef]

33. Patel, S.R.; Chitnis, V.R. Leptonic modelling of Ton 599 in flare and quiescent states. *Mon. Not. R. Astron. Soc.* **2020**, *492*, 72–78. [CrossRef]

34. Atwood, W.B.; Abdo, A.A.; Ackermann, M.; Althouse, W.; Anderson, B.; Axelsson, M.; Baldini, L.; Ballet, J.; Band, D.L.; Barbiellini, G.; et al. The Large Area Telescope on the Fermi Gamma-Ray Space Telescope Mission. *Astrophys. J.* **2009**, *697*, 1071–1102. [CrossRef]

35. Mattox, J.R.; Bertsch, D.L.; Chiang, J.; Dingus, B.L.; Digel, S.W.; Esposito, J.A.; Fierro, J.M.; Hartman, R.C.; Hunter, S.D.; Kanbach, G.; et al. The Likelihood Analysis of EGRET Data. *Astrophys. J.* **1996**, *461*, 396. [CrossRef]

36. Vaughan, S.; Edelson, R.; Warwick, R.S.; Uttley, P. On characterizing the variability properties of X-ray light curves from active galaxies. *Mon. Not. R. Astron. Soc.* **2003**, *345*, 1271–1284. [CrossRef]

37. Pandey, A.; Gupta, A.C.; Wiita, P.J. X-ray Flux and Spectral Variability of Six TeV Blazars with NuSTAR. *Astrophys. J.* **2018**, *859*, 49. [CrossRef]

38. Nolan, P.L.; Abdo, A.A.; Ackermann, M.; Ajello, M.; Allafort, A.; Antolini, E.; Atwood, W.B.; Axelsson, M.; Baldini, L.; Ballet, J.; et al. Fermi Large Area Telescope Second Source Catalog. *Astrophys. J. Suppl.* **2012**, *199*, 31. [CrossRef]

39. Paliya, V.S. Fermi-Large Area Telescope Observations of the Exceptional Gamma-Ray Flare from 3C 279 in 2015 June. *Astrophys. J. Lett.* **2015**, *808*, L48. [CrossRef]

40. Zhang, L.Z.; Fan, J.H.; Cheng, K.S. The Multiwavelength Doppler Factors for a Sample of Gamma-Ray Loud Blazars. *Publ. Astron. Soc. Jpn.* **2002**, *54*, 159–169. [CrossRef]

41. Foschini, L.; Ghisellini, G.; Tavecchio, F.; Bonnoli, G.; Stamerra, A. Search for the shortest variability at gamma rays in flat-spectrum radio quasars. *Astron. Astrophys.* **2011**, *530*, A77. [CrossRef]

42. Pushkarev, A.B.; Kovalev, Y.Y.; Lister, M.L.; Savolainen, T. Jet opening angles and gamma-ray brightness of AGN. *Astron. Astrophys.* **2009**, *507*, L33–L36. [CrossRef]

43. Ghisellini, G.; Maraschi, L. Bulk acceleration in relativistic jets and the spectral properties of blazars. *Astrophys. J.* **1989**, *340*, 181–189. [CrossRef]

44. Begelman, M.C.; Sikora, M.; Giommi, P.; Barr, P.; Garilli, B.; Gioia, I.M.; Maccacaro, T.; Maccagni, D.; Schild, R.E. Inverse Compton scattering of ambient radiation by a cold relativistic jet - A source of beamed, polarized continuum in blazars? *Astrophys. J.* **1987**, *322*, 650–661. [CrossRef]

45. Sikora, M.; Begelman, M.C.; Rees, M.J. Comptonization of diffuse ambient radiation by a relativistic jet: The source of gamma rays from blazars? *Astrophys. J.* **1994**, *421*, 153–162. [CrossRef]

46. Sbarrato, T.; Foschini, L.; Ghisellini, G.; Tavecchio, F. Study of the variability of blazars gamma-ray emission. *Adv. Space Res.* **2011**, *48*, 998–1003. [CrossRef]

47. Abdo, A.A.; Ackermann, M.; Ajello, M.; Atwood, W.B.; Axelsson, M.; Baldini, L.; Ballet, J.; Barbiellini, G.; Bastieri, D.; Bechtol, K.; et al. Spectral Properties of Bright Fermi-Detected Blazars in the Gamma-Ray Band. *Astrophys. J.* **2010**, *710*, 1271–1285. [CrossRef]

48. Paliya, V.S.; Sahayanathan, S.; Stalin, C.S. Multi-Wavelength Observations of 3C 279 During the Extremely Bright Gamma-Ray Flare in 2014 March-April. *Astrophys. J.* **2015**, *803*, 15. [CrossRef]

49. Rajput, B.; Stalin, C.S.; Sahayanathan, S.; Rakshit, S.; Mandal, A.K. Temporal correlation between the optical and γ-ray flux variations in the blazar 3C 454.3. *Mon. Not. R. Astron. Soc.* **2019**, *486*, 1781–1795. [CrossRef]

50. Coogan, R.T.; Brown, A.M.; Chadwick, P.M. Localizing the γ-ray emission region during the 2014 June outburst of 3C 454.3. *Mon. Not. R. Astron. Soc.* **2016**, *458*, 354–365. [CrossRef]

51. Cerruti, M.; Dermer, C.D.; Lott, B.; Boisson, C.; Zech, A. Gamma-Ray Blazars near Equipartition and the Origin of the GeV Spectral Break in 3C 454.3. *Astrophys. J. Lett.* **2013**, *771*, L4. [CrossRef]

52. Costamante, L.; Cutini, S.; Tosti, G.; Antolini, E.; Tramacere, A. On the origin of gamma-rays in Fermi blazars: Beyond the broad-line region. *Mon. Not. R. Astron. Soc.* **2018**, *477*, 4749–4767. [CrossRef]

53. Tramacere, A.; Giommi, P.; Perri, M.; Verrecchia, F.; Tosti, G. Swift observations of the very intense flaring activity of Mrk 421 during 2006. I. Phenomenological picture of electron acceleration and predictions for MeV/GeV emission. *Astron. Astrophys.* **2009**, *501*, 879–898. [CrossRef]

54. Dermer, C.D.; Yan, D.; Zhang, L.; Finke, J.D.; Lott, B. Near-equipartition Jets with Log-parabola Electron Energy Distribution and the Blazar Spectral-index Diagrams. *Astrophys. J.* **2015**, *809*, 174. [CrossRef]

galaxies

Article

The Radiative Newtonian $1 < \gamma \leq 1.66$ and the Paczyński–Wiita $\gamma = 5/3$ Regime of Non-Isothermal Bondi Accretion onto a Massive Black Hole with an Accretion Disc

Jose M. Ramírez-Velásquez [1,*], Leonardo Di G. Sigalotti [2], Ruslan Gabbasov [3], Jaime Klapp [4] and Ernesto Contreras [5]

[1] School of Physical Sciences and Nanotechnology, Yachay Tech University, Hacienda San José, Urcuqui 100119, Ecuador

[2] Área de Física de Procesos Irreversibles, Departamento de Ciencias Básicas, Universidad Autónoma Metropolitana-Azcapotzalco (UAM-A), Av. San Pablo 180, Ciudad de México 02200, Mexico; ldgsd@azc.uam.mx

[3] Departamento de Sistemas, Universidad Autónoma Metropolitana-Azcapotzalco (UAM-A), Av. San Pablo 180, Ciudad de México 02200, Mexico; gabbasov@azc.uam.mx

[4] Departamento de Física, Instituto Nacional de Investigaciones Nucleares (ININ), Carretera México-Toluca km. 36.5, La Marquesa, Ocoyoacac 52750, Mexico; jaime.klapp@inin.gob.mx

[5] Departamento de Física, Colegio de Ciencias e Ingeniería, Campus Cumbayá, Universidad San Francisco de Quito, Quito 170901, Ecuador; econtreras@usfq.edu.ec

* Correspondence: jmramirez@yachaytech.edu.ec

Citation: Ramírez-Velásquez, J.M.; Sigalotti, L.D.G.; Gabbasov, R.; Klapp, J.; Contreras, E. The Radiative Newtonian $1 < \gamma \leq 1.66$ and the Paczyński–Wiita $\gamma = 5/3$ Regime of Non-Isothermal Bondi Accretion onto a Massive Black Hole with an Accretion Disc. *Galaxies* **2021**, *9*, 55. https://doi.org/10.3390/galaxies 9030055

Academic Editor: Dimitris M. Christodoulou

Received: 11 May 2021
Accepted: 9 August 2021
Published: 11 August 2021

Publisher's Note: MDPI stays neutral with regard to jurisdictional claims in published maps and institutional affiliations.

Abstract: We investigate the non-isothermal Bondi accretion onto a supermassive black hole (SMBH) for the unexplored case when the adiabatic index is varied in the interval $1 < \gamma \leq 1.66$ and for the Paczyński–Wiita $\gamma = 5/3$ regime, including the effects of X-ray heating and radiation force due to electron scattering and spectral lines. The X-ray/central object radiation is assumed to be isotropic, while the UV emission from the accretion disc is assumed to have an angular dependence. This allows us to build streamlines in any desired angular direction. The effects of both types of radiation on the accretion dynamics is evaluated with and without the effects of spectral line driving. Under line driving (and for the studied angles), when the UV flux dominates over the X-ray heating, with a fraction of UV photons going from 80% to 95%, and γ varies from 1.66 to 1.1, the inflow close to the gravitational source becomes more supersonic and the volume occupied by the supersonic inflow becomes larger. This property is also seen when this fraction goes from 50% to 80%. The underestimation of the Bondi radius close to the centre increases with increasing γ, while the central overestimation of the accretion rates decreases with increasing γ, for all the six studied cases.

Keywords: black hole evolution; supermassive black hole; accretion of matter; galaxies: evolution; galaxies: nuclei

1. Introduction

Observations of giant elliptic galaxies provide firm evidence of the presence of supermassive black holes (SMBHs) in their centres [1,2], which accrete matter from the surroundings and liberate enormous amounts of energy that affect their environments from pc to Mpc scales (see [3] for spherically symmetric black holes with quantum corrections). In particular, the mass accretion rate onto these SMBHs, which is a key quantity to understand the galactic evolution, is usually estimated using Bondi accretion theory [4]. However, semi-analytic calculations and numerical simulations based on the Bondi accretion model do not provide information about the flow transport down to pc and sub-pc scales and the mass accretion rates (e.g., [5–8]). Since active galactic nuclei (AGNs) evolve concomitantly with their host galaxies, they affect each other. For example, good evidence for this kind of feedback has been provided by observations of AGN-starbursts [9]. On the other hand, the energy released during accretion onto a SMBH can hinder further accretion and drive the gas away, which in turn self-regulates the galaxy growth [10].

An earlier study on the structure of X-ray-irradiated accretion discs in AGNs was considered by the authors of [11]. They found that the irradiated region above and below the disc consists of a region which is supported by the radiation pressure where the UV flux is created and a warmer thin layer above this region, which is optically thin to the UV radiation. In the last years, studies of flows in AGNs with the inclusion of radiation source terms have become progressively more consistent [12–17]. All of these studies were able to compute cooling and heating functions, which are important to produce the correct opacities in these environments, and new methodologies were proposed to properly include the coupling between matter and radiation. Winds and accretion processes are of primary importance to improve our understanding of the feedback between the galaxy and its guest SMBH [18–23] and therefore on the galactic evolution [24–27]. Photoionization calculations of radiative forces due to spectral lines (i.e., [15]) have also shown the importance of a proper treatment of the coupling between matter and radiation along with the influence of the non-LTE effects (i.e., [28]).

Much effort has been devoted to study the dynamical evolution of the system in terms of the accretion rates onto a SMBH from the numerical solution of the hydrodynamics equations, where a common assumption has been to fix the boundary conditions at infinity as it is indeed required by the classical Bondi solution. However, an inconvenience with this approach is that when exploring the dynamics close to the black hole, this boundary conditions may fail to represent the finiteness of the spatial region under study. Several calculations of accretion processes with an ideal equation of state with different values of the adiabatic index (γ; in full symmetric and axisymmetric configurations in relativistic contexts e.g.; [29–35]) show it playing an important role in the estimation of the mass supply in cosmological simulations [21,36,37], and so does the exploration of the effects of varying the adiabatic index on the AGN accretion dynamics. In this paper we extend the analysis of [38] and present solutions for the radial Mach number and density profiles along with the critical points of transonic solutions for the radiative, radial accretion of matter in the potential well of a SMBH for adiabatic indices in the range $1.1 \leq \gamma \leq 1.66$. It is found that variations of the adiabatic index may change the whole dynamics of the system as well as the mass supply. The methodology employed is detailed in Section 2 and the results are described in Section 3. A catalogue of pure absorption lines for $1.1 \leq \gamma \leq 1.66$ is given in Section 4. Estimates of the Bondi radius and mass accretion rate as functions of the adiabatic index are given in Section 5. Finally, Section 6 summarizes the relevant conclusions.

2. Non-Isothermal Radial Bondi Accretion

Under the effects of irradiation by X rays, the accretion flow onto the central SMBH is described by the mass and momentum conservation laws

$$\frac{d\rho}{dt} = -\rho \nabla \cdot \mathbf{v}, \tag{1}$$

$$\frac{d\mathbf{v}}{dt} = -\frac{1}{\rho}\nabla p + \mathbf{g} + \mathbf{F}^{\text{rad}}, \tag{2}$$

where ρ denotes the density, $\mathbf{v}$ the velocity field, p the gas pressure, $\mathbf{g}$ the gravitational acceleration due to the SMBH, $\mathbf{F}^{\text{rad}} = (F_r, F_\theta = 0, F_\phi = 0)$ the radiation force per unit mass, and $d/dt = \partial/\partial t + \mathbf{v} \cdot \nabla$ the material time derivative. Using spherical coordinates and assuming that the angular components of the velocity vanish (i.e., $v_\theta = v_\phi = 0$), Equation (2) can be written as

$$\frac{\partial v_r}{\partial t} + v_r \frac{\partial v_r}{\partial r} = -\frac{1}{\rho}\frac{\partial p}{\partial r} - \frac{\partial}{\partial r}\psi_{\text{grav}}(r) + \frac{\partial}{\partial r}\psi_{\text{rad}}(r), \tag{3}$$

for the radial velocity component, where $v_r = v_r(r,\theta,\phi)$, $p = p(r,\theta,\phi)$, M_{BH} is the mass of the SMBH, and $\psi_{\text{rad}}(r)$ is a function defined below. If we further assume azimuthal symmetry (i.e., $\partial/\partial\phi = 0$) and perform the analysis for a fixed angle θ_0, such that $v_r(r,\theta,\phi) = v_r(r,\theta_0) = v_r$ and $p(r,\theta,\phi) = p(r,\theta_0) = p$, the above equation becomes

$$\frac{d\mathcal{H}(r)}{dr} = p\frac{d}{dr}\left(\frac{1}{\rho}\right) - \frac{\partial v_r}{\partial t}, \tag{4}$$

where

$$\mathcal{H}(r) = \frac{p}{\rho} + \frac{1}{2}v_r^2 - \psi_{\text{grav}}(r) - \psi_{\text{rad}}(r), \tag{5}$$

is the Bernoulli function $\mathcal{H}(r)$, $\psi_{\text{grav}} = (GM_{\text{BH}})/r$ and $\psi_{\text{rad}} = -C(r)/r$ are the gravitational and radiation potentials, respectively. Compared to the analysis of [38], a radial dependence is now assumed for the term C in the definition of the radiation potential. For simplicity the accretion disc is assumed to be flat, Keplerian, geometrically thin and optically thick. The pressure is related to the density by means of an ideal equation of state

$$p = \frac{k_B \rho T}{\mu m_p} = p_\infty \tilde{\rho}^\gamma, \tag{6}$$

where k_B is the Boltzmann constant, T is the gas temperature, μ is the mean molecular weight, and m_p is the proton mass. Here γ is varied in the range $1 \leq \gamma < 5/3$ and $\tilde{\rho} \equiv \rho/\rho_\infty$, where p_∞ and ρ_∞ are the values of the pressure and density at infinity, respectively. The sound speed is defined according to

$$c_{\text{s}}^2 = \frac{\gamma p}{\rho}. \tag{7}$$

The radiative contributions in Equations (2) and (4) are here implemented using the strategies used by [13,39–41] (hereafter P07). Although the radiation field from both the disc and the central black hole are modelled as in P07 by adding a radiation force in the momentum equation, there are important differences with P07 that are worth commenting. For instance, in P07 the analysis is based on time-dependent, axisymmetric simulations with $\gamma = 5/3$, while our analysis consists of steady-state calculations along radial streamlines for fixed θ and varying values of the adiabatic index in the interval $1 \lesssim \gamma < 5/3$. These calculations have been designed to serve as initial conditions for fully three-dimensional simulations of accretion discs [42]. As in P07, the disc is assumed to emit only UV light (i.e., $f_{\text{disc}} = f_{\text{UV}}$) and the central SMBH to emit only X-rays (i.e., $f_\star = f_{\text{X}} = 1 - f_{\text{disc}}$) so that the disc and central luminosities are given by $L_{\text{disc}} = f_{\text{disc}}L$ and $L_\star = f_\star L$, respectively, where L is the total accretion luminosity. The radial component of the radiation force is approximated according to the relation (e.g., P07)

$$F_{r_{\text{rad}}}(r,\theta_0) = \frac{\sigma_T L}{4\pi r^2 c m_p}[f_\star + 2\cos\theta_0 f_{\text{disk}}(1 + M(r,t))], \tag{8}$$

where c is the speed of light in vacuum, σ_T/m_p is the mass scattering coefficient for free electrons, θ_0 is some fixed and constant polar angle measured from the rotational axis of the disk, and t is the optical depth

$$t = \frac{(\sigma_T/m_p)\rho v_{\text{th}}}{|dv_r/dr|}, \tag{9}$$

where v_{th} is the thermal velocity (set using a temperature of 25,000 K), dv_r/dr is the velocity gradient along the radial direction, and $M(r,t)$ is the so-called force multiplier [43], defined by

$$M(r,t) = kt^{-\alpha}\left[\frac{(1 + \tau_{\text{max}})^{(1-\alpha)} - 1}{\tau_{\text{max}}^{(1-\alpha)}}\right], \tag{10}$$

where $\alpha = 0.6$ is the ratio of optically thick to optically thin lines. According to [44], the parameter k is given by the minimun between

$$k = 0.03 + 0.385 \exp\left(-1.4\xi^{0.6}\right), \tag{11}$$

and

$$\log k = \begin{cases} -0.383, & \text{for } \log T \leq 4, \\ -0.630 \log T + 2.138, & \text{for } 4 < \log T \leq 4.75, \\ -0.870 \log T + 17.528, & \text{for } \log T > 4.75, \end{cases} \tag{12}$$

which is based on detailed photoionization calculations performed using the XSTAR code (P07), while $\tau_{\max} = t\eta_{\max}$ with

$$\log \eta_{\max} = \begin{cases} 6.9 \exp\left(0.16\xi^{0.4}\right), & \text{if } \log\xi \leq 0.5, \\ 9.1 \exp\left(-7.96 \times 10^{-3}\xi\right), & \text{if } \log\xi > 0.5, \end{cases} \tag{13}$$

where ξ is the photoionization parameter. The parameter $\eta_{\max}$ defined by Equation (13) is used to determined the maximum force multiplier, i.e., $M_{\max} = k(1-\alpha)\eta_{\max}^{\alpha}$. The local X-ray flux

$$\mathcal{F}_X = \frac{L_\star}{4\pi r^2} \exp(-\tau_X), \tag{14}$$

is used to estimate the photoionization parameter $\xi = 4\pi\mathcal{F}_X/n$, where

$$\tau_X = \int_0^r \kappa_X \rho \, dr, \tag{15}$$

is the X-ray optical depth, evaluated between the centre ($r = 0$) and a radius r in the accreting flow, and $n = \rho/(\mu m_p)$ is the number density of the accreting gas. The absorption coefficient κ_X is set to 0.4 g^{-1} cm^2 for all values of ξ.

The $\gamma = 5/3$ regime of non-isothermal Bondi accretion is modelled using the Paczyński–Wiita (PW) potential [45,46]

$$\psi_{\mathrm{grav}} = \frac{GM_{\mathrm{BH}}}{r - R_s}, \tag{16}$$

where $R_s = 2GM_{\mathrm{BH}}/c^2$ is the gravitational radius of the black hole. While this pseudo-Newtonian potential does not obey the Poisson equation, it has become a standard tool because it accurately models the general relativistic accretion of matter onto a nonrotating SMBH. In particular, this form of the potential reproduces the radii of a marginally stable Keplerian orbit ($r = 3R_s$) and of a marginally bound orbit ($r = 2R_s$) as predicted by Einstein's gravity in the Schwarzschild metric [46].

Under the assumption of steady-state motion, Equations (4) and (5) can be combined to produce after a few algebraic steps the integral equation

$$\int \frac{d}{dr}\left(\frac{\mathcal{M}^2\tilde{c}_s^2}{2} + \frac{\tilde{\rho}^{(\gamma-1)}}{\gamma-1} - \frac{1}{x} + \frac{l_{\mathrm{tot}}^{\mathrm{rad}}|_{\theta=\theta_0}(x)}{x}\right) dr = 0, \tag{17}$$

where

$$x \equiv \frac{r}{r_B}, \quad \tilde{c}_s \equiv \frac{c_s}{c_\infty} = \tilde{\rho}^{(\gamma-1)/2}, \quad \mathcal{M} \equiv \frac{v_r}{c_s}, \tag{18}$$

and

$$l_{\mathrm{tot}}^{\mathrm{rad}}|_{\theta=\theta_0}(x) = l_{\mathrm{Edd}}^{\mathrm{rad}} f_{\mathrm{rad}}|_{\theta=\theta_0}(x). \tag{19}$$

Here $C|_{\theta=\theta_0}(x) = GM_{\mathrm{BH}} l_{\mathrm{tot}}^{\mathrm{rad}}|_{\theta=\theta_0}(x)$,

$$r_B = \frac{GM_{\mathrm{BH}}}{c_\infty^2}, \tag{20}$$

is the Bondi radius, $c_\infty^2 = \gamma p_\infty/\rho_\infty$, $\mathcal{M}$ is the Mach number, $l_{\mathrm{Edd}}^{\mathrm{rad}} = L/L_{\mathrm{Edd}}$ with $L_{\mathrm{Edd}} = 4\pi c G M_{\mathrm{BH}} m_p/\sigma_T$ being the Eddington luminosity, $\sigma_T = 6.6524 \times 10^{-25}$ cm^2 is the Thomson cross section, and $f_{\mathrm{rad}}|_{\theta=\theta_0}(x)$ is the radiative force parameter given by the relation

$$f_{\mathrm{rad}}|_{\theta=\theta_0}(x) = f_\star + 2\cos\theta_0 f_{\mathrm{disk}}[1 + M(x,t)]. \tag{21}$$

Although the force multiplier is an explicit function of x and t, for simplicity hereafter we omit the x-dependence and write it as $M(t = \tau)$.

Equation (17) can be integrated to give

$$\tilde{\rho}^{(\gamma-1)}\left(\frac{\mathcal{M}^2}{2} + \frac{1}{\gamma-1}\right) = \frac{1}{x} - \frac{l_{\mathrm{tot}}^{\mathrm{rad}}|_{\theta=\theta_0}(x)}{x} + \frac{1}{\gamma-1}. \tag{22}$$

This equation describes the non-isothermal ($\gamma > 1$), steady-state accretion along a radial streamline for a fixed angle $\theta = \theta_0$ when the effects of radiation emission due to electron scattering and spectral discrete lines with appropriate boundary conditions at infinity and ionization changes due to temperature corrections are considered. A sketch showing the geometry of the system is displayed in Figure 1, where a radial streamline at a meridional angle θ_0 is shown. In terms of the above assumptions and the normalized parameters (18), Equation (1) becomes

$$x^2 \mathcal{M} \tilde{\rho}^{(\gamma+1)/2} = \lambda, \tag{23}$$

where λ is the accretion parameter given by

$$\lambda = \frac{\dot{M}_B}{(4\pi f_{\mathrm{solid}})r_B^2 \rho_\infty c_\infty}, \tag{24}$$

which determines the accretion rate for given boundary conditions and mass of the SMBH. In this expression, $4\pi f_{\mathrm{solid}} = \int \sin\theta d\theta d\phi$ is the solid angle covered by the streamline at the polar angle θ_0 (see Figure 1). When $f_{\mathrm{solid}} = 1$ (full solid angle), Equation (24) reduces to the classical accretion parameter. However, for a θ_0-dependent force $f_{\mathrm{solid}} \ll 1$. This dependence is important only for the final calculation of $\dot{M}_B = \lambda(4\pi f_{\mathrm{solid}})r_B^2 \rho_\infty c_\infty$, which is fixed once the value of θ_0 is chosen. Replacing $\tilde{\rho}$ from Equation (23) into Equation (22), the radiative-radial Bondi problem reduces to solving the equation

$$g(\mathcal{M}) = \Lambda f(x), \quad \text{with} \quad \Lambda = \lambda^{2(1-\gamma)/(\gamma+1)}, \tag{25}$$

where $\chi_{\mathrm{tot}}^{\mathrm{rad}}|_{\theta=\theta_0} = 1 - l_{\mathrm{tot}}^{\mathrm{rad}}|_{\theta=\theta_0}$, $\lambda = \chi_{\mathrm{tot}}^{\mathrm{rad}}|_{\theta=\theta_0}^2 \lambda_{\mathrm{cr}}$, and

$$g(\mathcal{M}) = \mathcal{M}^{2(1-\gamma)/(\gamma+1)}\left(\frac{\mathcal{M}^2}{2} + \frac{1}{\gamma-1}\right), \tag{26}$$

$$f(x) = x^{4(\gamma-1)/(\gamma+1)}\left(\frac{\chi_{\mathrm{tot}}^{\mathrm{rad}}|_{\theta=\theta_0}}{x} + \frac{1}{\gamma-1}\right), \tag{27}$$

$$\lambda_{\mathrm{cr}} = \frac{1}{4}\left(\frac{2}{5-3\gamma}\right)^{(5-3\gamma)/[2(\gamma-1)]}. \tag{28}$$

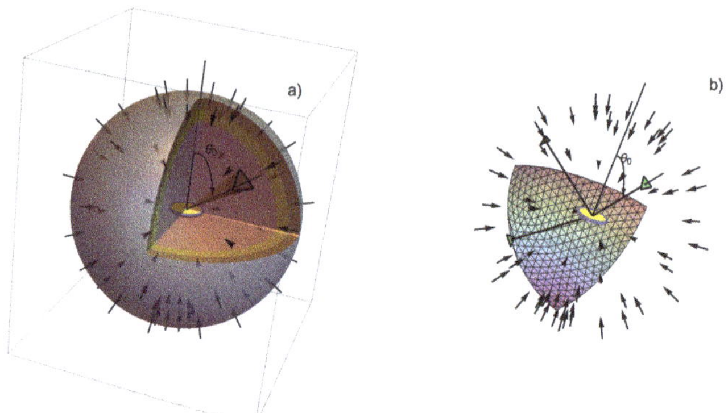

Figure 1. Model geometry. (**a**) Integration is performed along a radial streamline for a fixed meridional angle θ_0. (**b**) The full angular dependence can be obtained by integrating along different radial streamlines.

For this case Equations (17)–(27) remain the same with the only change being the substitution of x by $x - x_s$, where $x_s = R_s/r_B$. Here we take $R_s/r_B = 1.75 \times 10^{-5}$ for the calculations of Figure 2 and 2.18×10^{-6} for the rest of the calculations. The radiative Bondi-like problem then reduces to solving Equations (25)–(28) for the Mach number, $\mathcal{M}$, as a function of radius. On the other hand, it must be noticed that the PW potential, in addition to reproducing the innermost stable circular orbit and the marginally bound orbit, also accounts for a non-vanishing critical radius for $\gamma = 5/3$, in contrast to the purely Newtonian radial accretion, for which

$$x_{\mathrm{crit}} = x_s + \sqrt{\frac{2\chi_{\mathrm{total}}^{\mathrm{rad}} x_s}{3}}, \tag{29}$$

where now the factor $\chi_{\mathrm{total}}^{\mathrm{rad}}$ (>0) appears, which is not present in [38,47].

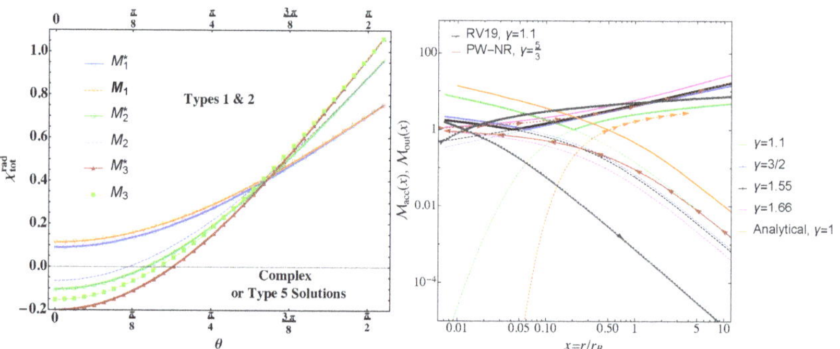

Figure 2. (**Left**) Angular dependence of the UV emission for models $M_{1,2,3}$ and $M_{1,2,3}^{\star}$. The horizontal black solid line at $\chi_{\mathrm{tot}}^{\mathrm{rad}} = 0$ mark the transition from type 5 to types 1 and 2 solutions. (**Right**) Mach number as a function of radius for model $M_1^{\star}$ and $\gamma = 1$ (analytical solution; orange curves), $\gamma = 1.1$ (numerical solution from [38] with $\psi_{\mathrm{rad}} = -C/r$; gray curves, and this work with $C = C(r)$; green curves), $\gamma = 3/2$ (blue curves), $\gamma = 1.55$ (black curves), $\gamma = 1.66$ (magenta curves) and $\gamma = 5/3$ (non-radiative (NR), with $x_s = 2.5 \times 10^{-3}$; red curves). The thick solid curves represent supersonic inflow ($x < x_s$) and outflow ($x > x_s$) solutions, while the dashed lines represent inflow and outflow subsonic solutions.

The angular variation of the UV emission, $\chi_{\rm tot}^{\rm rad}$, for models with and without the force multiplier is shown in the top panel of Figure 2. Model M_1 has $f_\star = 0.5$ and $f_{\rm disc} = 0.5$, model M_2 has $f_\star = 0.2$ and $f_{\rm disc} = 0.8$, while model M_3 has $f_\star = 0.05$ and $f_{\rm disc} = 0.95$, with $M(\tau) = 0$ so that both types of radiation can be evaluated in detail. Models $M_1^\star$, $M_2^\star$, and $M_3^\star$ differ from the preceding models in that $M(\tau) \neq 0$. For different fractions of $f_\star$, $f_{\rm disc}$, and position x, we estimate the critical angles for which a transition from type 5 to type 1 and 2 solutions is allowed[1]. This plot, which corresponds to $\tilde{\rho} = 0.001$, $x = 3$, and $\gamma = 1.55$, is an example of the complexity of the system. In particular, $M_2^\star$ (green line with empty squares) and $M_3^\star$ (red line with full triangles) have physical transonic solutions for $\theta > \pi/8$, while model $M_1^\star$ (blue line with empty circles) does not admit solutions for which $\chi_{\rm tot}^{\rm rad} < 0$, and so this flow can be seen in any angular direction. The transition for $M_2^\star$ occurs at a lower angle than for $M_3^\star$, meaning that we should find more collimated streamlines for the $M_2^\star$ flows than for the $M_3^\star$ flows. It is interesting to see that model M_1, for which only electron scattering is present ($M(\tau) = 0$) and $f_\star - f_{\rm disc} - 0.5$, behaves like model $M_1^\star$ with no transition to the unphysical region where $\chi_{\rm tot}^{\rm rad} < 0$. As the fraction of disc photons increases (i.e., from $f_{\rm disc} = 0.8$ to 0.95), critical angles for which unphysical type 5 solutions occur (for $\theta_{M_{2,3}}^{\rm phase} < \pi/8$) are found even when only electron scattering is dominating.

In both sets of models $M_{1,2,3}$ and $M_{1,2,3}^\star$, the radiation acceleration depends on the ionization parameter defined as

$$\xi = \frac{(f_\star L) \exp\left(-\tau_X\right)}{nr^2},\tag{30}$$

where $L = 7.45 \times 10^{45}$ erg s^{-1}, which is appropriate for a $10^8 M_\odot$ SMBH accreting at an efficiency of 8%. The gas density is $n(x) = (\rho_\infty \tilde{\rho}(x))/(\mu m_p)$ cm^{-3}, with $\rho_\infty = 10^{-20}$ gr cm^{-3}, $\mu = 0.7$, and the optical depth τ_X is integrated using Equation (15) with $\kappa_X = 0.4$ g^{-1} cm^2. The physical parameters for all models computed in this study are listed in Table 1 (a full analysis of the angular dependence is beyond the scope of the present work, and will be presented in a forthcoming paper). In all cases we choose $\theta_0 = \theta_{\rm acc}^{\rm phase}$. The fractions $f_\star$ and $f_{\rm disc}$ in Table 1 are exactly the same used by P07, which comply with observational results from [49,50].

Table 1. Parameters of the accretion models.

Run	Lines [a]	$f_\star$	$f_{\rm disc}$	γ [b]	$\theta^{\rm phase}$ [c] (rad)	$\theta_{\rm acc}^{\rm phase}$ (rad)
$M_1^\star$	yes	0.50	0.50	-	$\frac{3}{8}\pi$	1.1
$M_2^\star$	yes	0.20	0.80	-	$\frac{3}{8}\pi$	1.6
$M_3^\star$	yes	0.05	0.95	-	$\frac{3}{8}\pi$	1.7
M_1	no	0.50	0.50	-	$\frac{3}{8}\pi$	1
M_2	no	0.20	0.80	-	$\frac{3}{8}\pi$	1.5
M_3	no	0.05	0.95	-	$\frac{3}{8}\pi$	1.6

[a] $M(t) \neq 0$ (yes) and $M(t) = 0$ (no). [b] The values of $\gamma = \{1.1, 1.2, 1.3, 13/9, 3/2, 1.53, 1.55, 1.63, 1.65, 1.66\}$. [c] The angle $\theta_{\rm acc}^{\rm phase}$ is given in terms of $\theta^{\rm phase}$, i.e., $\theta_{\rm acc}^{\rm phase} = 1.1\theta^{\rm phase}$. All models $M_\star^\star$ and $M_\star$ shared the same angle.

The bottom panel of Figure 2 shows the dependence of the Mach number of the inflow and outflow transonic solutions with the normalized radial distance from the black hole for model $M_1^\star$ and various characteristic values of the adiabatic index, namely $\gamma = 1$ (full analytical solution; orange curves), $\gamma = 1.1$ (numerical solution from [38] with $\psi_{\rm rad} = -C/r$; gray curves, and this paper with $\psi_{\rm rad} = -C(r)/r$; green curves), $\gamma = 3/2$ (blue curves), $\gamma = 1.55$ (black curves), $\gamma = 1.66$ (magenta curves), and $\gamma = 5/3$ (red curves). Note that the latter curves correspond to the PW model with no radiation, i.e., with $f_\star = f_{\rm disc} = 0$. The intersection of curves of the same colour define the position of the sonic point (x_s). The thick solid curves for $x < x_s$ represent supersonic inflow solutions, while those at larger radii (i.e., $x > x_s$) represent supersonic outflow solutions. The dashed curves at radii $x > x_s$ are subsonic inflow solutions and those for $x < x_s$ are subsonic outflow solutions. The Mach number profiles are shown up to radial distances of $\approx 10^{-3.1} r_B$ from the centre. It is evident from this figure that isothermal ($\gamma = 1$) inflow occurs at slightly higher values of $\mathcal{M}_{\rm acc}(r)$ than its non-isothermal counterparts. As the adiabatic index is increased the accretion rate occurs at relatively smaller values of the Mach number everywhere and closer to the central source so that the region of supersonic accretion involves progressively smaller central volumes as γ increases. In contrast, the isothermal outflow takes place at much lower Mach numbers compared to the non-isothermal cases. These differences are clearly expected to influence the rates for non-isothermal accretion. As the adiabatic index is increased above $\gamma = 3/2$ the outflow close to $x = 10^{-3}$ becomes supersonic only when $\gamma = 1.66$ and keeps subsonic for all values <1.66. In the interval $0.01 \lesssim x \lesssim 0.1$, the outflow proceeds supersonically for all non-isothermal cases ($\gamma > 1$) with a small dependence of the Mach number on γ, while towards larger $x (\gtrsim 0.2)$ the outflow becomes more supersonic as the adiabatic index is increased.

3. Results

The Mach number as a function of distance from the source is shown in the top panel of Figure 3 for model $M_1^\star$ and varied γ. The solid curves depict the inflow solutions, while the dotted lines with arrows correspond to the outflow solutions. In all cases, the complete subsonic and supersonic solutions are shown. The position of the sonic point, $x_{\rm crit}$, is defined by the intersection between the inflow and the outflow solutions. When matching the subsonic portion of the solution at $x \leq x_{\rm crit}$ with the supersonic one at $x > x_{\rm crit}$, we then obtain the "Bondi-outflow" solution. The difference with the Parker solution is the location of the integration limits in the Bernoulli equation. In the Bondi problem they are located at infinity, while in the Parker problem they are located at the base of the wind. For all models, the Bondi outflows are always depicted by dotted lines with small arrows pointing out of the system. On the other hand, when matching the subsonic part of the solution at $x > x_{\rm crit}$ with the supersonic one at $x \leq x_{\rm crit}$, we then obtain the "Bondi-inflow" solution. Increased values of γ produce slower inflows and faster outflows everywhere, while the position of the sonic point is shifted towards smaller radii. The left panel of Figure 4 shows a close comparison of the Mach number profiles between the Newtonian $\gamma = 1.66$ case with spectral line forces (gray curves) and the $\gamma = 5/3$ case with $M(\tau) = 0$ (orange curves). For $M(\tau) \neq 0$ the inflow and outflow Mach number profiles essentially overlap those for the Newtonian $\gamma = 1.66$ case with $M(\tau) \neq 0$ regardless of the radiation field. In spite of differences in the value of $M(\tau)$, the inflow and outflow profiles for both cases look almost identical.

Figure 3 (right panel) displays the Mach number as a function of radius for model $M_2^\star$ and varied γ. Compared to the left panel of Figure 3, we see that increasing the fraction of UV emission (from $f_\star = 0.5$ to 0.2) increases the inflow Mach number everywhere, while the position of the sonic point is shifted towards larger radii. Consequently, for given γ, when the ionizing flux decreases faster inflows are produced everywhere. In particular, at small radii $x \lesssim 0.01$ the inflow becomes more supersonic. A similar trend is observed for model $M_3^\star$ with the weakest X-ray heating ($f_\star = 0.05$) and the strongest UV emission ($f_{\rm disc} = 0.95$), where the sonic point is moved farther away compared to model $M_2^\star$ and therefore the supersonic inflow close to the centre occupies a larger volume. A comparison of the right panel of Figure 4 with that of left panel shows that as the UV heating dominates over the X-ray luminosity, the ($\gamma = 5/3$) inflow becomes faster than the $\gamma = 1.66 - M_1^\star$ inflow. It is interesting to note that the increase of the UV emission from $f_{\rm disc} = 0.8$ to $f_{\rm disc} = 0.95$ has only mild effects on the inflow and outflow Mach number profiles as well. The Mach number profiles for models M_1 and M_2 with no force multiplier (i.e., $M(\tau) = 0$) and varied γ are now displayed in Figure 3 (right), respectively. The profiles for model M_3 are very similar to those shown in Figure 3 for model M_2. For model M_1, where the central X-ray heating is the strongest and no line driving is present, the solution does not admit type 5 flows regardless of the incident angle. As shown in Figure 3 (bottom) the critical points occur far from the central source between $x \approx 0.2$ (for $\gamma = 1.1$) and $x \approx 0.007$ (for $\gamma = 1.66$). These distances are relatively similar to those found for model $M_1^\star$ with force multiplier. For $\theta_0 = \frac{3}{8}\pi$ and close to the SMBH ($x \approx 10^{-3}$), the Mach number ranges from ≈ 1 (for $\gamma = 1.66$) to ≈ 8 (for $\gamma = 1.1$). When the fraction of UV emission is increased to $f_{\rm disc} = 0.8$, the position of the sonic point is shifted outwards so that the flow becomes supersonic in a larger central volume as can be seen by comparing bottom with top. However, model M_2 is characterized by smaller supersonic volumes than model M_3. Similarly to the cases with force multiplier, as the UV emission is increased, the flow close to the central black hole becomes more supersonic. Moreover, for $\gamma = 1.66$ models M_2 and M_3 both exhibit outflows that are slower at large radii ($x > 1$) than their Bondi-inflow counterparts close to the central source, while the inverse is seen to occur for larger γ.

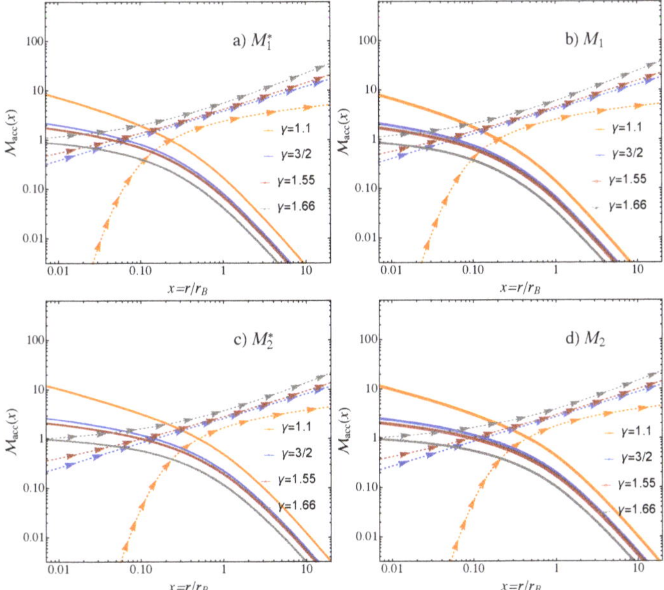

Figure 3. (Top) Radial Mach number profiles of the inflow (solid lines) and outflow (dotted lines with arrows) solutions for model **(a)** $M_1^\star$ with $M(\tau) \neq 0$ and **(b)** M_1 with $M(\tau) = 0$, $f_\star = 0.5$, $f_{\rm disc} = 0.5$, and varied γ in the interval $1.1 \leq \gamma \leq 1.66$. **(Bottom)** **(c)** $M_2^\star$ with $M(\tau) \neq 0$ and **(d)** M_2 with $M(\tau) = 0$, $f_\star = 0.2$, $f_{\rm disc} = 0.8$.

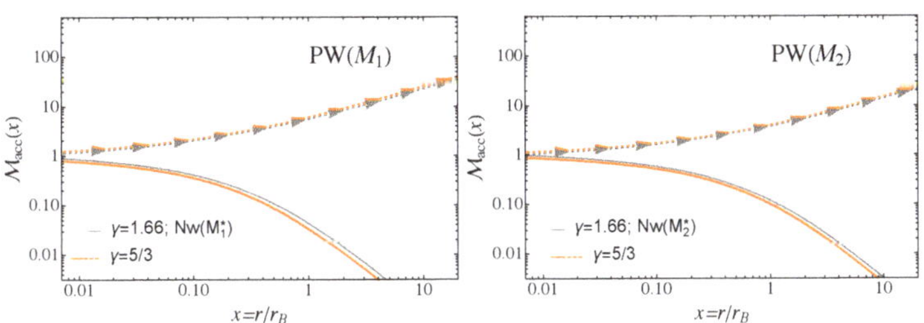

Figure 4. Comparison of the Mach number profiles between the Newtonian ($\gamma = 1.66$) case with spectral line forces and the ($\gamma = 5/3$) case with $M(\tau) = 0$.

For all models the density profiles are similar and almost independent of the adiabatic index. In all cases the density increases for $x \lesssim 0.1$ from $\approx(1-7)$ to values as high as $\approx 10^3$ when the UV emission rises from $f_{\mathrm{disc}} = 0.5$ to $f_{\mathrm{disc}} = 0.95$ for models with $M(\tau) \neq 0$. In contrast, when $M(\tau) = 0$ the highest central densities are seen to occur for model M_1 with $\tilde{\rho} \approx 5 \times 10^2$ at $x = 10^{-3}$, while models M_2 and M_3 with stronger UV emission and weaker central X-ray heating reach central densities of ≈ 1000 at the same radial distance from the SMBH, respectively. Additionally, in these cases the density profiles show a very little dependence on the adiabatic index. It is clear from the present results that the flow dynamics are sensitive to the radiative process that dominates the environment.

4. Catalogue of Pure Absorption Spectral Line Shapes for $1.1 \leq \gamma \leq 1.66$

As resolution improves with the emergence of more powerful telescopes, it becomes mandatory to predict the shape of absorption spectral lines as seen by a distant observer. This can be done by imagining a radially falling atom onto the SMBH according to the predictions of models $M^\star_{1,2,3}$, which would absorb photons emitted by the inner region at 10 in the rest-frame. Then we use the Doppler-shift formula [51]:

$$w = \frac{w_0}{\gamma L \left(1 - \frac{\mathrm{v}(x)}{c} \cos \phi\right)},\tag{31}$$

where w is the angular frequency of an emitted photon with rest-frame frequency w_0 as measured by the observer, γL is the Lorentz factor, $\mathrm{v}(x)$ is the velocity of the absorbing atom, and ϕ is the angle between the streamline and the line-of-sight towards the observer, which we set to $\phi = 0$. For falling particles we set $\mathrm{v}(x) = -\mathcal{M}(x)c_s$, where c_s is given by Equation (7)^2, for values of ≈ 360, 420, 430, 440, and 420 km s^{-1}. A simple description of the absorption spectrum is given by

$$F_\lambda(x) = \exp\left\{ -A(x) \exp\left[-\frac{(\lambda(x) - \lambda_0)^2}{\sigma^2} \right] \right\},\tag{32}$$

where $v_0\lambda_0 = w_0/(2\pi)\lambda_0 = c$, σ is given by $0.1\%\lambda(x)$, c is the speed of light, and λ_0 is the photon wavelength in the rest frame. The intensity of the absorption is modelled by setting $A(x) = \rho(x)/\rho_{\mathrm{max}}$, where $\rho(x) = \tilde{\rho}(x)\rho_\infty$ and $\rho_{\mathrm{max}} = \tilde{\rho}(x = 3 \times 10^{-4})\rho_\infty$. As the particle is getting closer and closer to the black hole, the density increases and the shape of the spectral line tends to have a deeper deep shifted towards the red. Figure 5 depicts the predicted accretion absorption line shape for models $M^\star_1$, $M^\star_2$, and $M^\star_3$ with varied values of γ. From this figure, it is clear that the most asymmetrical lines are for model $M^\star_3$ corresponding to $f_\star = 0.05$, for which the high-energy radiation is weaker making the particle to attain higher velocities and therefore more asymmetry. Telescopes with higher resolutions will be needed to resolve the asymmetry of these lines. As the UV emission dominates over the X-ray heating the absorption lines become broader, while they become narrower with increasing γ regardless of the radiation field. For $\gamma = 1.1$ the symmetry is enough to distinguish which source of radiation is dominant with a resolution of ≈ 2500 km s^{-1}. However, as γ increases this property is lost as changes in the fraction of UV heating from $f_{\mathrm{disc}} = 0.5$ to 0.95 cannot be easily distinguished from the shape of the spectral absorption lines.

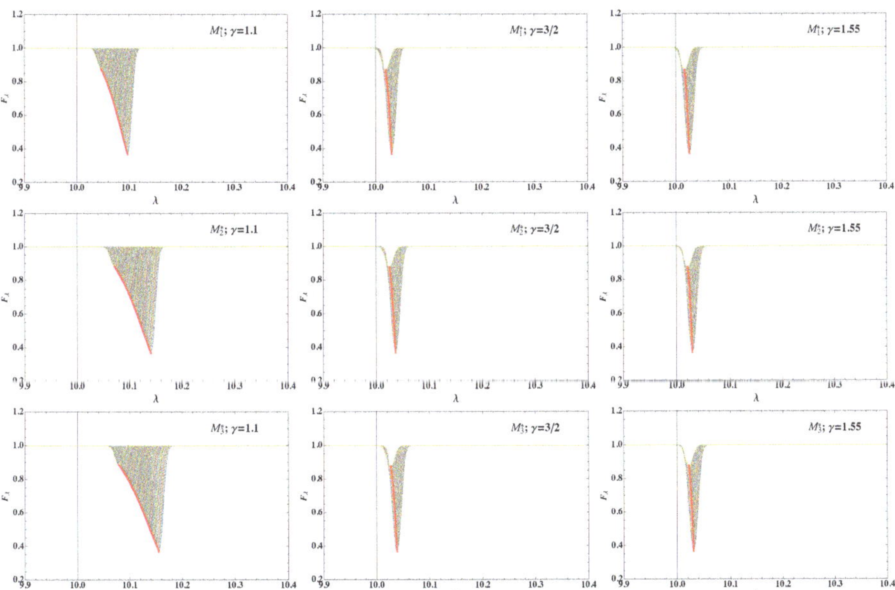

Figure 5. Absorption line shapes as predicted for models $M_1^\star$, $M_2^\star$, and $M_3^\star$ with varied adiabatic indices in the range $1.1 \leq \gamma \leq 1.66$. The solid red lines depict the profiles of a typical absorption line for an atom falling off onto a black hole.

5. Estimated Bondi Radius and Mass Accretion with Respect to Conditions at Infinity

We are now in the position to quantify the ratio between the estimated (r_e) and the true (r_B) values of the Bondi radius and that between the estimated ($\dot{M}_e$) and the true ($\dot{M}_{\rm rad}$) accretion rates when the boundary conditions are not at infinity. These ratios are given as functions of the radial distance from the SMBH by the following relations

$$\frac{r_{e;\{{\rm acc,out}\}}(x)}{r_B} = \left(\frac{x^2 \mathcal{M}_{\{{\rm acc,out}\}}}{\lambda} \right)^{2(\gamma-1)/(\gamma+1)}, \tag{33}$$

$$\frac{\dot{M}_{e;\{{\rm acc,out}\}}(x)}{\dot{M}_{\rm rad}} = \frac{1}{\chi_{\rm tot}^{\rm rad}|_{\theta=\theta_0}^2} \left[\frac{r_{e;\{{\rm acc,out}\}}(x)}{r_B} \right]^{-(5-3\gamma)/[2(\gamma-1)]}, \tag{34}$$

respectively, where $\chi_{\rm tot}^{\rm rad}(\theta, x)$ is the angle- and space-dependent radiative factor. Figure 6 (left) shows the estimated Bondi radii for model $M_1^\star$ for adiabatic indices between $\gamma = 1.1$ and 1.66. At small radial distances ($x = 8 \times 10^{-3}$), model $M_1^\star$ with $\gamma = 1.1$ has an estimated Bondi radius of $r_e/r_B \approx 0.50$. As the adiabatic index is increased, the ratio r_e/r_B decreases reaching values slightly less than 0.5 for $\gamma = 1.1$ and as low as $\approx$0.03 for $\gamma = 1.66$ at $x = 10^{-2}$. The same trends are also seen for models $M_2^\star$ (see Figure 6 (right)) and $M_3^\star$ as γ is increased, except that now the estimated Bondi radii achieve slightly lower values as the radiation field is dominated by the UV emission from the accretion disc. The profiles for model $M_3^\star$ are very similar to those displayed in Figure 6 with the ratios r_e/r_B achieving slightly smaller values close to the centre. For all models regardless of the adiabatic index the ratio $r_e/r_B \rightarrow 1$ at large radii ($x = 3$).

Figure 7 shows the estimated accretion rates for models $M_1^\star$, and $M_2^\star$, for varied values of γ. When the UV emission is increased from $f_{\rm disc} = 0.8$ to 0.95 for model $M_3^\star$ the trends are almost the same as those in Figure 7 (right) for model $M_2^\star$. In all cases, the radiative effects lead to an overestimation of the accretion rates for $x \lesssim 0.1$ (model $M_1^\star$) and $x \lesssim 0.12$ (models $M_2^\star$ and $M_3^\star$). The differences between $\dot{M}_e(x)$ and $\dot{M}_{\rm rad}$ grow rapidly as we get closer to the central source. However, the rapid growth of the ratio $\dot{M}_e(x)/\dot{M}_{\rm rad}$ towards the centre becomes progressively less steep as γ is increased, with $\dot{M}_e(x)/\dot{M}_{\mathrm{rad}}$ becoming almost flat for $\gamma = 1.66$. For instance, model $M_1^\star$ has an accretion rate at $x = 10^{-2}$ for $\gamma = 1.1$ that is about 100 times larger than for $\gamma = 1.66$. A similar difference between both values of γ happens when comparing models $M_2^\star$ and $M_3^\star$, where the UV emission from the disc dominates over the incident X-ray flux. The observed large overestimations in the estimated accretion rates make sense because as the closer the accreting matter is to the source of gravitation, the less accurate are the determination of ρ and c_{s} from their values at infinity. However, the authors of [38] found that the level of overestimation of the accretion rates compared to the classical Bondi problem is slightly attenuated when the effects of line driving are ignored (model M_1 with $\gamma = 1.1$). The same is true for higher values of γ.

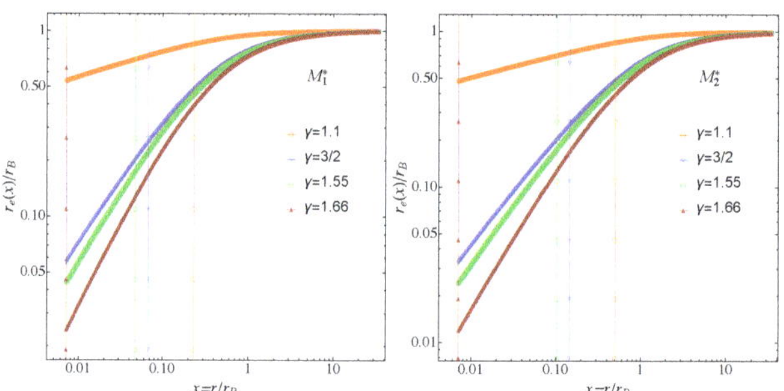

Figure 6. (**Left**) Estimated Bondi radii for model $M_1^\star$ with $M(\tau) \neq 0$. (**Right** $M_2^\star$ with $M(\tau) \neq 0$), $f_\star = 0.5(0.2)$ and $f_{\mathrm{disc}} = 0.5(0.8)$ for varied values of the adiabatic index in the range $1.1 \leq \gamma \leq 1.66$. The vertical solid lines mark the position of the sonic point and the symbols on them identify to which inflow solution they correspond to.

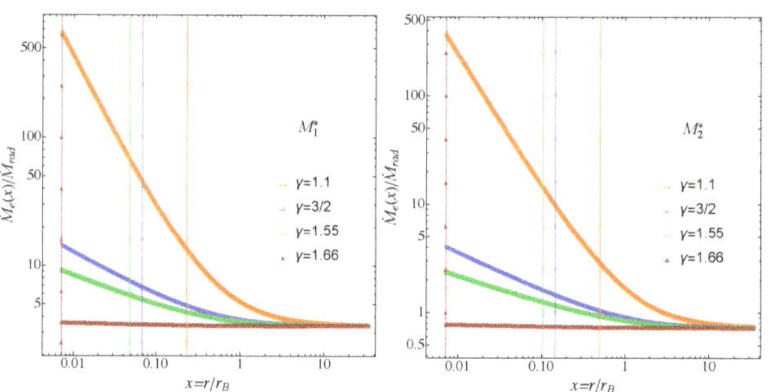

Figure 7. (**Left**) Estimated accretion rates for model $M_1^\star$ with $M(\tau) \neq 0$ (**Right** $M_2^\star$ with $M(\tau) \neq 0$), $f_\star = 0.5$ and $f_{\mathrm{disc}} = 0.5$ for varied values of the adiabatic index in the range $1.1 \leq \gamma \leq 1.66$. The vertical solid lines mark the position of the sonic point and the symbols on them identify to which inflow solution they correspond to.

6. Conclusions

In this paper, we have investigated the non-isothermal Bondi accretion onto a supermassive black hole (SMBH) when the adiabatic index is varied in the interval $1.1 \leq \gamma \leq 1.66$ by including the effects of X-ray heating and UV emission. The radiation field is modelled using the prescriptions introduced by [13] (P07), where a central SMBH is considered as a source of X-ray heating and a surrounding optically thick, geometrically thin, standard accretion disc as a source of UV radiation. The radiative, non-isothermal radial accretion is solved semi-analytically using a method similar to that developed by the authors of [41]. The X-ray heating from the SMBH is assumed to be isotropic, while the UV emission from the accretion disc is assumed to have an angular dependence. The effects of both types of radiation are studied for different flux fractions ($f_\star$ and f_{disc}) and an incident angle

$\theta_0 = \theta_{\mathrm{acc}}^{\mathrm{phase}}$ in the presence and absence of the effects of spectral line driving. The $\gamma = 5/3$ regime is also explored using the Paczyński–Wiita (PW) potential for the case when the effects of line driving are ignored.

The relevant conclusions are summarized as follows:

- Mass inflows with $1.1 \leq \gamma \leq 1.66$ occur at lower Mach numbers everywhere compared with pure isothermal ($\gamma = 1$) inflows, while isothermal outflows take place at much lower Mach numbers than their non-isothermal counterparts. In addition, for the isothermal accretion the sonic radius occurs at larger distances from the centre compared to the non-isothermal inflows, implying that in the former case the supersonic flow occupies a much larger central volume.
- When the UV flux dominates over the X-ray heating in models with spectral line driving, the inflow for given γ becomes faster everywhere. In particular, at radial distances as close as $10^{-3} r_B$ from the central source the inflow is always supersonic regardless of γ, where r_B is the Bondi radius. However, as the adiabatic index is increased from $\gamma = 1.1$ to $\gamma = 1.66$, the inflow of matter occurs at lower Mach numbers.
- The position of the sonic point depends on both the radiation field and the adiabatic index. In particular, when line driving is allowed and γ is increased from 1.1 to 1.66 the sonic point is at smaller radial distances from the centre. Hence, the central volume occupied by the supersonic inflow decreases in radius with increasing γ. For given γ, as the UV emission dominates over the X-ray heating the sonic point is shifted towards larger radii and, as a consequence, the supersonic inflow will occupy larger volumes.
- As long as the fraction of X-ray heating is comparable to that of UV emission from the accretion disc, the outflow at large radii from the central source becomes more supersonic than the inflow close to the black hole. Independently of the radiation field, the faster outflows always occur when $\gamma = 1.66$.
- With no line driving, the inflow becomes less supersonic as the UV emission dominates over the X-ray heating and the values of γ increases. In fact, as the UV radiation becomes stronger, the central volume occupied by the supersonic inflow becomes larger as the sonic points are shifted towards larger radii from the gravitational source. In contrast, the outflows become more supersonic as the UV emission becomes stronger and γ is increased.
- At distances of $10^{-3} r_B$ from the central source, the ratio of the estimated to the true Bondi radius is always below one. Independently of the dominant type of radiation, the deviations between the estimated and true Bondi radius increase with increasing γ. For given γ between 1.1 and 1.66, this ratio drops faster as the radiation field is dominated by the UV emission.
- Under the effects of line driving, the radiative effects lead to an overestimation of the accretion rates close to the centre. The deviation between the estimated and the true accretion rate increases with decreasing γ. For given γ, the deviation decreases as the UV emission becomes stronger.
- The models predict broader absorption lines when the UV emission dominates over the high-energy (X-ray) heating, when going from $f_{\mathrm{disk}} = 0.8 \rightarrow 0.95$, which in turn become narrower as the value of γ is increased. For low $\gamma (= 1.1)$ the lines are so asymmetric, e.g., [52,53], that they can be used to infer which source of radiation is dominant with resolutions of ≈ 3000 km s^{-1}. As γ is increased and the lines become narrower, their shape can no longer be used to distinguish the dominant source of heating.
- Compared to the Newtonian $\gamma = 1.66$ case, the PW model with $\gamma = 5/3$ and no spectral line driving exhibits almost identical inflow Mach numbers everywhere. As the UV emission dominates over the X-ray heating, i.e., when f_{disk} increases from 0.5 to 0.8, both the inflow and outflow become slightly faster for the Newtonian $\gamma = 1.66$ case.

Author Contributions: Conceptualization, J.M.R.-V. and L.D.G.S.; methodology, J.M.R.-V., L.D.G.S. and E.C.; software, J.M.R.-V.; formal analysis, J.M.R.-V., L.D.G.S. and J.K.; validation, J.M.R.-V., L.D.G.S., J.K., R.G. and E.C.; writing—review and editing, J.M.R.-V., L.D.G.S., J.K., R.G. and E.C. All authors have read and agreed to the published version of the manuscript.

Funding: This work has been partially supported by Yachay Tech under the project for the use of the supercomputer QUINDE I and by UAM-A through internal funds. One of us (J. K.) acknowledges financial support from CONACYT under grant 283151.

Institutional Review Board Statement: Not applicable.

Informed Consent Statement: Not applicable.

Data Availability Statement: The code to reproduce the average $\langle \zeta \rangle$ calculations underlying this article are available in Zenodo at doi:10.5281/zenodo.4075323. For the new varying ζ, it will be available under request to the corresponding author.

Acknowledgments: IMPETUS is a collaboration project between the ABACUS-Centro de Matemática Aplicada y Cómputo de Alto Rendimiento of Cinvestav-IPN, the School of Physical Sciences and Nanotechnology of the Yachay Tech University, and the Área de Física de Procesos Irreversibles of the Departamento de Ciencias Básicas of the Universidad Autónoma Metropolitana–Azcapotzalco (UAM-A) aimed at the SPH modelling of astrophysical flows.

Conflicts of Interest: The authors declare no conflict of interest.

Notes

[1] The structure of our solutions is in X, in the same terms as in [48]. Types 5 and 6 solutions are doubled valued in $\mathcal{M}$ for a given position x, and we exclude them as physical solutions.

[2] With $T = 10^7$ K and $\mu = 0.7$.

References

1. Kormendy, J.; Richstone, D. Inward bound—The search for supermassive black holes in galactic nuclei. *Annu. Rev. Astron. Astrophys.* **1995**, *33*, 581–624. [CrossRef]
2. Ferrarese, L.; Ford, H. Supermassive black holes in galactic nuclei: Past, present and future research. *Space Sci. Rev.* **2005**, *116*, 523–624. [CrossRef]
3. Contreras, E.; Rincón, Á.; Ramírez-Velásquez, J.M. Relativistic dust accretion onto a scale-dependent polytropic black hole. *Eur. Phys. J. C* **2019**, *79*, 53. [CrossRef]
4. Bondi, H. On spherically symmetrical accretion. *Mon. Not. R. Astron. Soc.* **1952**, *112*, 195–204. [CrossRef]
5. Salpeter, E.E. Accretion of interstellar matter by massive objects. *Astrophys. J.* **1964**, *140*, 796–800. [CrossRef]
6. Fabian, A.C. The obscured growth of massive black holes. *Mon. Not. R. Astron. Soc.* **1999**, *308*, L39–L43. [CrossRef]
7. Barai, P. Large-scale impact of the cosmological population of expanding radio galaxies. *Astrophys. J. Lett.* **2008**, *682*, L17–L20. [CrossRef]
8. Germain, J.; Barai, P.; Martel, H. Anisotropic active galactic nucleus outflows and enrichment of the intergalactic medium. I. Metal disribution. *Astrophys. J.* **2009**, *704*, 1002–1020. [CrossRef]
9. Croston, J.H.; Hardcastle, M.J.; Birkinshaw, M.; Worrall, D.M.; Laing, R.A. An XMM-Newton study of the environments, particle content and impact of low-power radio galaxies. *Mon. Not. R. Astron. Soc.* **2008**, *386*, 1709–1728. [CrossRef]
10. DiMatteo, T.; Springel, V.; Hernquist, L. Energy input from quasars regulates the growth and activity of black holes and their host galaxies. *Nature* **2005**, *433*, 604–607. [CrossRef]
11. Sincell, M.W.; Krolik, J.H. The vertical structure and ultraviolet spectrum of X-ray-irradiated accretion disks in active galactic nuclei. *Astrophys. J.* **1997**, *476*, 605–619. [CrossRef]
12. Proga, D. Winds from accretion disks driven by radiation and magnetocentrifugal force. *Astrophys. J.* **2000**, *538*, 684–690. [CrossRef]
13. Proga, D. Dynamics of accretion flows irradiated by a quasar. *Astrophys. J.* **2007**, *661*, 693–702. [CrossRef]
14. Proga, D.; Waters, T. Cloud formation and acceleration in a radiative environment. *Astrophys. J.* **2015**, *804*, 137. [CrossRef]
15. Ramírez-Velasquez, J.M.; Klapp, J.; Gabbasov, R.; Cruz, F.; Sigalotti, L.D. Impetus: New cloudy's radiative tables for accretion onto a galaxy black hole. *Astrophys. J. Suppl. Ser.* **2016**, *226*, 22. [CrossRef]
16. Dyda, S.; Dannen, R.; Waters, T.; Proga, D. Irradiation of astrophysical objects—SED and flux effects on thermally driven winds. *Mon. Not. R. Astron. Soc.* **2017**, *467*, 4161–4173. [CrossRef]

17. Ramírez-Velasquez, J.M.; Klapp, J.; Gabbasov, R.; Cruz, F.; Sigalotti, L.D. The IMPETUS project: Using ABACUS for the high performance computation of radiative tables for accretion onto a galaxy black hole. In *Recent Developments in High Performance Computing*; Barrios Hernández, C.J., Gitler, I., Klapp, J., Eds.; Springer: Berlin/Heidelberg, Germany, 2017; Volume 697, pp. 374–386.
18. Silk, J.; Rees, M.J. Quasars and galaxy formation. *Astron. Astrophys.* **1998**, *331*, L1–L4.
19. Furlanetto, S.R.; Loeb, A. Metal absorption lines as probes of the intergalactic medium prior to the reionization epoch. *Astrophys. J.* **2003**, *588*, 18–34. [CrossRef]
20. Haiman, Z.; Bryan, G.L. Was star formation suppressed in high-redshift minihalos? *Astrophys. J.* **2006**, *65*, 7–11. [CrossRef]
21. Hopkins, P.F.; Hernquist, L.; Cox, T.J.; Robertson, B.; Springel, V. Determining the properties and evolution of red galaxies from the quasar luminosity function. *Astrophys. J. Suppl. Ser.* **2006**, *163*, 50–79. [CrossRef]
22. Ciotti, L.; Ostriker, J.P.; Proga, D. Feedback from central black holes in elliptical galaxies. III. Models with both radiative and mechanical feedback. *Astrophys. J.* **2010**, *717*, 708–723. [CrossRef]
23. McCarthy, I.G.; Schaye, J.; Ponman, T.J.; Bower, R.G.; Booth, C.M.; DallaVecchia, C.; Crain, R.A.; Springel, V.; Theuns, T.; Wiersma, R.P.C. The case for AGN feedback in galaxy groups. *Mon. Not. R. Astron. Soc.* **2010**, *406*, 822–839. [CrossRef]
24. Ostriker, J.P.; Choi, E.; Ciotti, L.; Novak, G.S.; Proga, D. Momentum driving: Which physical processes dominate active galactic nucleus feedback? *Astrophys. J.* **2010**, *722*, 642–652. [CrossRef]
25. Fabian, A.C. Observational evidence of active galactic nuclei feefback. *Annu. Rev. Astron. Astrophys.* **2012**, *50*, 455–489. [CrossRef]
26. Faucher-Giguère, C.-A.; Quataert, E.; Murray, N. A physical model of FeLoBALs: Implications for quasar feedback. *Mon. Not. R. Astron. Soc.* **2012**, *420*, 1347–1353. [CrossRef]
27. Choi, E.; Naab, T.; Ostriker, J.P.; Johansson, P.H.; Moster, B.P. Consequences of mechaical and radiative feedback from black holes in disc galaxy mergers. *Mon. Not. R. Astron. Soc.* **2014**, *442*, 440–453. [CrossRef]
28. Dannen, R.C.; Proga, D.; Kallman, T.R.; Waters, T. Photoionization calculations of the radiation force due to spectral lines in AGNs. *arXiv* **2019**, arXiv:1812.01773v2.
29. Palit, I.; Janiuk, A.; Sukova, P. Effects of adiabatic index on the sonic surface and time variability of low angular momentum accretion flows. *Mon. Not. R. Astron. Soc.* **2019**, *487*, 755–768. [CrossRef]
30. Chakrabarti, S.K.; Das, S. Model dependence of transonic properties of accretion flows around black holes. *Mon. Not. R. Astron. Soc.* **2001**, *327*, 808–812. doi:10.1046/j.1365-8711.2001.04758.x. [CrossRef]
31. Das, T.K. On some transonic aspects of general relativistic spherical accretion on to Schwarzschild black holes. *Mon. Not. R. Astron. Soc.* **2002**, *330*, 563-566. [CrossRef]
32. Meliani, Z.; Sauty, C.; Tsinganos, K.; Vlahakis, N. Relativistic Parker winds with variable effective polytropic index. *Astron. Astrophys.* **2004**, *425*, 773–781.:20035653. [CrossRef]
33. Fukue, J. Transonic disk accretion revisited. *Publ. Astron. Soc. Jpn.* **1987**, *39*, 309-327.
34. Chattopadhyay, I.; Ryu, D. Effects of fluid composition on spherical flows around black holes. *Astrophys. J.* **2009**, *694*, 492. [CrossRef]
35. Sarkar, S.; Chattopadhyay, I.; Laurent, P. Two-temperature solutions and emergent spectra from relativistic accretion discs around black holes. *Astron. Astrophys.* **2020**, *642*, A209. [CrossRef]
36. Fabian, A.C.; Rees, M.J. The accretion luminosity of a massive black hole in an elliptical galaxy. *Mon. Not. R. Astron. Soc.* **1995**, *277*, L55–L58. [CrossRef]
37. Volonteri, M.; Rees, M.J. Rapid growth of high-redshift black holes. *Astrophys. J.* **2005**, *633*, 624–629. [CrossRef]
38. Ramírez-Velasquez, J.M.; Sigalotti, L.D.; Gabbasov, R.; Klapp, J.; Contreras, E. Bondi accretion for adiabatic flows onto a massive black hole with an accretion disc. The one dimensional problem. *Astron. Astrophys.* **2019**, *631*, A13. [CrossRef]
39. Ciotti, L.; Pellegrini, S. Isothermal Bondi accretion in Jaffe and Hernquist galaxies with a central black hole: Fully analytical solutions. *Astrophys. J.* **2017**, *848*, 29. [CrossRef]
40. Ciotti, L.; ZiaeeLorzad, A. Two-component Jaffe models with a central black hole—I. The spherical case. *Mon. Not. R. Astron. Soc.* **2018**, *473*, 5476–5491. [CrossRef]
41. Korol, V.; Ciotti, L.; Pellegrini, S. Bondi accretion in early-type galaxies. *Mon. Not. R. Astron. Soc.* **2016**, *460*, 1188–1200. [CrossRef]
42. Ramírez-Velasquez, J.M.; Sigalotti, L.D.; Gabbasov, R.; Cruz, F.; Klapp, J. IMPETUS: Consistent SPH calculations of 3D spherical Bondi accretion on to a black hole. *Mon. Not. R. Astron. Soc.* **2018**, *477*, 4308–4329. [CrossRef]
43. Castor, J.I.; Abbott, D.C.; Klein, R.I. Radiation-driven winds of of stars. *Astrophys. J.* **1975**, *195*, 157–174. [CrossRef]
44. Stevens, I.R.; Kallman, T.R. X-ray illuminated stellar winds: Ionization effects in the radiative driving of stellar winds in massive X-ray binary systems. *Astrophys. J.* **1990**, *365*, 321–331. [CrossRef]
45. Paczyński, B.; Wiita, P.J. Thick accretion disks and supercritical luminosities. *Astron. Astrophys.* **1980**, *88*, 23–31.
46. Abramowicz, M.A. The Paczyński-Wiita potential. A step-by-step derivation. *Astron. Astrophys.* **2009**, *500*, 213–214. [CrossRef]
47. Proga, D.; Begelman, M.C. Accretion of low angular momentum material onto black holes: Two-dimensional hydrodynamical inviscid case. *Astrophys. J.* **2003**, *582*, 69–81. [CrossRef]
48. Frank, J.; King, A.; Raine, D.J. *Accretion Power in Astrophysics*, 3rd ed.; Cambridge University Press: Cambridge, UK, 2002.
49. Laor, A.; Fiore, F.; Elvis, M.; Wilkes, B.J.; McDowell, J.C. The soft X-ray properties of a complete sample of optically selected quasars. II. Final results. *Astrophys. J.* **1997**, *477*, 93–113. [CrossRef]

50. Zheng, W.; Kriss, G.A.; Telfer, R.C.; Grimes, J.P.; Davidsen, A.F. A composite HST spectrum of quasars. *Astrophys. J.* **1997**, *475*, 469–478. [CrossRef]
51. Rybicki, G.B.; Lightman, A.P. *Radiative Processes in Astrophysics*; Wiley-Interscience: New York, NY, USA, 1979.
52. Ramírez, J.M.; Bautista, M.; Kallman, T. Line asymmetry in the Seyfert galaxy NGC 3783. *Astrophys. J.* **2005**, *627*, 166–176. [CrossRef]
53. Ramírez, J.M. Kinematics from spectral lines for AGN outflows based on time-independent radiation-driven wind theory. *Rev. Mex. Astron. Astrofísica* **2011**, *47*, 385–399.

Article

Geodesic Circular Orbits Sharing the Same Orbital Frequencies in the Black String Spacetime

Sanjar Shaymatov [1,2,3,4,5,*] **and Farruh Atamurotov** [3,6]

1 Institute for Theoretical Physics and Cosmology, Zheijiang University of Technology, Hangzhou 310023, China
2 College of Engineering, Akfa University, Kichik Halqa Yuli Street 17, Tashkent 100095, Uzbekistan
3 Ulugh Beg Astronomical Institute, Astronomicheskaya 33, Tashkent 100052, Uzbekistan; atamurotov@yahoo.com
4 Physics Faculty, National University of Uzbekistan, Tashkent 100174, Uzbekistan
5 Tashkent Institute of Irrigation and Agricultural Mechanization Engineers, Kori Niyozly, 39, Tashkent 100000, Uzbekistan
6 School of Computer and Information Engineering, Inha University in Tashkent, Ziyolilar 9, Tashkent 100170, Uzbekistan
* Correspondence: sanjar@astrin.uz

Abstract: We consider isofrequency pairing of geodesic orbits that share the same three orbital frequencies associated with $\Omega^{\hat{r}}$, $\Omega^{\hat{\phi}}$, and $\Omega^{\hat{\omega}}$ in a particular region of parameter space around black string spacetime geometry. We study the effect of a compact extra spatial dimension on the isofrequency pairing of geodesic orbits and show that such orbits would occur in the allowed region when particles move along the black string. We find that the presence of the compact extra dimension leads to an increase in the number of the isofrequency pairing of geodesic orbits. Interestingly we also find that isofrequency pairing of geodesic orbits in the region of parameter space cannot be realized beyond the critical value $\mathcal{J}_{cr} \approx 0.096$ of the conserved quantity of the motion arising from the compact extra spatial dimension.

Keywords: isofrequency; geodesic orbits; black string

Citation: Shaymatov, S.; Atamurotov, F. Geodesic Circular Orbits Sharing the Same Orbital Frequencies in the Black String Spacetime. *Galaxies* **2021**, *9*, 40. https://doi.org/10.3390/galaxies9020040

Received: 21 April 2021
Accepted: 25 May 2021
Published: 28 May 2021

1. Introduction

In general relativity, black holes are regarded as one of the most interesting and fascinating objects due to their geometric and remarkable gravitational properties. However, the point to be noted is that their existence had been predicted as a potential tool in explaining observational phenomena until the discovery of the gravitational waves detected by LIGO and Virgo scientific collaborations [1,2] and as well as the first direct shadow of supermassive black hole at the center of galaxy 87 by the Event Horizon Telescope (EHT) collaboration [3,4]. These observations opened a qualitatively new stage to reveal black hole's unknown properties and test remarkable nature of the background geometry around black hole's horizon irrespective of the fact that there are still fundamental problems that general relativity faces, i.e., the occurrence of singularity, spacetime quantization, etc. In this framework, the motion of test particles in the strong gravitational field regime has been a productive field of study for several years [5–19]. On the other hand, there is an extensive body of work devoted to understand the nature of radiative inspirals around black holes as a source of gravitational waves and binary systems [20–28].

It is worth noting that experiments and observations always allow to test the nature of the geometry through alternative theories of modified gravity in the strong field regime. The spactime geometry that can depart from the spherically symmetric Schwarzschild black hole was considered by Grunau and Khamesra [29]. This spacetime metric describes a five dimensional black string due to the compact extra dimension added to the well known Schwarzschild metric. Thus, it is one of reasons why this solution is interesting object to investigate its properties. The motion of test particles in the different gravity string models has been investigated thoroughly by several authors [30–35]. Furthermore, the geodesic motion of test particles in the black string spacetime for both rotating and non-rotating cases has been studied in detail [29]. The geodesic motion of colliding particles was also studied in the vicinity of black string spacetime [36].

The theoretical investigation of isofrequency pairing of geodesic circular orbits has been considered in the vicinity of Schwarzschild and Kerr black holes, i.e., the region of the parameter space where such geodesic orbits occur has been discussed by providing an intuitive explanation of their occurrence. The above investigation was addressed by taking an external asymptotically uniform magnetic field in the Schwarzschild spacetime [25]. It turns out that the surface of allowed region where isofrequency pairing orbits occur decreases as a consequence of the presence of the external magnetic field. It was later extended to the spinning particles in the Schwarzschild-de Sitter spacetime [26]. In the present paper, we consider isofrequency pairing of geodesic orbits around black string spacetime geometry, as shown by the line element in [29]. We then study the effect of the compact extra spatial dimension ω on the isofrequency pairing of geodesic spiral orbits sharing the same three orbital frequencies.

In this paper, we consider a new frequency $\Omega^{\hat{\omega}}$ arisen from the compact extra spatial dimension associated with the motion in the extra direction along the black string. Generally, the bound orbits are confined to the interior of a compact special spiral given by $h_{z1}(e = 0) \leq h_z \leq h_{z2}(e = 1)$; frequencies, in particular, Ω^{r} and Ω^{θ} are considered as "libration"-type frequencies associated with the radial and longitudinal periods, while $\Omega^{\hat{\phi}}$ is a "rotation"-type frequency. In the case of black string, new frequency $\Omega^{\hat{\omega}}$ can be defined by a "libration-rotation" type frequency which involves two kinds of type frequencies, as described above. For the sake of clarity, we will further focus on the motion of test particles at the equatorial plane (i.e., $\theta = \pi/2$), and hence we omit $\Omega^{\hat{\theta}}$ as it loses its meaning. Then bound geodesic orbits around black string spacetime are respectively characterized by three orbital frequencies, i.e., Ω^{r}, $\Omega^{\hat{\phi}}$, and $\Omega^{\hat{\omega}}$ associated with the periodic motions. Thus, these orbits are considered as triperiodic orbits around the black string spacetime. In doing so we predict that an infinite number of pairs of such orbits may occur in a region of parameter space around the black string spacetime where physically distinct orbits possessing $\mathcal{E}$, $\mathcal{L}$, and $\mathcal{J}$ share the same orbital frequencies associated with $\Omega^{\hat{\phi}}$, Ω^{r}, and $\Omega^{\hat{\omega}}$, thereby referring to the "isofrequency" bound geodesic orbits synchronized in phase φ while passing periastra at the same time. Isofrequency pairing of geodesic orbits occuring in the vicinity of black string spacetime may in principle play an important role in an astrophysical scenario in understanding the nature of radiative inspirals around black holes.

The paper is organised as follows: In Section 2 we briefly describe the metric for black string and provide the geodesic equations of motion. We further study the effect of the compact extra dimension on the isofrequency pairing of geodesic spiral orbits in Section 3. We end up with conclusion in Section 4. Throughout the paper we use a system of geometric units in which $G = c = 1$.

2. The Metric and the Geodesic Equations of Motion

The metric describing five dimensional black string spacetime with the compact extra spatial dimension ω added to the Schwarzschild metric in the Boyer-Lindquist coordinates $(t, r, \theta, \varphi, \omega)$ is given by [29]

$$ds^2 = -\left(1 - \frac{2M}{r}\right)dt^2 + \left(1 - \frac{2M}{r}\right)^{-1}dr^2$$
$$+ r^2(d\theta^2 + \sin^2 d\varphi^2) + d\omega^2 \tag{1}$$

with M being the total mass of black string and ω corresponding to the compact extra spatial dimension.

The radial motion of geodesic test particles in the equatorial plane of black string spacetime satisfies the following equation

$$\dot{r}^2 = \mathcal{E}^2 - V_{eff}(r; \mathcal{L}, \mathcal{J}),$$ (2)

where

$$V_{eff}(r; \mathcal{L}, \mathcal{J}) = \left(1 - \frac{2M}{r}\right)\left(1 + \mathcal{J}^2 + \frac{\mathcal{L}^2}{r^2}\right),$$ (3)

with $\mathcal{E} = E/m$, $\mathcal{L} = L/m$, and $\mathcal{J} = J/m$ being the conserved constants per unit mass and related to the particle's specific energy, angular momentum, and conserved quantity arising from ω, respectively.

It is well known that bound orbits exist in the case when $\mathcal{L} > 2\sqrt{3}M$ with $2\sqrt{2}/3 < \mathcal{E} < 1$ for the Schwarzschild black hole. Whereas, in the case of black string, bound orbits only exist when $\mathcal{L} > 2\sqrt{3}(1 + \mathcal{J}^2)^{1/2}M$ with $2\sqrt{2}(1 + \mathcal{J}^2)^{1/2}/3 < \mathcal{E} < (1 + \mathcal{J}^2)^{1/2}$ is satisfied, and the motion of particles is then restricted by the turning points being labeled as the periastron r_p and the apastron r_a, respectively.

We now consider the innermost stable circular orbit (ISCO) for test particles orbiting around black string spacetime. To find radii of the ISCO one needs to solve the following equation

$$V''_{eff}(r; \mathcal{L}, \mathcal{J}) = 0.$$ (4)

In our case r_{ISCO} can be determined implicitly from the condition

$$6\mathcal{L}^2(r - 4M) - 4(1 + \mathcal{J}^2)Mr^2 = 0,$$ (5)

with $\mathcal{L}$ obtained on solving $V'_{eff}(r; \mathcal{L}, \mathcal{J}) = 0$. Thus, we have the ISCO radius $r_{ISCO} = 6M$ for black string spacetime. It is worth noting that the ISCO is the same in contrast for the Schwarzschild black hole.

The standard condition $V_{eff}(r_p, \mathcal{L}, \mathcal{J}) = V_{eff}(r_a, \mathcal{L}, \mathcal{J}) = \mathcal{E}^2$ leads to the specific energy $\mathcal{E}$ and angular momentum $\mathcal{L}$ of the particle, given in terms of p and e. Here, $r_p = Mp/(1 + e)$ and $r_a = Mp/(1 - e)$ respectively refer to the periastron and the apastron [25]. In between these two turning points lies bound orbits with corresponding specific energy and angular momentum. Note that parameters p and e respectively measure the size of the orbit and its degree of noncircularity [37]. Then we have

$$\mathcal{E}^2 = \frac{(p - 2 - 2e)(p - 2 + 2e)}{p(p - 3 - e^2)}(1 + \mathcal{J}^2),$$

$$\mathcal{L}^2 = \frac{p^2 M^2}{p - 3 - e^2}(1 + \mathcal{J}^2).$$ (6)

Following Cutler et al. [38], we turn to the integration of the geodesic equations in terms of $t(r)$, $\varphi(r)$, and $\omega(r)$ coordinates

$$t(r) = \mathcal{E} \int \frac{r\,dr}{(r - 2M)\left[\mathcal{E}^2 - V_{eff}(r; \mathcal{L}, \mathcal{J})\right]^{1/2}},$$ (7)

$$\varphi(r) = \mathcal{L} \int \frac{dr}{r^2\left[\mathcal{E}^2 - V_{eff}(r; \mathcal{L}, \mathcal{J})\right]^{1/2}},$$ (8)

and

$$\omega(r) = \mathcal{J} \int \frac{dr}{\left[\mathcal{E}^2 - V_{eff}(r;\mathcal{L},\mathcal{J})\right]^{1/2}}. \tag{9}$$

Using Equations (6)–(9) and substituting $r(\chi) = Mp/(1 + e\cos\chi)$ into Equation (3), we then have the following expressions together with χ related to the coordinates t, φ, and ω as

$$\frac{dt}{d\chi} = \frac{Mp^2[(p-2)^2 - 4e^2]^{1/2}(p-6-2e\cos\chi)^{-1/2}}{(p-2-2e\cos\chi)(1+e\cos\chi)^2}, \tag{10}$$

$$\frac{d\varphi}{d\chi} = \frac{p^{1/2}}{(p-6-2e\cos\chi)^{1/2}}, \tag{11}$$

$$\frac{d\omega}{d\chi} = \frac{\mathcal{J}Mp^{3/2}(p-3-e^2)^{1/2}}{(1+\mathcal{J}^2)^{1/2}(p-6-2e\cos\chi)^{1/2}(1+e\cos\chi)^2}. \tag{12}$$

The radial period and frequency are defined by [24,25,38]

$$T^{\hat{r}} = \int_0^{2\pi} \frac{dt}{d\chi}\,d\chi, \qquad \Omega^{\hat{r}} = \frac{2\pi}{T^{\hat{r}}}. \tag{13}$$

The azimuthal frequency of the orbit will then have the form as

$$\Omega^{\hat{\varphi}} = \frac{1}{T^{\hat{r}}} \int_0^{T^{\hat{r}}} \frac{d\varphi}{dt}\,dt = \frac{\Delta\varphi}{T^{\hat{r}}}, \tag{14}$$

with azimuthal phase $\Delta\varphi$, which is given by

$$\begin{aligned}
\Delta\varphi &= \int_0^{2\pi} \frac{d\varphi}{d\chi}\,d\chi = \\
&= \int_0^{2\pi} \frac{\sqrt{p}}{\sqrt{p-6-2e\cos\chi}}\,d\chi = 4\sqrt{\frac{p}{\epsilon}}\,K\left(-\frac{4e}{\epsilon}\right),
\end{aligned} \tag{15}$$

where $\epsilon = p - p_s(e) = p - 6 - 2e$ and $K(x) = \int_0^{\pi/2} d\theta(1 - x\sin^2\theta)^{-1/2}$ is the complete elliptic integral of the first kind.

As a consequence of the presence of the compact extra dimension ω, the new frequency can be defined by

$$\Omega^{\hat{\omega}} = \frac{1}{T^{\hat{r}}} \int_0^{T^{\hat{r}}} \frac{d\omega}{dt}\,dt = \frac{\Delta\omega}{T^{\hat{r}}}, \tag{16}$$

where $\Delta\omega$ is the new extra phase accumulated over time interval $T^{\hat{r}}$.

3. Isofrequency Pairing Orbits in Black String Spectime Geometry

Now we investigate the isofrequency and bound orbits of the particles in the vicinity of the black string spacetime. We consider orbits lying extremely close to the separatrix ϵ in order to estimate the divergent quantities of the azimuthal and the new extra phases, the radial period, and their ratio in Equations (14) and (16). Considering the near-separatrix analytic expansions we obtain the corresponding expression for $\Delta\omega$ as

$$\Delta\omega = \frac{\mathcal{J}M(3+2e-e^2)^{1/2}}{(1+\mathcal{J}^2)^{1/2}(1+2e)}\sqrt{\frac{(6+2e)^3}{e}}$$
$$\times\left[1+\mathcal{O}\left(\frac{\epsilon}{4e}\right)\right]\ln\left(\frac{64e}{\epsilon}\right). \tag{17}$$

The previous investigations suggest that the occurrence of isofrequency pairing of geodesic orbits is strongly related to the presence of boundary regions referred as *separatrix* and *circular-orbit duals* (COD), and each and every circular orbit in the open range between the separatrix and singular curve and between the singular curve and COD as well has a dual isofrequency pairing. Note that the existence of the *separatrix* plays an important role for occurrence of these dual orbits in this isofrequency scenario. Considering the singular curve and the circular orbit duals in the particular region of the gravitational object is particularly important, and the COD can also play an important role due to the fact that as a boundary region allowing to keep pairs of isofrequency orbits that exist in this region [24–26].

Let us then study the effect of the compact extra dimension on the occurrence of the isofrequency pairing of geodesic orbits in black string spacetime. Based on the above discussions and Equations (13)–(15) we keep the result for which the range of frequencies of any pairs of orbits on $\Omega^{\hat\phi} = (M/r_b^3)^{1/2}$ in the black string is given by

$$\tilde\Omega^{\hat\phi}_{e=0} = 0.062 \ < \ \tilde\Omega^{\hat\phi} \ < \ \tilde\Omega^{\hat\phi}_{e=1} = 0.125, \tag{18}$$

where we have defined $\tilde\Omega^{\hat\omega} = M\Omega^{\hat\phi}$ as a dimensionless parameter.

Further, we determine the values of radial and azimuthal frequencies numerically and tabulate their values in Table 1.

Table 1. Values of $\Omega^{\hat{r}}$ and $\Omega^{\hat\phi}$ are tabulated for different values of eccentricity e for isofrequency pairing of geodesic spiral orbits around the black string.

e	0.1	0.3	0.4	0.5	0.6
$\Omega^{\hat\phi}$	0.063	0.068	0.073	0.079	0.091
$\Omega^{\hat{r}}$	0.0069	0.0124	0.0152	0.0181	0.0226

e	0.7	0.8	0.89	0.96	1
$\Omega^{\hat\phi}$	0.103	0.109	0.115	0.121	0.125
$\Omega^{\hat{r}}$	0.0272	0.0304	0.0334	0.0362	0.0380

In Table 2, we show numerical values of the frequency $\tilde\Omega^{\hat\omega}$ of the bound orbits around the black string. As can be seen from Table 2 the value of the frequency $\tilde\Omega^{\hat\omega}$ arisen from the compact extra spatial dimension is increasing with an increase in the value of the forth conserved quantity $\mathcal{J}$, but the height of the horizontal direction along black string is decreasing for given values of eccentricity e. However, numerical analysis leads to the fact that when the value of the fourth conserved quantity attains $\mathcal{J}_{cr} \approx 0.096$ geodesic spiral orbits are not allowed all through to occur around the black string. Thus, the pairs of geodesic circular spiral orbits of particles with the same radial $\tilde\Omega^{\hat{r}}$, azimuthal $\tilde\Omega^{\hat\phi}$, and new $\tilde\Omega^{\hat\omega}$ frequencies occur in the particular region of the gravitational object under this critical value irrespective of the fact that these orbits are physically distinct with different values of conserved quantities (i.e., $\mathcal{E}$, $\mathcal{L}$, and $\mathcal{J}$). Then the pairs of these spiral orbits of particles with three frequencies associated with ($\tilde\Omega^{\hat{r}}$, $\tilde\Omega^{\hat\phi}$, $\tilde\Omega^{\hat\omega}$) would move and oscillates in a cylindrical form along the black string.

It turns out that as a consequence of the decrease in the value of $\mathcal{J}$ the above mentioned cylindrical region becomes larger around the black string. It could then play an important role in allowing the particles to move along the black string, thus increasing the amount of isofrequency pairs of geodesic spiral orbits in the region between the *separatrix, circular-orbit duals* and *horizontal length* h_z, i.e., in the particular region of the black string. Hence, one can keep in mind that there would exist an infinite number of pairs of circular spiral orbits having the same three orbital frequencies around the black string regardless of physically distinct obits with three conserved quantities. As a consequence of the presence of compact extra spatial dimension, the occurrence of isofrequency pairing of spiral orbits would be of particularly importance in understanding and explaining the behavior of black string spacetime and the nature of the radiative inspirals as a source of gravitational waves or binary systems.

Table 2. The values of $\Omega^{\hat{\omega}}$ are tabulated for different values of the fourth conserved parameter $\mathcal{J}$ for isofrequency pairing of geodesic spiral orbits around the black string. The point to be noted here is that h_z corresponds to the horizontal length and will be valid only for the non vanishing compact extra dimension, i.e., $\omega \neq 0$.

	$\mathcal{J}$	0.01	0.03	0.05	0.07	0.09	0.096
$e = 0$	$\tilde{\Omega}^{\hat{\omega}}$	0.007	0.021	0.035	0.049	0.063	-
	r	27.17	13.05	9.29	7.43	6.29	
	h_z	26.51	11.59	7.09	4.38	1.89	
$e = 1$	$\tilde{\Omega}^{\hat{\omega}}$	0.006	0.019	0.033	0.046	0.059	-
	r	28.25	13.57	9.66	7.73	6.55	
	h_z	27.53	11.98	7.25	4.36	1.45	

4. Conclusions

We considered isofrequency pairing of geodesic spiral orbits in the black string spacetime, which recovers the Schwarzschild one in the case of vanishing ω. Thus, in an astrophysical scenario, it is worth investigating these geodesic orbits in the vicinity of black string as it allows to test the effects arising from the five dimensions. Our analysis suggests that the compact extra dimension ω leads to an increase in the number of pairs of circular spiral orbits. These pairs of orbits having the same three orbital frequencies (i.e., $\Omega^{\hat{r}}$, $\Omega^{\hat{\phi}}$, and $\Omega^{\hat{\omega}}$) occur in the particular region and oscillate in the cylindrical form around the black string in spite of the fact that they are physically distinct orbits possessing different conserved quantities (i.e., $\mathcal{E}$, $\mathcal{L}$, and $\mathcal{J}$). Furthermore, we showed that pairs of geodesic spiral orbits are not allowed all through to occur in the black string vicinity in the case when the fourth conserved quantity attains $\mathcal{J} \approx 0.096$.

The obtained results suggest that an infinite number of pairs of such orbits may occur in the particular region around the black string and it would be particularly important in explaining not only the possibility of occurrence of pairs of geodesic spiral orbits around the black holes, but also the nature of the black string spacetime. These theoretical results and discussions would be useful to the possible interpretation of astrophysical observations, and they could help to provide details of the validity of alternative models to black holes and also explain the nature of the radiative inspirals since isofrequency paring of geodesic spiral orbits can give the same information as such inspiral objects.

In a future work, we would be considering the possible extensions of recent results to the case of rotating black string and studying the effects of the rotation and a compact extra dimension on the specific isofrequency pairing of geodesic orbits. This would be of primary astrophysical importance.

Author Contributions: Conceptualization, S.S. and F.A.; investigation, S.S. and F.A.; methodology, S.S. and F.A.; validation, S.S. and F.A.; formal analysis, S.S. Both authors have read and agreed to the published version of the manuscript.

Funding: This research received no external funding.

Institutional Review Board Statement: Not applicable.

Informed Consent Statement: Not applicable.

Data Availability Statement: Data sharing not applicable—no new data generated.

Acknowledgments: S.S. acknowledges the support of the Uzbekistan Ministry for Innovative Development. F.A. acknowledges the support of INHA University in Tashkent.

Conflicts of Interest: The authors declare no conflict of interest.

References

1. Abbott, B.P.; Abbott, R.; Abbott, T.D.; Abernathy, M.R.; Acernese, F.; Ackley, K.; Adams, C.; Adams, T.; Addesso, P.; Adhikari, R.X.; et al. Observation of Gravitational Waves from a Binary Black Hole Merger. *Phys. Rev. Lett.* **2016**, *116*, 061102. [CrossRef]
2. Abbott, B.P.; Abbott, R.; Abbott, T.D.; Abernathy, M.R.; Acernese, F.; Ackley, K.; Adams, C.; Adams, T.; Addesso, P.; Adhikari, R.X.; Adya, V.B.; et al. Properties of the Binary Black Hole Merger GW150914. *Phys. Rev. Lett.* **2016**, *116*, 241102. [CrossRef]
3. Akiyama, K.; Alberdi, A.; Alef, W.; Asada, K.; Azulay, R.; Baczko, A.; Ball, D.; Baloković, M.; Barrett, J.; Bintley, D.; et al. First M87 Event Horizon Telescope Results. I. The Shadow of the Supermassive Black Hole. *Astrophys. J.* **2019**, *875*, L1. [CrossRef]
4. Akiyama, K.; Alberdi, A.; Alef, W.; Asada, K.; Azulay, R.; Baczko, A.; Ball, D.; Baloković, M.; Barrett, J.; Bintley, D.; et al. First M87 Event Horizon Telescope Results. VI. The Shadow and Mass of the Central Black Hole. *Astrophys. J.* **2019**, *875*, L6. [CrossRef]
5. Herrera, L.; Paiva, F.M.; Santos, N.O.; Ferrari, V. Geodesics in the γ Spacetime. *Int. J. Mod. Phys. D* **2000**, *9*, 649–659. [CrossRef]
6. Herrera, L.; Magli, G.; Malafarina, D. Non-spherical sources of static gravitational fields: Investigating the boundaries of the no-hair theorem. *Gen. Relativ. Gravit.* **2005**, *37*, 1371–1383. [CrossRef]
7. Kovář, J.; Stuchlík, Z.; Karas, V. Off-equatorial orbits in strong gravitational fields near compact objects. *Class. Quantum Gravity* **2008**, *25*, 095011. [CrossRef]
8. Kovář, J.; Kopáček, O.; Karas, V.; Stuchlík, Z. Off-equatorial orbits in strong gravitational fields near compact objects II: halo motion around magnetic compact stars and magnetized black holes. *Class. Quantum Gravity* **2010**, *27*, 135006. [CrossRef]
9. Shaymatov, S.; Ahmedov, B.; Stuchlík, Z.; Abdujabbarov, A. Effect of an external magnetic field on particle acceleration by a rotating black hole surrounded with quintessential energy. *Int. J. Mod. Phys. D* **2018**, *27*, 1850088. [CrossRef]
10. Dadhich, N.; Tursunov, A.; Ahmedov, B.; Stuchlík, Z. The distinguishing signature of magnetic Penrose process. *Mon. Not. R. Astron. Soc.* **2018**, *478*, L89–L94. [CrossRef]
11. Pavlović, P.; Saveliev, A.; Sossich, M. Influence of the vacuum polarization effect on the motion of charged particles in the magnetic field around a Schwarzschild black hole. *Phys. Rev. D* **2019**, *100*, 084033. [CrossRef]
12. Shaymatov, S. Magnetized Reissner–Nordström black hole restores cosmic censorship conjecture. *Int. J. Mod. Phys. Conf. Ser.* **2019**, *49*, 1960020. [CrossRef]
13. Bini, D.; Boshkayev, K.; Geralico, A. Tidal indicators in the spacetime of a rotating deformed mass. *Class. Quantum Gravity* **2012**, *29*, 145003. [CrossRef]
14. Toshmatov, B.; Malafarina, D. Spinning test particles in the γ spacetime. *Phys. Rev. D* **2019**, *100*, 104052. [CrossRef]
15. Haydarov, K.; Abdujabbarov, A.; Rayimbaev, J.; Ahmedov, B. Magnetized Particle Motion around Black Holes in Conformal Gravity: Can Magnetic Interaction Mimic Spin of Black Holes? *Universe* **2020**, *6*, 44. [CrossRef]
16. Stuchlík, Z.; Kološ, M.; Kovář, J.; Slaný, P.; Tursunov, A. Influence of Cosmic Repulsion and Magnetic Fields on Accretion Disks Rotating around Kerr Black Holes. *Universe* **2020**, *6*, 26. [CrossRef]
17. Shaymatov, S.; Vrba, J.; Malafarina, D.; Ahmedov, B.; Stuchlík, Z. Charged particle and epicyclic motions around 4 D Einstein-Gauss-Bonnet black hole immersed in an external magnetic field. *Phys. Dark Universe* **2020**, *30*, 100648. [CrossRef]
18. Shaymatov, S.; Malafarina, D.; Ahmedov, B. Effect of perfect fluid dark matter on particle motion around a static black hole immersed in an external magnetic field. *arXiv* **2020**, arXiv:2004.06811.
19. Narzilloev, B.; Rayimbaev, J.; Shaymatov, S.; Abdujabbarov, A.; Ahmedov, B.; Bambi, C. Dynamics of test particles around a Bardeen black hole surrounded by perfect fluid dark matter. *Phys. Rev. D* **2020**, *102*, 104062. [CrossRef]
20. Carter, B. Global Structure of the Kerr Family of Gravitational Fields. *Phys. Rev.* **1968**, *174*, 1559–1571. [CrossRef]
21. Bardeen, J.M.; Press, W.H.; Teukolsky, S.A. Rotating Black Holes. Locally Nonrotating Frames, Energy Extraction, and Scalar Synchrotron Radiation. *Astrophys. J.* **1972**, *178*, 347–370. [CrossRef]
22. Drasco, S.; Hughes, S.A. Rotating black hole orbit functionals in the frequency domain. *Phys. Rev. D* **2004**, *69*, 044015. [CrossRef]
23. Barack, L.; Sago, N. Beyond the geodesic approximation: Conservative effects of the gravitational self-force in eccentric orbits around a Schwarzschild black hole. *Phys. Rev. D* **2011**, *83*, 084023. [CrossRef]
24. Warburton, N.; Barack, L.; Sago, N. Isofrequency pairing of geodesic orbits in Kerr geometry. *Phys. Rev. D* **2013**, *87*, 084012. [CrossRef]
25. Shaymatov, S.; Atamurotov, F.; Ahmedov, B. Isofrequency pairing of circular orbits in Schwarzschild spacetime in the presence of magnetic field. *Astrophys. Space Sci.* **2014**, *350*, 413–419. [CrossRef]
26. Kunst, D.; Perlick, V.; Lämmerzahl, C. Isofrequency pairing of spinning particles in Schwarzschild-de Sitter spacetime. *Phys. Rev. D* **2015**, *92*, 024029. [CrossRef]
27. Osburn, T.; Warburton, N.; Evans, C.R. Highly eccentric inspirals into a black hole. *Phys. Rev. D* **2016**, *93*, 064024. [CrossRef]
28. Lewis, A.G.M.; Zimmerman, A.; Pfeiffer, H.P. Fundamental frequencies and resonances from eccentric and precessing binary black hole inspirals. *Class. Quantum Gravity* **2017**, *34*, 124001. [CrossRef]

29. Grunau, S.; Khamesra, B. Geodesic motion in the (rotating) black string spacetime. *Phys. Rev. D* **2013**, *87*, 124019. [CrossRef]
30. Gal'Tsov, D.V.; Masar, E. Geodesics in spacetimes containing cosmic strings. *Class. Quantum Gravity* **1989**, *6*, 1313–1341. [CrossRef]
31. Chakraborty, S.; Biswas, L. Motion of test particles in the gravitational field of cosmic strings in different situations. *Class. Quantum Gravity* **1996**, *13*, 2153–2161. [CrossRef]
32. Özdemir, N. Gravitomagnetic effects and cosmic strings. *Class. Quantum Gravity* **2003**, *20*, 4409–4417. [CrossRef]
33. Özdemir, F.; Özdemir, N.; Kaynak, B.T. Multi-Black Holes Solution with Cosmic Strings. *Int. J. Mod. Phys. A* **2004**, *19*, 1549–1557. [CrossRef]
34. Hackmann, E.; Hartmann, B.; Lämmerzahl, C.; Sirimachan, P. Test particle motion in the space-time of a Kerr black hole pierced by a cosmic string. *Phys. Rev. D* **2010**, *82*, 044024. [CrossRef]
35. Hackmann, E.; Hartmann, B.; Lämmerzahl, C.; Sirimachan, P. Complete set of solutions of the geodesic equation in the space-time of a Schwarzschild black hole pierced by a cosmic string. *Phys. Rev. D* **2010**, *81*, 064016. [CrossRef]
36. Tursunov, A.; Kološ, M.; Abdujabbarov, A.; Ahmedov, B.; Stuchlík, Z. Acceleration of particles in spacetimes of black string. *Phys. Rev. D* **2013**, *88*, 124001. [CrossRef]
37. Darwin, C. The Gravity Field of a Particle. II. *Proc. R. Soc. Lond. Ser. A* **1961**, *263*, 39–50. [CrossRef]
38. Cutler, C.; Kennefick, D.; Poisson, E. Gravitational radiation reaction for bound motion around a Schwarzschild black hole. *Phys. Rev. D* **1994**, *50*, 3816–3835. [CrossRef]

Article

Study of Accretion Flow Dynamics of V404 Cygni during Its 2015 Outburst

Arghajit Jana [1,2,*], **Jie-Rou Shang** [3], **Dipak Debnath** [2], **Sandip K. Chakrabarti** [2], **Debjit Chatterjee** [2,4] **and Hsiang-Kuang Chang** [5]

1 Physical Research Laboratory, Navrangpura, Ahmedabad 380009, India
2 Indian Centre for Space Physics, Garia Station Road, Kolkata 700084, India; dipakcsp@gmail.com (D.D.); sandipchakrabarti9@gmail.com (S.K.C.); debjitchatterjee92@gmail.com (D.C.)
3 Institute of Astronomy, National Tsing Hua University, Hsinchu 30013, Taiwan; bandyshang@gmail.com
4 Indian Institute of Astrophysics, Koramangala, Bangalore 560034, India
5 Department of Physics, National Tsing Hua University, Hsinchu 30013, Taiwan; hkchang@mx.nthu.edu.tw
* Correspondence: argha@prl.res.in or argha0004@gmail.com

Abstract: The 2015 Outburst of V404 Cygni is an unusual one with several X-ray and radio flares and rapid variation in the spectral and timing properties. The outburst occurred after 26 years of inactivity of the black hole. We study the accretion flow properties of the source during its initial phase of the outburst using *Swift*/XRT and *Swift*/BAT data in the energy range of 0.5–150 keV. We have done spectral analysis with the two component advective flow (TCAF) model fits file. Several flow parameters such as two types of accretion rates (Keplerian disk and sub-Keplerian halo), shock parameters (location and compression ratio) are extracted to understand the accretion flow dynamics. We calculated equipartition magnetic field B_{eq} for the outburst and found that the highest $B_{eq} \sim 900$ Gauss. Power density spectra (PDS) showed no break, which indicates no or very less contribution of the Keplerian disk component, which is also seen from the result of the spectral analysis. No signature of prominent quasi-periodic oscillations (QPOs) is observed in the PDS. This is due to the non-satisfaction of the condition for the resonance shock oscillation as we observed mismatch between the cooling timescale and infall timescale of the post-shock matter.

Keywords: X-rays: binaries—stars individual: (V404 Cygni)—stars:black holes—accretion; accretion disks—shock waves—radiation:dynamics

Citation: Jana, A.; Shang, J.-R.; Debnath, D.; Chakrabarti, S.K.; Chatterjee, D.; Chang, H.-K. Study of Accretion Flow Dynamics of V404 Cygni during Its 2015 Outburst. *Galaxies* **2021**, *9*, 39. https://doi.org/10.3390/galaxies9020039

Academic Editors: Yosuke Mizuno and Emilio Elizalde

Received: 10 April 2021
Accepted: 21 May 2021
Published: 25 May 2021

Publisher's Note: MDPI stays neutral with regard to jurisdictional claims in published maps and institutional affiliations.

1. Introduction

Transient black hole candidates (BHCs) have two phases in their lives: the quiescence phase and the outbursting phase. They spend most of their lifetimes in the quiescence phase. A sudden rise in viscosity leads to an outburst when the X-ray intensity rises by a factor of thousands or more that of the quiescence phase. Matter from the companion star accretes to the central black hole, and in this process, gravitational potential energy is converted to heat and radiation. A black hole (BH) X-ray spectrum generally consists of two components: a multi-color blackbody and a hard power-law tail. The multi-color blackbody part is believed to originate from a Shakura-Sunyaev type standard thin disk [1]. The power-law tail is believed to originate from a Compton corona [2,3]. In the two-component advective flow (TCAF) solution, the CENBOL or CENtrifugal pressure supported BOundary Layer [4,5] replaces the Compton corona used in other models, such as disk-corona model [6,7] or evaporated disk in ADAF [8]. In this paper, we used the TCAF solution to study the accretion flow dynamics of V404 Cygni during its first outburst in 2015 after a long quiescent of $\sim$26 years.

V404 Cygni is one of the most studied black hole X-ray binary systems. It is also known as GS 2023+338. It was first identified as an optical nova in 1938. In 1956, another nova outburst was reported in this system [9]. In 1989, V404 Cygni went through another outburst. The 1989 outburst was discovered by the all sky monitor onboard *Ginga*. It is located at RA = 306°.01 and Dec = 33°.86. After spending $\sim$26 years in the quiescent state, V404 Cygni went through a short but violent outburst on 15 June 2015. In December 2015, another short activity was observed [10]. The binary system V404 Cygni harbors a black hole of mass 9–12 $M_\odot$ at the center with a K-III type companion of mass $\sim$1 $M_\odot$ [11–13]. The inclination angle of the binary system is $\sim$67° [12,13]. The orbital period of the system is 6.5 days [11]. V404 Cygni is located at a distance of 2.39 kpc, measured by parallax method [14]. V404 Cygni has a high spinning black hole with spin parameters $a^* > 0.92$ [15].

The 2015 outburst of V404 Cygni was discovered on 15 June simultaneously by *Swift*/BAT [10] and *MAXI*/GSC [16]. This outburst of the source was extensively observed in multi-wavelength bands, such as in radio [17], optical [18] and X-ray [19,20]. *INTE-GRAL* observation reported multiple X-ray flares during the outburst [19]. Several radio flares were also observed [17]. Extreme variable and rotating jet is observed during the outburst [21–23]. The source showed rapid changes in the spectral properties in a very short time [24]. *INTEGRAL* observation detected $e^- - e^+$ pair annihilation on 20 June 2015 [25]. *FERMI*/LAT detected high energy γ-ray jet in the source on 26 June 2015 [26]. King et al. [27] reported detection of emission lines with Chandra-HETG, indicating strong disc wind emission. Motta et al. [24] argued that a slim disk is formed at the inner accretion flow due to the super-Eddington accretion in the source.

In this *paper*, we study the timing and the spectral properties of V404 Cygni with combined *Swift*/XRT and *Swift*/BAT data in the broad energy range of 0.5–150 keV during the initial phase of the 2015 outburst. We have done spectral analysis with the TCAF model-based *fits* file to extract physical flow parameters. The nature of these model-fitted accretion flow parameters allowed us to investigate the physical reasons behind the origin of the several flares and their variability and turbulent features. We have also calculated the equipartition magnetic field and cooling time during the outburst of the source.

The paper is organized in the following way. In Section 2, we briefly discuss the disk structure prescribed by TCAF and the way flow parameters decide on the spectral shape. In Section 3, we discuss the observations and the data analysis procedure. In Section 4, we present the results of our analysis. In Section 5, we make a discussion based on our result, and finally, in Section 6, we summarize our findings.

2. TCAF Solution

TCAF configuration is based on the solution of a set of equations which govern viscous, transonic flows around a black hole [28]. In the TCAF solution, accreting flow consists of two components: (i) high viscous, high angular momentum, optically thick and geometrically thin Keplerian disk flow ($\dot{m}_{\rm d}$), (ii) low viscous, low angular momentum, optically thin sub-Keplerian halo flow ($\dot{m}_{\rm h}$). The Keplerian disk accretes on the equatorial plane and is immersed within the sub-Keplerian flow. Centrifugal force rises as flow moves close to the black hole and becomes comparable with the gravitational force. Low viscous halo matter is fast-moving and forms an axisymmetric shock at the centrifugal barrier [29]. The post-shock region i.e., CENBOL 'puffed-up', and becomes 'hot'. The Keplerian disk truncates at the shock location. The multi-color black body part of the observed spectrum that is formed by the soft photons is emitted from the Keplerian disk. A fraction of these soft photons are intercepted by the CENBOL. Depending on the temperature and size of the CENBOL, soft photons become hard photons through repeated inverse-Comptonization scattering at the CENBOL. Conversely, a fraction of Comptonized photons is reprocessed at the Keplerian disk and produces a reflection hump. Thus in the TCAF solution, the reflection component is self-consistently incorporated. However, a Gaussian line may be required to add if an iron emission line is present. CENBOL is also considered to be the base of the jets or outflows [30]. Toroidal magnetic flux tubes are responsible for the collimation of jet [31,32]. Oscillation of CENBOL can be triggered when the cooling and heating times inside CENBOL are similar, and the emerging photons produce the quasi-periodic oscillations (QPOs) [33–35] hereafter C15.

Transient BHCs generally show different spectral states during their outbursts. In TCAF, these observed spectral states are controlled by the flow parameters (two types of accretion rates and two shock parameters). A typical outbursting BHC generally goes through spectral state transitions to form a hysteresis loop as follows: hard state (HS) → hard-intermediate state (HIMS) → soft-intermediate state (SIMS) → soft state (SS) → soft-intermediate state (SIMS) → hard-intermediate state (HIMS) → hard state (HS) [36,37]. In the upper panel of Figure 1 (adopted from [38]), we show a cartoon diagram of the above four spectral states under the TCAF solution. In the lower panel, typical spectra of each spectral state correspond to the diagrams are shown. In the cartoon diagrams, brown, light green, dark green, and grey regions represent Keplerian disk, sub-Keplerian halo, CENBOL and jet, respectively. It is to be noted that the cartoon diagram and corresponding spectra are only for illustrative purposes. TCAF model fits file assumes the shock location as the truncation radius of the Keplerian disk ; however, it may not occur all the time. For example, in the HS, a disk may not be formed at all, or in the SS, the shock may not be formed with the negligible sub-Keplerian flow. However, these scenarios are the example of extreme cases. In general, both Keplerian disk and sub-Keplerian halo matters are present in the accretion flow, and the TCAF model fits file covers the most probable parameter space of the system.

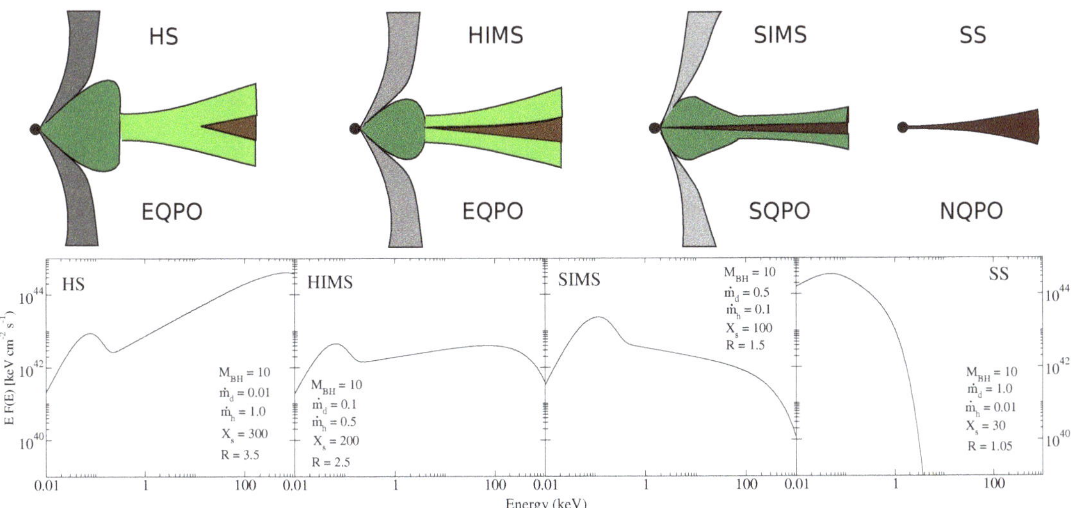

Figure 1. In upper panel, cartoon diagrams of four commonly observed spectral states under TCAF solution are shown (adopted from [38]). Brown, light green, dark green and grey region represent Keplerian disk, sub-Keplerian halo, CENBOL and jet, respectively. An outbursting black hole evolves as HS → HIMS → SIMS → SS → SIMS → HIMS → HS. Generally both rising and declining HS and HIMS show monotonically evolving type-C (low frequency) QPOs, and SIMS shows sporadic type-B or type-A QPO. In soft sate these QPOs are absent. Here, EQPO means evolving QPO; SQPO means sporadic QPO; NQPO means no QPO. In the bottom panel, theoretical spectra corresponding to the top paneled spectral states are shown. These spectra are generated using five input parameters (M_{BH} in $M_\odot$; $\dot{m}_d$ in $\dot{M}_{Edd}$; $\dot{m}_h$ in $\dot{M}_{Edd}$; X_s in r_s, R), whose values are marked inset. The cartoon diagram and corresponding spectra are only for illustrative purpose.

When an outburst is triggered due to lower viscosity and angular momentum, the sub-Keplerian matter moves with free-fall velocity, whereas the Keplerian flow moves in viscous time. Thus, the sub-Keplerian flow dominates the accretion process since it moves faster than the flow in the Keplerian disk. The Keplerian disk is truncated very far away by a large CENBOL. A strong shock (higher R) is formed at hundreds of Schwarzschild radius (r_s) away from the BH. Thus, it is difficult to cool the CENBOL by the Keplerian component. Hard X-ray flux dominates, and a hard state is observed. Compact jet is launched in this state from the CENBOL [30]. Evolving type-C QPO is produced in this state due to the resonance oscillation of the shock [33].

The source enters the HIMS after HS. The Keplerian disk accretion rate continues to rise and becomes comparable with the sub-Keplerian halo accretion rate. As a result, accretion rate ratio (ARR = $\dot{m}_h/\dot{m}_d$) decreases. Due to the rise of the Keplerian disk accretion rate, CENBOL shrinks and becomes cool as the shock moves farther inward. Shock strength decreases as the Compton cooling reduces the post-shock thermal pressure. Further, in this intermediate state, in the presence of jets, mass outflow rate to inflow rate ratio becomes maximum. In this state, evolving type-C QPOs may also be observed.

In the SIMS, the Keplerian rate keeps on increasing, although the sub-Keplerian flow rate started to decrease. This is because more and more sub-Keplerian flow becomes Keplerian by viscous transport. The shock becomes weak in this state. The shock further moves in, and the CENBOL becomes small. The soft X-ray flux increases, and the hard X-ray flux decreases in this state due to rapid rise in $\dot{m}_d$ and slow decrease in $\dot{m}_h$. Generally, type-A or type-B QPOs are observed sporadically in this state due to weak oscillation of the CENBOL (type-B) or due to oscillation of shock-less centrifugal barrier (type-A) [34,35]. The time (day) difference between $\dot{m}_h$ and $\dot{m}_d$ peaks gives us rough estimation of the viscous time scale of the source [39,40]. In the SS, the Keplerian disk dominates and completely cools down the CENBOL. The soft X-ray flux dominates over the hard X-ray flux. No shock is formed. As a result, the jet is completely quenched in this state [30,41]. No QPO is produced in the soft state.

The spectral states evolve in the reverse order during the declining phase of the outburst as flow parameters show opposite nature. Starting from the SS to the SIMS transition day, both the Keplerian disk accretion rate and the sub-Keplerian halo accretion rate decreases, although the Keplerian disk rate decreases faster. As a result, ARR increases. As in the SIMS of the rising phase, one may see sporadic type-B or A QPOs in the declining SIMS. In the declining HIMS and HS, evolving type-C QPOs could be seen. Similar to the rising phase, one could observe a compact jet in the HIMS and HS in the declining phase.

TCAF solution is implemented in XSPEC [42] as an additive table model to analyze the spectral properties around the black holes [43,44]. TCAF model has four input flow parameters ($\dot{m}_d$, $\dot{m}_h$, X_s, R) other than mass of the black hole (M_{BH}) and normalization (N). In the recent years, accretion flow dynamics around several black holes are studied quite successfully using TCAF model [45–48]. Frequencies of the dominating QPOs are predicted from the TCAF model fitted shock parameters [49]. Masses of the black holes are estimated quite successfully from spectral analysis with the TCAF model [49–51]. Jet contribution in the X-rays are also calculated using TCAF solution [52–54].

While fitting with the TCAF model, one must take care of degeneracy. Like any other spectral model, TCAF model is degenerate between different parameters. Thus, one must be very careful while fitting with TCAF model. One must always remember the physical condition of the system. For example, a single spectrum can be fitted with two different set of parameters: ($\dot{m}_d = 0.1\ M_{Edd}$, $\dot{m}_h = 0.5\ M_{Edd}$, $M_{BH} = 10\ M_\odot$, $X_s = 100\ r_s$, $R = 3.5$) and ($\dot{m}_d = 0.2\ M_{Edd}$, $\dot{m}_h = 0.8\ M_{Edd}$, $M_{BH} = 10\ M_\odot$, $X_s = 10\ r_s$, $R = 4$). Clearly, the second set of parameters is unphysical (the Xs can not be low while the the shock is strong and mh is higher than md). Also, one needs to find the parameters with the steppar command to avoid the degeneracy.

3. Observation and Data Analysis

We analyzed *Swift* data for 19 observations between 2015 June 15 (MJD 57,188.77) and 26 June (MJD 57,199.52). We studied the source in 0.5–150 keV energy band with combined XRT and BAT data for five observations (MJD 57,191.01, 57,194.54, 57,197.21, 57,197.33, and 57,198.02). The 15–150 keV BAT data was used for four observations (MJD 57,188.77, 57,193.56, 57,198.15 and 57,199.52) when only BAT observations were available. For the rest of the ten observations, we studied 0.5–10 keV using XRT data.

We used WT mode data for XRT observation. Cleaned event files were generated for XRT using `xrtpipeline` command. To reduce pile-up effects, we used grade-0 data. For pile-up correction, we chose an annular region around the source. We chose an outer radius of 30 pixels and a varying inner region, depending on the count rate. We extracted spectra for the different inner regions and tested for the pile-up. The piled-up spectra are harder than expected. When increasing the inner region did not alter the spectra, the pile-up had been excluded. A background region is chosen far away from the source at the same axis as the source with an annular of same radii as the source. Then, we obtained *source*, and *background* files using cleaned event files in `XSELECT v2.4`. A scaling factor was applied to the source and background with `backscal` task. For an annular of outer radius of r_2 and inner radius of r_1, we used the scaling factor $r_2 - r_1$. The spectra were re-binned to 20 counts per bin using `grppha` command. The 0.5–10 keV 0.01 s lightcurves were generated in `XSELECT v2.4` using cleaned source and background event files. We followed standard procedures to generate BAT spectra and lightcurves. Detector plane images (dpi) were generated using the task `batbinevt`. For appropriate detector quality, we used `batdetmask` task. Noisy detectors were found, and a quality map was obtained using `bathotpix`. Then `batmaskwtevt` was run to apply mask weighting to the event mode data. A systematic error was applied to the BAT spectra using `batphasyserr`. Ray-tracing was corrected using `batupdatephakw` task. Then a response matrix for the spectral file was generated using `batdetmask`.

Here, we used the TCAF model-based *fits* file for the spectral analysis. We also used combined `diskbb` (DBB) and `powerlaw` (PL) models to get rough estimation about the thermal and the non-thermal fluxes where a reflection component is often required to find the best fits. TCAF model-based fits do not require any additional component for reflection since the reflection component is already incorporated in the model. We required a `Gaussian` model to incorporate Fe Kα emission line. We used `phabs` model for interstellar absorption and `pcfabs` model for partial absorption. With the TCAF, we extracted physical parameters such as the mass of the black hole (M_{BH}) in solar mass ($M_\odot$), the Keplerian disk rate ($\dot{m}_d$) in Eddington rate ($\dot{M}_{\mathrm{Edd}}$), the sub-Keplerian halo rate ($\dot{m}_d$) in Eddington rate ($\dot{M}_{\mathrm{Edd}}$), the shock location (X_s) (i.e., size of the Compton cloud) in Schwarzschild radius (r_s) and the shock compression ratio ($R = \rho_+/\rho_-$ with ρ_+ and ρ_- are post- and pre-shock density, respectively). In TCAF, N depends on the source's distance and inclination angle and is just a constant factor between the emitted spectra and observed spectra by a given instrument. However, it can vary if any physical processes are present other than the accretion. Since the current version of the TCAF model fits file does not include jets, for instance, a variation of normalization is observed if they are present [52,54]. We first analyzed the spectra after keeping the mass of the black hole as a free parameter. We obtained the mass of the black hole in the range of 9.5–11.5 $M_\odot$ or the average value of 10.6 $M_\odot$. This measured mass range agrees very well with previously reported values by many authors [11,13]. Then, we refitted all the spectra after keeping the mass of the black hole frozen at 10.6 $M_\odot$. The result based on the later analysis is presented here.

We achieved best-fittings using `steppar` command. After obtaining a best-fit based on $\chi^2_{red}(\sim 1)$ with TCAF, we ran `steppar` to verify fitted parameter values. The 'steppar' command ran for pair of parameters $\dot{m}_d - \dot{m}_h$ and $X_s - R$. We also calculated uncertainties with the `steppar` with 90% confidence. In Figures 2 and 3, 2D-contour plots for two observations (MJD 57,188.77 and 57,189.62) are shown.

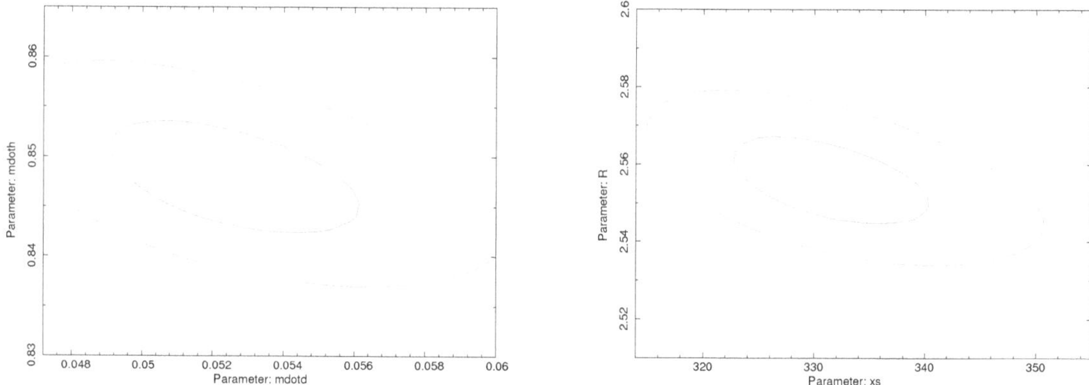

Figure 2. 2D confidence contour plot for MJD 57,188.77, for $\dot{m}_\mathrm{d}-\dot{m}_\mathrm{h}$ in the left panel and $X_\mathrm{s}-R$ in the right panel.

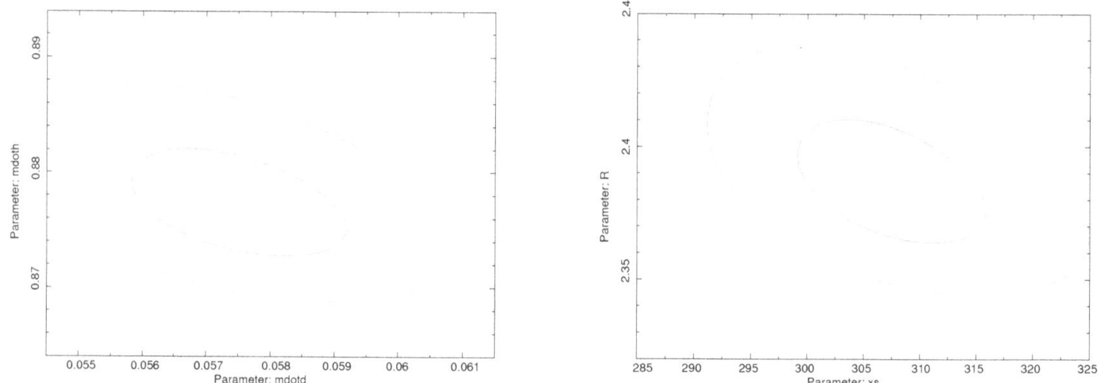

Figure 3. 2D confidence contour plot for MJD 57189.62, for $\dot{m}_\mathrm{d}-\dot{m}_\mathrm{h}$ in the left panel and $X_\mathrm{s}-R$ in the right panel.

4. Results

We present the results of spectral and temporal analysis of the source in 0.5–150 keV energy band using combined XRT+BAT or only BAT (in 15–150 keV) or only XRT (in 0.5–10 keV) data. In Figure 4, we show two XRT PDS for observations on MJD 57,191.01 (18 June 2015) and MJD 57,194.54 (21 June 2015). In Figure 5, TCAF model fitted spectrum of combined XRT plus BAT data in the broad energy range of 0.5–150 keV is shown for the observation on MJD 57,191.01.

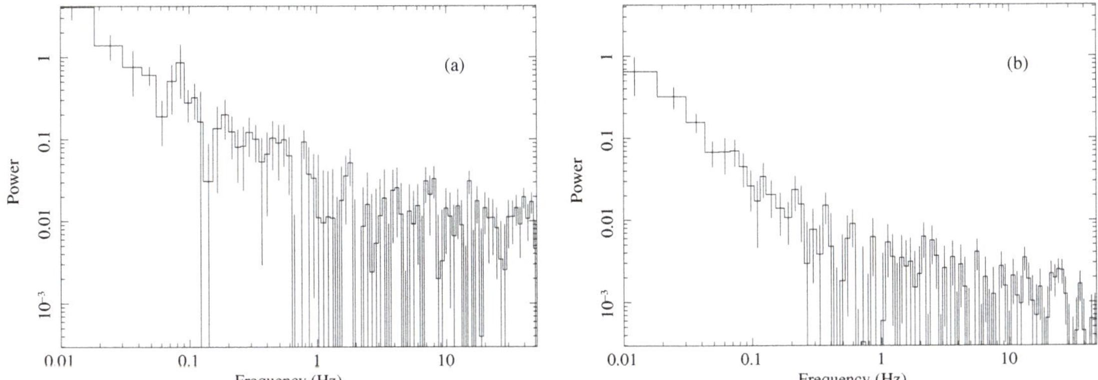

Figure 4. Power density spectra (PDS) of 0.01 s time binned lightcurves of 0.5–10 keV XRT data. The observation date of these PDS are: (**a**) 18 June 2015 (MJD 57,191.01) and (**b**) 21 June 2015 (MJD 57,194.54).

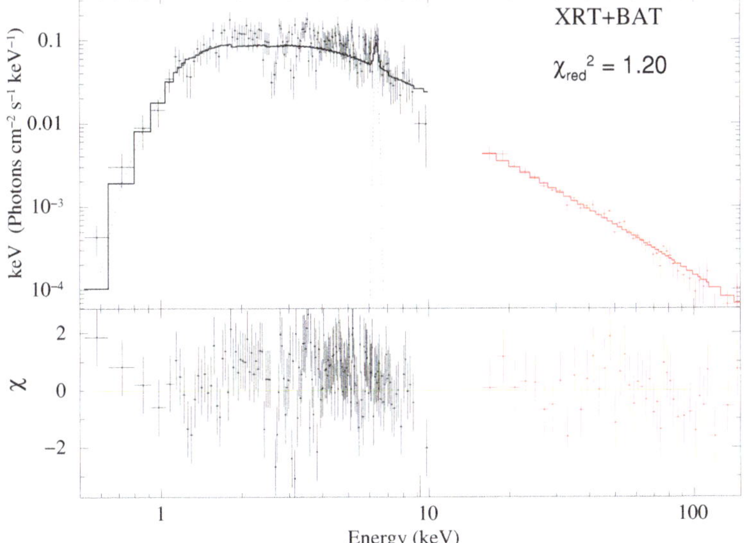

Figure 5. TCAF model fitted combined XRT+BAT spectra in the energy range of 0.5–150 keV, observed on MJD 57,191.03.

In Figure 6a, we show the evolution of BAT and XRT fluxes. In Figure 6b–c, we show the variation of the Keplerian disk rate ($\dot{m}_\mathrm{d}$), the sub-Keplerian halo rate ($\dot{m}_\mathrm{h}$) with day (in MJD). In Figure 6d, we show the evolution of accretion rate ratio (ARR = $\dot{m}_\mathrm{h}/\dot{m}_\mathrm{d}$). In Figure 7a, we show the variation of the equipartition magnetic field (B_eq) with the day. In Figure 7b–d, we show the variation of the shock location (X_s), the shock compression ratio (R), and TCAF model normalization with the day. In Figure 8, the time-resolved *Swift*/BAT spectra in the energy range of 15–150 keV for the observation on 15 June 2015 (MJD 57,188.77) are shown. The three spectra are marked as (a) (online green), (b) (online black) and (c) (online red) with exposures of 220 s, 680 s and 160 s, respectively.

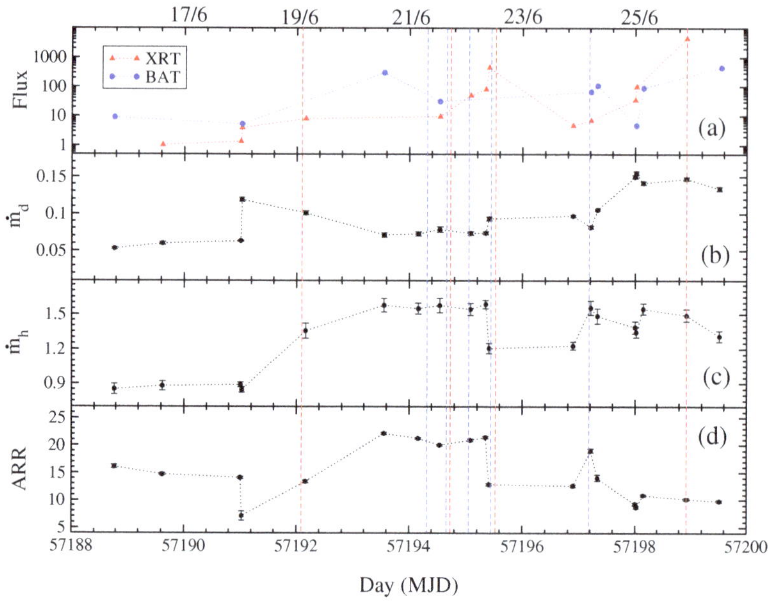

Figure 6. Panel (**a**) shows variation of 0.5–10 keV XRT and 15–150 keV BAT flux with day (MJD). The red (online) triangle represents XRT flux while blue (online) circle represents BAT flux. The fluxes are in the unit of $10^{-9}\,\mathrm{erg\,cm^{-2}\,s^{-1}}$. The variation of (**b**) $\dot{m}_\mathrm{d}$ in $\dot{M}_\mathrm{Edd}$ and (**c**) $\dot{m}_\mathrm{h}$ in $\dot{M}_\mathrm{Edd}$ are shown with day (MJD). In panel (**d**) the variation of ARR ($=\dot{m}_\mathrm{h}/\dot{m}_\mathrm{d}$) is shown. Blue and red dotted lines represent reported X-ray and Radio flaring activities, respectively.

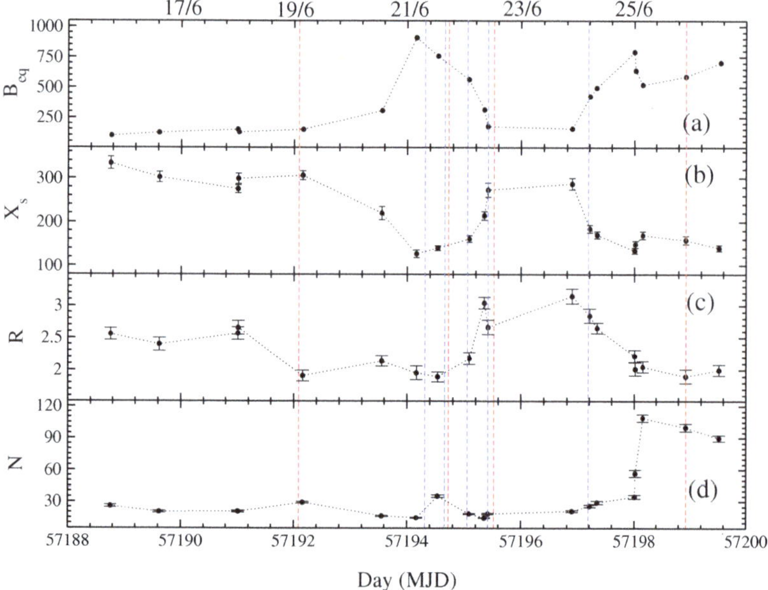

Figure 7. The variation (**a**) equipartition magnetic field (B_eq) in *Gauss*, (**b**) shock location (X_s) in r_s, (**c**) shock compression ratio (R) and (**d**) normalization (N) are shown with day (MJD).

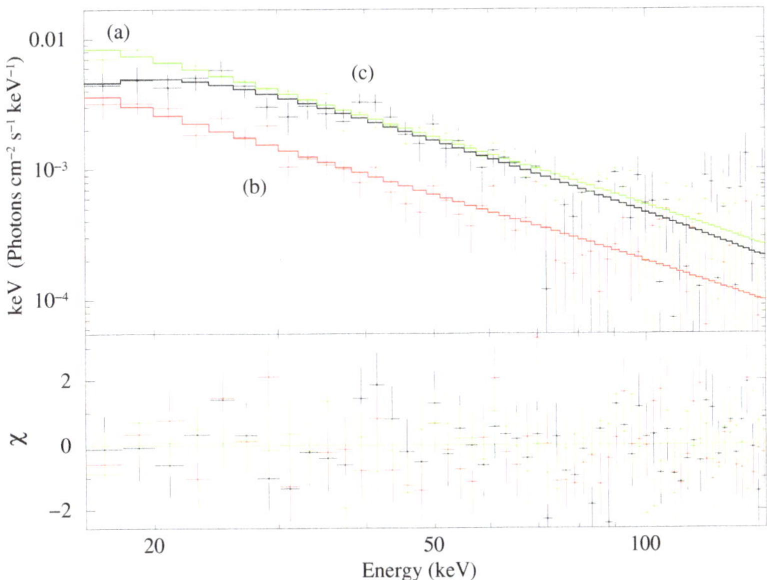

Figure 8. Time resolved BAT spectra for observation on MJD 57,188.77 is shown. The spectra correspond to exposures of (**a**) 220 s, (**b**) 680 s and (**c**) 160 s.

4.1. Temporal Evolution

The 2015 outburst of V404 Cygni is different from other regular outbursts of transient BHCs. It showed rapid changes in a very short timescale in both XRT and BAT count rates. The flux changed significantly in minutes to hours timescales. Several radio and X-ray flares were observed during this epoch. *INTEGRAL* observation revealed 18 X-ray flares during the outburst [19]. Radio flares were also reported during the outburst period [17]. We find that the luminosity was super-Eddington in some observations. On 22 June 2015 (MJD 57,195.41), the XRT flux increased rapidly to 4.61×10^{-7} erg cm^{-2} s^{-1} from 8.2×10^{-8} erg cm^{-2} s^{-1} of the previous observation. This corresponds to the source luminosity, $L \simeq 3.13 \times 10^{38}$ erg s^{-1}. On 25 June 2015 (MJD 57,198.93), the XRT flux reached at 4.59×10^{-6} erg cm^{-2} s^{-1} which corresponds to the source luminosity, $L \simeq 3.12 \times 10^{39}$ erg s^{-1}. BAT count rate also increased rapidly within our analysis period. It became maximum on 26 June 2015 (MJD 57,198.93) at 4.68×10^{-7} erg cm^{-2} s^{-1}, i.e., luminosity was $L \simeq 3.18 \times 10^{38}$ erg s^{-1}.

We studied power density spectra (PDS) of 0.01 s binned lightcurve of *Swift*/XRT. We did not find any clear evidence of QPOs during the initial phase of the outburst in *Swift*/XRT PDSs. A weak signature of QPO was observed in the PDS on 18 June 2015 (MJD 57,191.03) (see Figure 4a). The XRT PDS showed an almost flat spectral slope with broad-band noise. The noise decreased as the outburst progressed, during which we found two slopes in the PDS, a steep power-law slope in the lower frequency and a flat slope at a higher frequency. In some PDS, we observed that power diminished very rapidly. We did not find any break in the PDS.

4.2. Evolution of the Spectral Properties

We have done the spectral analysis using 0.5–150 keV combined *Swift*/XRT and *Swift*/BAT data between 15 and 26 June 2015. Radhika et al. (2016) analyzed the same data set using phenomenological `diskbb` and `powerlaw` models [20]. We analyzed the data with combined `diskbb`, and `power-law` models and have found similar results as in [20]. In general, we used `phabs*pcf*(diskbb+powerlaw+gaussian)` model to estimate thermal and non-thermal fluxes. While analyzing with phenomenological models, we did not require a diskbb component regularly. Diskbb component was required only in 9 observations out of a total of 19 observations. During the entire period, PL photon index varied between 0.60 and 2.43, although the PL flux dominated over the DBB flux. We also required a Gaussian for Fe Kα line emission along with the diskbb+powerlaw model. Detailed results of the phenomenological model fitted spectra are given in Table 1.

In the present paper, our main goal is to study the accretion flow dynamics of the source from spectral analysis with the physical TCAF model. For this purpose, we used phabs*pcf*(TCAF+gaussian) model. Detailed results using this model is presented in Table 2. From the spectral analysis in the initial few observations, we obtained very low values of the Keplerian disk rate ($\dot{m}_\mathrm{d}$) while the sub-Keplerian halo rates ($\dot{m}_\mathrm{h}$) were found to be high (>0.85$\dot{M}_\mathrm{Edd}$). On 19 June 2015 (MJD 57,192.16), $\dot{m}_\mathrm{h}$ increased to 1.37 $\dot{M}_\mathrm{Edd}$ from its previous day value of 0.85 $\dot{M}_\mathrm{Edd}$. After that, it varied within 1.24–1.58 $\dot{M}_\mathrm{Edd}$ until the end of our analysis period. On 25 June 2015 (MJD 57,198.02), we observed sudden rise in $\dot{m}_\mathrm{d}$ from its previous day, i.e., from 0.11 $\dot{M}_\mathrm{Edd}$ to 0.15 $\dot{M}_\mathrm{Edd}$. Before that, $\dot{m}_\mathrm{d}$ varied in the range of 0.05–0.12 $\dot{M}_\mathrm{Edd}$. After MJD 57,198.02, $\dot{m}_\mathrm{d}$ was obtained in a narrow range of 0.13–0.16 $\dot{M}_\mathrm{Edd}$ (see Figure 6b).

A strong shock ($R = 2.56$) was found far away from the black hole ($X_\mathrm{s} = 334\ r_\mathrm{s}$) on the first day (MJD 57,188.77) of our observation (see Figure 7b,c). The shock remained strong for the next five days. After that, the shock was found to move closer to the black hole as the Keplerian disk rate increased. The shock was found at 126 r_s on MJD 57,194.16 with $R = 1.92$. The shock did not move closer than this. After that, shock moved away from the black hole. Again we found that the shock was moving inward after MJD 57,197.21. On the last day of our observation, we found the shock to be located at 143 r_s.

We added a Gaussian profile and the TCAF solution to incorporate the contribution of the *Fe* emission line in XRT data. Fe-line varied within 6.08 keV and 6.97 keV. In some observations, we required the Fe line width of >1 keV. The source was reported to be partially obscured due to the disk wind and outflow [24,27,55]. We used phabs models for the absorption. We did not freeze n_H at a particular value. Rather, we kept it free. In our analysis, we observed it to vary between 0.54 $\times$ 10^{22} to 1.49 $\times$ 10^{22} cm^{-2}. The variable absorption is unlikely due to the interstellar medium, rather it is likely to be associated with the intrinsic absorption of the source. In 11 observations out of total of 19 observations, we required to use pcfabs model to incorporate partial absorption in the XRT data. In some observations, the covering required as high as 95%. In general, it varied between 50% and 95%. For covering absorption, n_H varied between 2.2 $\times$ 10^{22} and 28.9 $\times$ 10^{22} cm^{-2}.

4.3. Time Resolved BAT Spectra

To study the evolution of spectral nature in short time intervals, we analyzed the time-resolved BAT spectra. Rapid variation of the accretion rates and other physical flow parameters from observation to observation motivated us to make this study. Here, we analyzed time-resolved BAT spectra for four observations on 15 June 2015 (MJD 57,188.77), 18 June 2015 (MJD 57,191.03), 20 June 2015 (MJD 57,193.56) and 26 June 2015 (MJD 57,199.52). We found rapid variation in the BAT spectra within a very short time, even in one observation. In Table 3, TCAF model fitted parameters for time-resolved spectra are presented. For example, on 15 June 2015, within total BAT exposure of 1202 s (Figure 8), we observed the variations of $\dot{m}_\mathrm{d}$ between 0.041 and 0.057 $\dot{M}_\mathrm{Edd}$, and $\dot{m}_\mathrm{h}$ between 0.82 and 0.88 $\dot{M}_\mathrm{Edd}$. The shock was observed to vary in between $\sim$326–336 r_s. Similar rapid variation of the flow parameters ($\dot{m}_\mathrm{d}$ and $\dot{m}_\mathrm{h}$) were also observed for the remaining three observations.

Table 1. DBB+PL Model Fitted Spectral Analysis Results.

Obs ID	Day (MJD)	XRT exp (sec)	BAT exp (sec)	XRT Flux*	BAT Flux*	DBB Flux*	PL Flux*	n_{H}^{1} (10^{22})	n_{H}^{2} (10^{22})	CF	T_{in} (keV)	Γ	χ^2/dof
(1)	(2)	(3)	(4)	(5)	(6)	(7)	(8)	(9)	(10)	(11)	(12)	(13)	(14)
00643949000	57,188.77	—	1202	—	9.01 ± 0.09	—	9.01 ± 0.09	0.94 ± 0.07	—	—	—	1.59 ± 0.13	71/53
00031403035	57,189.62	1970	—	1.04 ± 0.04	—	—	1.04 ± 0.04	1.12 ± 0.09	32.7 ± 0.2	0.46 ± 0.06	—	1.09 ± 0.15	714/944
00031403038	57,191.01	610	—	1.36 ± 0.06	—	—	1.36 ± 0.06	0.91 ± 0.09	1.61 ± 0.06	0.82 ± 0.10	—	0.60 ± 0.07	1013/946
00644520000	57,191.03	217	1202	3.98 ± 0.07	5.37 ± 0.08	—	9.35 ± 0.15	0.59 ± 0.08	10.5 ± 0.4	0.73 ± 0.12	—	1.83 ± 0.18	394/250
00031403042	57,192.16	1262	—	8.05 ± 0.03	—	—	8.05 ± 0.03	1.28 ± 0.05	11.7 ± 0.6	0.95 ± 0.10	—	1.39 ± 0.18	1339/931
00645176000	57,193.56	—	542	—	306.7 ± 2.3	—	306.7 ± 2.3	0.51 ± 0.03	—	—	—	1.20 ± 0.15	56/51
00031403048	57,194.16	4208	—	4.89 ± 0.08	—	1.42 ± 0.03	3.47 ± 0.04	0.49 ± 0.03	—	—	0.19 ± 0.03	1.65 ± 0.17	970/889
00031403046	57,194.54	240	7	9.77 ± 0.10	31.8 ± 0.7	—	41.6 ± 0.3	0.66 ± 0.07	—	—	—	1.07 ± 0.19	1437/882
00031403045	57,195.08	930	—	50.8 ± 1.0	—	24.0 ± 0.4	26.7 ± 0.6	0.54 ± 0.05	5.97 ± 0.14	0.88 ± 0.12	0.15 ± 0.03	0.44 ± 0.06	1192/928
00031403049	57,195.35	2977	—	82.1 ± 1.5	—	16.9 ± 1.5	65.2 ± 2.2	0.62 ± 0.04	0.44 ± 0.02	0.69 ± 0.07	0.95 ± 0.12	0.55 ± 0.06	1282/702
00031403047	57,195.41	1903	—	461.9 ± 3.4	—	—	461.9 ± 3.4	0.71 ± 0.05	—	—	—	1.25 ± 0.14	975/713
00031403052	57,196.89	275	—	4.84 ± 0.09	—	2.67 ± 0.21	2.16 ± 0.16	0.75 ± 0.08	5.83 ± 0.15	0.95 ± 0.05	0.24 ± 0.07	1.48 ± 0.29	1061/941
00031403054	57,197.21	682	5	7.21 ± 0.15	67.8 ± 1.2	1.04 ± 0.08	74.0 ± 1.1	0.96 ± 0.11	29.4 ± 1.2	0.77 ± 0.11	0.21 ± 0.06	1.43 ± 0.22	1409/979
00031403053	57,197.33	1052	518	7.33 ± 0.12	109.7 ± 1.5	—	117.0 ± 1.5	0.48 ± 0.09	—	—	—	2.43 ± 0.36	1107/998
00031403055	57,198.00	1028	—	35.9 ± 0.4	—	19.7 ± 0.8	16.2 ± 0.8	1.08 ± 0.18	0.91 ± 0.11	0.61 ± 0.09	0.76 ± 0.10	1.67 ± 0.24	1250/926
00031403056	57,198.02	818	713	101.6 ± 1.8	4.82 ± 0.22	101.6 ± 1.8	4.82 ± 0.33	0.58 ± 0.08	9.15 ± 0.81	0.81 ± 0.11	0.17 ± 0.03	1.76 ± 0.31	1328/994
00031403057	57,198.15	—	745	—	91.4 ± 2.4	—	91.4 ± 2.4	1.25 ± 0.24	—	—	—	1.67 ± 0.25	39/50
00031403058	57,198.93	1312	—	4592 ± 12	—	1.34 ± 0.11	4590 ± 12	0.95 ± 0.09	0.47 ± 0.05	0.86 ± 0.09	0.98 ± 0.12	1.70 ± 0.35	1322/928
00646721000	57,199.52	—	965	—	468.9 ± 4.1	—	468.9 ± 4.1	1.22 ± 0.17	—	—	0.58 ± 0.08	1.82 ± 0.27	41/49

In Col. 2, UT dates of the year 2015 are mentioned in dd-mm format. In Cols. 4 & 5, XRT and exposures are mentioned. In Cols. 6 and 7, XRT (0.5–10 keV) and BAT (15–150 keV) fluxes are mentioned. In Cols. 8 and 9, model fitted disk blackbody (DBB) and powerlaw (PL) fluxes are mentioned. All fluxes are in the units of 10^{-9} erg cm^{-2} s^{-1}. n_{H}^{1} and n_{H}^{2} are in the unit of 10^{22} cm^{-2}. n_{H}^{1} is Hydrogen column density for interstellar absorption. n_{H}^{2} is Hydrogen column density for partial covering absorption. The Fe emission line energy and σ are mentioned in Col. 15 and 16. Best fitted values of χ^2 and degrees of freedom are mentioned in Col. 17 as χ^2/dof.

Table 2. TCAF Model Fitted Spectral Analysis Results.

Obs ID	UT Date (DD-MM)	Day (MJD)	n_H^1 (10^{22})	n_H^2 (10^{22})	CF	$\dot{m}_d$ ($\dot{M}_{Edd}$)	$\dot{m}_h$ ($\dot{M}_{Edd}$)	ARR ($\dot{m}_h/\dot{m}_d$)	X_S (r_s)	R	N	Fe.Line (keV)	LW (keV)	χ^2/dof
(1)	(2)	(3)	(4)	(5)	(6)	(7)	(8)	(9)	(10)	(11)	(12)	(13)	(14)	(15)
00643949000	15-06	57,188.77	1.49 ± 0.17	—	—	0.05 ± 0.01	0.85 ± 0.04	17.0 ± 1.0	334 ± 15	2.56 ± 0.10	15.5 ± 1.2	—	—	69/51
00031403035	16-06	57,189.62	1.86 ± 0.22	28.8 ± 1.8	0.41 ± 0.03	0.06 ± 0.01	0.88 ± 0.04	15.2 ± 0.8	305 ± 13	2.40 ± 0.10	10.2 ± 0.9	—	—	708/899
00031403038	18-06	57,191.01	0.87 ± 0.10	2.2 ± 0.1	0.93 ± 0.10	0.06 ± 0.01	0.90 ± 0.03	15.3 ± 0.4	275 ± 10	2.60 ± 0.12	17.6 ± 0.8	6.57 ± 0.11	0.89 ± 0.08	929/937
00644520000	18-06	57,191.03	1.27 ± 0.14	9.3 ± 0.4	0.63 ± 0.07	0.12 ± 0.02	0.85 ± 0.03	6.9 ± 0.2	300 ± 11	2.59 ± 0.12	19.7 ± 0.9	6.41 ± 0.15	1.10 ± 0.17	338/281
00031403042	19-06	57,192.16	1.20 ± 0.11	12.2 ± 0.8	0.91 ± 0.04	0.10 ± 0.01	1.37 ± 0.07	13.4 ± 0.7	302 ± 12	1.91 ± 0.08	24.0 ± 1.1	6.44 ± 0.15	0.74 ± 0.13	932/787
00645176000	20-06	57,193.56	0.58 ± 0.05	—	—	0.07 ± 0.01	1.58 ± 0.07	22.2 ± 1.2	220 ± 14	2.14 ± 0.09	14.4 ± 0.8	—	—	54/49
00031403048	21-06	57,194.16	0.57 ± 0.09	—	—	0.07 ± 0.01	1.53 ± 0.07	21.6 ± 1.1	126 ± 9	1.92 ± 0.11	12.4 ± 0.4	6.46 ± 0.29	0.45 ± 0.04	1086/899
00031403046	21-06	57,194.54	0.62 ± 0.10	—	—	0.08 ± 0.01	1.58 ± 0.08	20.2 ± 1.2	141 ± 7	1.88 ± 0.09	23.8 ± 1.5	6.39 ± 0.24	1.20 ± 0.22	1399/797
00031403045	22-06	57,195.08	0.55 ± 0.07	6.2 ± 0.2	0.86 ± 0.09	0.07 ± 0.01	1.54 ± 0.05	21.1 ± 0.9	162 ± 10	2.18 ± 0.11	18.3 ± 0.6	6.92 ± 0.22	1.40 ± 0.15	1117/928
00031403049	22-06	57,195.35	0.66 ± 0.06	18.0 ± 0.3	0.66 ± 0.05	0.08 ± 0.01	1.57 ± 0.05	20.3 ± 0.7	217 ± 12	3.01 ± 0.09	12.3 ± 0.5	6.91 ± 0.18	1.19 ± 0.20	1067/707
00031403047	22-06	57,195.41	0.63 ± 0.05	—	—	0.09 ± 0.01	1.24 ± 0.05	13.6 ± 0.6	269 ± 14	2.68 ± 0.12	18.9 ± 0.9	6.97 ± 0.23	0.86 ± 0.11	982/709
00031403052	23-06	57,196.89	0.84 ± 0.10	5.6 ± 0.5	0.95 ± 0.15	0.10 ± 0.01	1.26 ± 0.03	12.7 ± 0.3	290 ± 11	3.16 ± 0.11	22.1 ± 0.9	6.45 ± 0.21	0.30 ± 0.05	1001/937
00031403054	24-06	57,197.21	1.10 ± 0.14	28.9 ± 1.3	0.73 ± 0.07	0.08 ± 0.01	1.54 ± 0.04	19.5 ± 0.9	187 ± 9	2.88 ± 0.10	26.5 ± 1.2	6.64 ± 0.20	0.61 ± 0.14	1405/981
00031403053	24-06	57,197.33	0.54 ± 0.04	—	—	0.11 ± 0.01	1.44 ± 0.07	13.4 ± 0.6	168 ± 7	2.63 ± 0.10	20.3 ± 1.5	6.55 ± 0.18	0.29 ± 0.06	1213/991
00031403055	25-06	57,198.00	0.85 ± 0.07	5.2 ± 0.4	0.85 ± 0.06	0.15 ± 0.01	1.39 ± 0.05	9.0 ± 0.4	132 ± 8	2.22 ± 0.09	37.5 ± 2.7	6.08 ± 0.14	0.96 ± 0.13	937/727
00031403056	25-06	57,198.02	0.60 ± 0.04	8.2 ± 0.8	0.81 ± 0.06	0.15 ± 0.01	1.36 ± 0.04	8.9 ± 0.3	154 ± 7	2.04 ± 0.12	59.6 ± 3.6	6.30 ± 0.19	1.11 ± 0.16	1303/994
00031403057	25-06	57,198.15	1.20 ± 0.11	—	—	0.15 ± 0.01	1.57 ± 0.06	10.6 ± 0.4	174 ± 10	2.08 ± 0.07	93.4 ± 2.9	—	—	41/51
00031403058	25-06	57,198.93	0.87 ± 0.04	3.9 ± 0.5	0.92 ± 0.05	0.15 ± 0.01	1.45 ± 0.05	10.2 ± 0.4	160 ± 9	1.88 ± 0.10	89.5 ± 4.5	6.22 ± 0.22	0.90 ± 0.21	1423/943
00646721000	26-06	57,199.52	1.15 ± 0.16	—	—	0.13 ± 0.02	1.31 ± 0.05	9.8 ± 0.4	143 ± 8	1.99 ± 0.09	77.5 ± 3.3	—	—	36/51

In Col. 2, UT dates of the year 2015 are mentioned in dd-mm format. n_H^1 and n_H^2 are in the unit of 10^{22} cm^{-2}. n_H^1 is Hydrogen column density for interstellar absorption. n_H^2 is Hydrogen column density for partial covering absorption. TCAF model fitted/derived parameters are mentioned in Cols. 7–12. The Fe emission line energy and σ are mentioned in Col. 13 and 14. Best fitted values of χ^2 and degrees of freedom are mentioned in Col. 15 as χ^2/dof. Note: Mass of the black hole was kept frozen at 10.6 $M_\odot$ during spectral fitting with the TCAF model fits file. The average values of 90% confidence ± values obtained using `steppar` command in XSPEC, are placed as superscripts of fitted parameter values.

Table 3. Time Resolved BAT Spectra.

Obs ID	Day	Spectra	Exposures	$\dot{m}_\mathrm{d}$	$\dot{m}_\mathrm{h}$	ARR	X_S	R	N	χ^2/dof
	(MJD)		(sec)	$(\dot{M}_\mathrm{Edd})$	$(\dot{M}_\mathrm{Edd})$		(r_s)			
(1)	(2)	(3)	(4)	(5)	(6)	(7)	(8)	(9)	(10)	(11)
00643949000	57,188.77		1202	0.050 ± 0.0013	0.852 ± 0.046	17.04 ± 0.98	334 ± 15	2.55 ± 0.10	15.49 ± 1.19	69/51
		1	220	0.055 ± 0.0015	0.879 ± 0.043	15.98 ± 0.85	326 ± 13	2.63 ± 0.09	15.58 ± 1.22	66/51
		2	680	0.057 ± 0.0014	0.823 ± 0.045	14.44 ± 0.71	336 ± 13	2.37 ± 0.10	12.55 ± 1.09	72/51
		3	160	0.041 ± 0.0015	0.876 ± 0.043	21.36 ± 0.91	330 ± 13	2.55 ± 0.09	19.76 ± 1.51	59/51
00644520000	57,191.03		1202	0.124 ± 0.0022	0.851 ± 0.028	6.92 ± 0.22	300 ± 11	2.59 ± 0.10	19.69 ± 0.89	338/281
		1	200	0.131 ± 0.0022	0.854 ± 0.037	6.52 ± 0.39	317 ± 10	2.81 ± 0.11	21.54 ± 1.17	56/51
		2	220	0.109 ± 0.0026	0.847 ± 0.053	7.77 ± 0.45	269 ± 10	2.43 ± 0.12	15.76 ± 1.34	61/51
		3	50	0.096 ± 0.0025	0.909 ± 0.049	9.46 ± 0.51	322 ± 11	2.50 ± 0.13	22.42 ± 1.81	71/51
		4	150	0.124 ± 0.0037	0.822 ± 0.045	6.63 ± 0.42	309 ± 11	2.41 ± 0.11	20.36 ± 1.15	57/51
00645176000	57,193.56		542	0.071 ± 0.0037	1.577 ± 0.068	22.15 ± 1.17	220 ± 14	2.14 ± 0.09	14.43 ± 0.77	54/49
		1	150	0.071 ± 0.0044	1.582 ± 0.068	22.28 ± 1.19	218 ± 14	2.15 ± 0.09	14.04 ± 0.76	62/49
		2	260	0.071 ± 0.0036	1.553 ± 0.075	21.87 ± 1.22	230 ± 15	2.13 ± 0.09	14.93 ± 0.90	65/49
00646721000	57,199.52		965	0.134 ± 0.0207	1.311 ± 0.047	9.79 ± 0.17	143 ± 8	1.99 ± 0.09	77.55 ± 3.29	35/51
		1	530	0.136 ± 0.0212	1.312 ± 0.051	9.65 ± 0.22	142 ± 8	2.01 ± 0.10	74.15 ± 3.37	37/51
		2	40	0.133 ± 0.0194	1.310 ± 0.042	9.85 ± 0.18	144 ± 9	1.96 ± 0.09	89.22 ± 2.54	36/51

TCAF fitted extracted parameters for time resolved BAT spectra in the energy range of 15–150 keV. Note: First row in each spectra are TCAF model fitted spectral analysis results when entire data exposure including gaps are used. The mass of the BH is frozen at 10.6 $M_\odot$ during the fitting.

4.4. Estimation of Magnetic Field

Observation of presence of high magnetic field on 25 June 2015 by [56] motivated us to estimate magnetic field strength at the 'hot' Compton cloud region (here CENBOL) for the BHC V404 Cygni during its 2015 outburst. We made some simple assumptions as mentioned below to calculate the equipartition value of the magnetic field. At the shock location, energy conservation leads to the following equation [28],

$$na_+^2 + \frac{1}{2}u_+^2 = na_-^2 + \frac{1}{2}u_-^2,\tag{1}$$

where, n is the polytropic index of the flow, a is sound speed, u is particle velocity. '+' and '−' signs indicate the values at post- and pre-shock region respectively. The electron number density (n_e) is given by,

$$n_\mathrm{e} = \frac{\dot{m}_\mathrm{d} + \dot{m}_\mathrm{h}}{4\pi X_s u_+ H_\mathrm{shk} m_\mathrm{p}}.\tag{2}$$

For our calculation, we assume CENBOL shape as cylindrical. m_p is the mass of the proton. H_shk is the shock height. The shock height could be calculated from the TCAF model fitted shock parameters [44] using standard vertical equilibrium [29],

$$H_\mathrm{shk} = \sqrt{\frac{n(R-1)X_s^2}{R^2}}.\tag{3}$$

Now, we can calculate pressure at CENBOL using the following equation,

$$P_\mathrm{gas} = \frac{a_+^2 m_\mathrm{e} n_\mathrm{e}}{n}.\tag{4}$$

We consider equipartition magnetic field (B_eq) as,

$$\frac{B_\mathrm{eq}^2}{8\pi} = P_\mathrm{rad} + P_\mathrm{gas},\tag{5}$$

where, P_{rad} is the radiation pressure. The radiation pressure is negligible compared to the gas pressure. Using this equation, we calculated the value of the equipartition magnetic field for this source (see Table 4). We show the variation of the equipartition magnetic field in Figure 7a. We find $B_{\mathrm{eq}} = 97 \pm 3$ G on our first observation (MJD 57,188.77). Then we found that B_{eq} was increasing with the shock moving towards the black hole. B_{eq} was maximum on 2015 June 21 (MJD 57194.16) at 923 ± 44 G. On that day, the shock was closest to the black hole.

Table 4. Magnetic field and Resonance condition.

Obs ID	Day	n_{e}	B_{eq}	Λ_{syn}	t_{sync}	Λ_{Comp}	t_{Comp}	t_{inf}	$t_{\mathrm{sync}}/t_{\mathrm{inf}}$	$t_{\mathrm{Comp}}/t_{\mathrm{inf}}$
	(MJD)	(10^{16})	(*Gauss*)	$(10^{-9}\ \mathrm{erg\,s^{-1}})$	(s)	$(10^{-3}\ \mathrm{erg\,s^{-1}})$	$(10^{-3}\,\mathrm{s})$	(s)		
(1)	(2)	(3)	(4)	(5)	(6)	(7)	(8)	(9)	(10)	(11)
00643949000	57,188.77	1.07	97 ± 3	1.01	81.16	0.25	12.34	1.22	66.48	10.10
00031403035	57,189.62	1.51	124 ± 4	1.62	50.51	0.39	9.06	1.04	48.44	8.69
00031403038	57,191.01	2.05	149 ± 4	2.35	34.91	0.45	8.38	0.91	38.33	9.20
00644520000	57,191.03	1.59	125 ± 4	1.66	49.24	0.69	4.97	1.04	47.30	4.77
00031403042	57,192.16	2.28	153 ± 5	2.50	32.79	0.53	6.74	1.05	31.24	6.42
00645176000	57,193.56	6.67	308 ± 12	10.12	8.17	0.67	7.34	0.65	12.53	11.20
00031403048	57,194.16	34.38	923 ± 44	90.14	0.91	3.07	2.79	0.28	3.21	9.85
00031403046	57,194.54	25.57	752 ± 25	59.91	1.37	2.52	3.04	0.33	4.09	9.10
00031403045	57,195.08	16.55	564 ± 19	33.71	2.43	1.49	4.46	0.41	5.92	10.91
00031403049	57,195.35	7.28	305 ± 9	9.82	8.34	0.53	8.38	0.64	12.99	13.04
00031403047	57,195.41	3.03	182 ± 7	3.49	23.46	0.50	7.50	0.88	26.59	8.49
00031403052	57,196.89	2.56	154 ± 5	4.21	19.49	1.67	9.49	0.99	19.71	1.95
00031403054	57,197.21	11.29	414 ± 14	18.22	4.51	0.84	6.25	0.51	8.85	12.33
00031403053	57,197.33	14.49	504 ± 16	26.91	3.04	1.66	3.65	0.44	6.99	8.83
00031403055	57,198.00	28.70	820 ± 29	71.15	1.51	5.06	1.60	0.31	3.78	5.26
00031403056	57,198.02	17.84	602 ± 24	38.33	2.13	3.78	1.86	0.38	5.59	4.86
00031403057	57,198.15	13.94	500 ± 18	26.41	3.09	2.35	2.63	0.46	6.72	5.72
00031403058	57,198.93	16.65	568 ± 26	34.21	2.40	3.31	2.03	0.41	5.89	4.99
00646721000	57,199.55	21.38	685 ± 25	49.64	1.65	4.31	1.76	0.34	4.84	5.16

Electron number density (n_{e}) is in the unit of $\times 10^{16}$ cm^3.

4.5. Cooling and Infall Times

V404 Cygni did not show any strong QPO during its 2015 outburst. Although there are numerous suggestions for the origin of low-frequency QPOs, we believe that these QPOs are generated due to the oscillation of the shock formed in the TCAF solution. This shock oscillates when the resonance condition between cooling and infall time scales is satisfied or when Rankine-Hugoniot conditions are not satisfied to form a stable shock [33–35]. Thus to check if the resonance condition is satisfied or not, we have calculated both cooling and infall time scales during the outburst. Generally, in low mass X-ray binaries, Compton cooling is the primary process of cooling. We also considered the synchrotron cooling since the magnetic field was present.

We calculated both cooling time scales (synchrotron and Compton) during the entire period of our observations and compared that with the infall times. If cooling and infall timescales are roughly comparable, then we may say that the resonance condition for oscillation of the shock is satisfied. For simplicity, here we assume that the matter is moving radially at the post-shock region with speed u_+. This allowed us to calculate infall time using the following equation,

$$t_{\mathrm{inf}} = \frac{X_{\mathrm{s}}}{u_+} \tag{6}$$

Similar to the magnetic field calculation, here we also assume CENBOL as a cylindrical in shape. Now the total thermal energy of electron content within the CENBOL of volume V is,

$$E = \gamma_{\mathrm{e}} m_{\mathrm{e}} c^2 V n_{\mathrm{e}}, \tag{7}$$

where γ_e is the Lorentz factor. It is given by, $\gamma_e = 1/\sqrt{1 - \beta_e^2}$, where $\beta_e = u_e/c$, u_e being electron velocity and c is the velocity of light. The synchrotron cooling rate is [57],

$$\Lambda_{syn} = \frac{4}{3}\sigma_T c \beta_e \gamma_e^2 U_B, \tag{8}$$

where σ_T is Boltzman constant. U_B is magnetic energy density and given by, $U_B = B_{eq}^2/8\pi$. The synchrotron cooling time is given by,

$$t_{syn} = \frac{E}{\Lambda_{syn}}. \tag{9}$$

We calculated cooling time due to synchrotron using Equation (9). Note, we use $\gamma_e = 10$ for our calculation. We also checked the Compton cooling timescale (t_{Comp}). To calculate Compton cooling, we use the same method as described in C15. The Compton cooling timescale is given by Equation (6) of C15,

$$t_{Comp} = \frac{E}{\Lambda_{Comp}}. \tag{10}$$

Here, Λ_{Comp} is the Compton cooling rate (for more details, see, C15). Here we find that the Compton cooling is much faster than the synchrotron cooling (see Table 4). Thus, the inverse-Comptonization is the primary cooling process. Now if we compare t_{inf} with t_{syn} or with t_{Comp}, in both cases, the ratio deviates largely from unity. It is clear that the resonance condition was not satisfied during the 2015 outburst of V404 Cygni. This indeed verifies the observational fact that there was no prominent signature of type-C low-frequency QPOs during the outburst. The detailed results of our calculation of the magnetic field and two types of cooling time scales are presented in Table 4.

5. Discussion

We study the 2015 outburst of V404 Cygni using *Swift*/XRT and BAT data in the energy band of 0.5–150 keV. We used a total of 19 observations between 15 June 2015 (MJD 57,188.77) and 26 June (MJD 57,199.52) for the timing and spectral analysis. The TCAF model fitted spectral analysis allowed us to understand the nature of this violent outburst from a physical perspective. In the following sub-sections, we discuss our findings from the combined spectral and timing analysis.

5.1. Magnetic Field

A magnetic field is brought in by the accretion flow. The shear, convection, and advection amplify the predominantly toroidal flux tubes. They are expelled from the CENBOL due to the magnetic buoyancy [31,32] in the direction towards the pressure gradient force. If the buoyancy timescale is larger than the shear amplification timescale, then the magnetic flux tubes are amplified within the dynamical timescale; until they reach equipartition where gas pressure matches with the magnetic pressure. In general, the magnetic field is not found to contribute significantly to the accretion disk spectra of black holes as TCAF alone fits the spectral data very well. However, its role is important for the acceleration and collimation of jets and outflow.

Here, we calculated the equipartition magnetic field by equating magnetic pressure with the local gas pressure at the shock location. We found $B_{eq} = 97$ G on the first observation day. After that, it increased gradually with the increase of the accretion rate. We found the maximum value of the magnetic field to be $B_{eq} = 923$ G on 21 June 2015 (MJD 57,194.16). Then, it decreased to 154 G on MJD 57,196.89. Then, we observed that it was varying between 414 and 820 G within our observation. Dallilar et al. calculated the equipartition magnetic field for this source at hot Compton corona to be 461 ± 12 G on 25 June 2015 (MJD 57,198.18) [56]. We found $B_{eq} = 500 \pm 18$ G on the same observation day (MJD 57,198.15), which is consistent with the finding of Dallilar et al. (2017).

5.2. Power Density Spectra

No prominent QPOs were observed in the PDS of the Fourier transformed 0.01 c time binned XRT lightcurves in the 0.5–10 keV energy band. The presence of a significant magnetic field may cause turbulence in the accretion disk, which could be responsible for the white noise observed in the PDS. The PDSs were observed to have two slopes: a powerlaw component and a flat component. Powerlaw slope was observed in the lower frequency region while flat slope was observed in the higher frequency. The turbulence could be the reason behind the flat spectrum observed in the XRT PDS. Power in the PDS follows as, $P \sim 1/f^{\beta}$, where f is the frequency and β is powerlaw index. In the most PDS, we found $\beta \sim 0$, i.e., flat spectra. It indicates the presence of turbulence in the Keplerian disk. No break was found in power density spectra. This also indicates no or minimal contribution of the Keplerian disk accretion. Precisely, this is found from the spectral analysis.

5.3. Absence of QPOs

We believe that the oscillation of shock is responsible for the QPOs [33]. Generally strong type-C QPOs are observed if the resonance condition is satisfied, i.e., when the infall time of the post-shock matter roughly matches with the cooling time inside the CENBOL. The cooling process could be via inverse-Compton scattering, bremsstrahlung, or synchrotron emission. If the magnetic field is high enough, one can expect the dominance of the synchrotron cooling.

We did not observe any prominent type-C QPOs during the 2015 outburst of V404 Cygni. Non-satisfaction of the resonance condition could be responsible for it. To verify this assertion, we calculated synchrotron and Compton cooling times for this source. We found that the Compton cooling rate is much faster than the synchrotron cooling rate. Thus, the Compton cooling process was the dominant cooling process. We compared Compton cooling time and synchrotron cooling time with the infall time. We found that the infall time and cooling times were not comparable at all (see Table 4). C15 showed that QPO would be generated if the ratio of the infall and the cooling time is between 0.5 and 1.5, i.e., within 50% of unity either way. Since the resonance condition was not satisfied, type-C QPO was not produced.

Huppenkothen et al., 2017 reported of detection of mHz QPO with *Chandra*, *Swift*/XRT and *Fermi* observations [58]. They reported simultaneous detection of 18 mHz QPO with *Swift*/XRT and *Fermi*/GBM. They classified this QPO as a new type of low-frequency QPO. *Chandra*/ACIS observation revealed signatures of 73 mHz and 1.03 Hz QPOs, as well as a QPO at 136 mHz in a single Swift/XRT. They argued that the QPO at 136 mHz is standard type-C QPO. However, they did not seem to be type-C QPOs. However, Radhika et al. also argued that they did not find any clear signature of this QPO [20]. Thus the resonance condition was not behind the origin of these QPOs. Non-satisfaction of the Rankine-Hugoniot conditions at the shock front could be the reason for this type of QPO [34].

5.4. Spectral and Temporal Evolution

V404 Cygni showed complex behavior during the 2015 outburst. We required two absorption models to fit the spectra: `phabs` model for interstellar absorption and `pcfabs` for the partial covering absorption. The later absorption may be due to disk wind emission or outflow emission [20,27]. We studied V404 Cygni till MJD 57198. After that, variable dust scattering rings were observed, and the dust halo started to dominate the field of view [59].

Some observations of V404 Cygni could not be fitted with simple `disk blackbody` and `powerlaw` models since a significant amount of reflection was present. In the first six observations, the DBB component was not required to fit the spectra (till MJD 57,193.56). After that, it was required occasionally. PL photon indices varied randomly between 0.60 and 2.43. Overall, PL flux dominated over the thermal flux during the entire period of our observation. We often observed a 'hump' region after ~ 8 keV in the XRT data. This 'reflection hump' was extended up to $\sim$20 keV in the BAT data. Radhika et al. analyzed 0.5–150 keV combined XRT+BAT data using disk blackbody (diskbb) and PL models and found similar results [20]. Later, they used pexrav [60] model instead of 'PL' model for this reflection. In their analysis, the disk blackbody component was not required in a few observations. They found that photon indices varied randomly between 0.60 and 4.43. Refs. [19,61,62] analyzed *INTEGRAL*/IBIS and *INTEGRAL*/SPI data with Comptonization model in the broad energy range of 20–650 keV. They required an additional cut-off in the energy range of 400–600 keV. They found that the Comptonization temperature was $\sim$40 keV with seed photon temperature $\sim$7 keV. This is very high for disk emission. They concluded that this emission could be of different origins, such as synchrotron emission from the jet.

We extracted the physical parameters of the accretion flows from each fit. We first fitted each spectrum by keeping all model input parameters free. We found a variation of $M_{\rm BH}$ in a narrow range of 9.5–11.5 $M_\odot$ with an average value of $10.6^{+0.9}_{-1.1}$ $M_\odot$. This estimated mass of V404 Cygni agrees well with other reported values in the range of 9–12 $M_\odot$ [11–13]. We then re-fitted all the spectra by keeping $M_{\rm BH}$ frozen at mean value ($M_{\rm BH} = 10.6$ $M_\odot$) to extract values of other physical flow parameters during the outburst.

From the variation of the TCAF model fitted flow parameters (i.e., high dominance of $\dot{m}_h$ over $\dot{m}_d$, and presence of strong shock ($R > 2.3$) with a large CENBOL), we infer that the source was in the hard state at the beginning of the outburst. High ARR ($\sim$14–22) also indicates this except in one observation made on 18 June 2015 (MJD 57,191.03). From the phenomenological model fittings, the dominance of powerlaw flux also indicates this. On 25 June 2015 (MJD 57,198), we observed that $\dot{m}_d$ started to increase rapidly, although other parameters did not change much, leading to the softening of the spectra. At this last phase of our observation, we observed that ARR decreased with an increase in the Keplerian disk rate. During this phase, the source could be in the hard-intermediate state. In general, the spectral state classification is done based on the variation of ARR and the nature of QPO. Due to the absence of QPOs and the ARR's complex behavior, we could not assert the exact state transition day, but the source likely entered in the hard-intermediate state on MJD 57,197.21 when $\dot{m}_h$ was found to achieve its maximum rate. After this day, $\dot{m}_d$ was found to increase rapidly. However, our classification may not be valid since the source showed rapid fluctuation in a very short timescale [24]. Motta et al. analyzed the time-resolved 0.5–10 keV *Swift*/XRT spectra and found that spectra shape changed in as short as $\sim$10 s [24].

The 2015 outburst is somewhat similar to the 2003 outburst of H 1743-322. Both outbursts occurred after long quiescence periods. Similar to the 2015 outburst of V404 Cygni, the 2003 outburst of H 1743-322, which also occurred after about 26 years, showed peculiar behavior with several flares and outflow activity. Thus it is possible that a huge amount of matter accumulated at a large pileup radius over a long period of the quiescence phase. With a sudden enhancement of viscosity, the outburst is triggered and leads to a violent and non-settling disk-jet activity [63].

5.5. Evolution of the Spectral Properties with Flares

The present outburst of V404 Cygni did not behave like any other typical outburst of a classical transient black hole. V404 Cygni showed a strong jet associated with several flares. The flares were observed in multi-wavebands, from X-ray, optical, IR to radio [19,64]. Eighteen X-ray flares were reported with the *INTEGRAL* and *SWIFT*/BAT observations between 20 June 2015 (MJD 57,193) and 25 June 2015 (MJD 57198). The magnetic field was the strongest during this phase of the outburst. The magnetic field could be responsible for this flaring activity.

In general, we see a decreasing ARR in an observation immediately after a flare, indicating softening of the spectra (see Figure 6). For example, an X-ray flare was observed on MJD 57,194.31 [19]. This could be due to the high magnetic field on MJD 57,194.16. Immediately after the flare, on MJD 57,194.54, we found that the ARR decreased slightly from its previous observation (21.6 to 20.2). On 18 June 2015 (MJD 57,191) we observed that the Keplerian disk rate suddenly rose to 0.12 $\dot{M}_{Edd}$ from 0.06 $\dot{M}_{Edd}$ within $\sim$30 mins. This could be associated with the radio flare observed on MJD 57,191.09 with AMI-LA observation [17]. This is expected since a large amount of mass was ejected from the CENBOL during a flare, and inflowing matter rapidly moved inward to fill the vacant space. This led to the softening of the spectrum as the CENBOL size was reduced. However, this was not observed after every flare. It is possible that those flares were not localized, and the disk was unstable. Around MJD 57,198, we found that B_{eq} decreased sharply, although N increased very rapidly. This could be due to the high magnetic field that produced flare and outflow. This flare and the outflow was responsible for the rapid rise of N. Additionally, flaring events could also contribute to the rapid variability. On 21 June 21 2015 (MJD 57,194), the B_{eq} was observed to be at an increasing N. On the same day, the rapid variation of the spectral properties was observed [65], although the mass accretion rates did not change.

The X-ray jet flux can be calculated based on the deviation of the constancy of the TCAF model normalization [52–54]. However, to calculate jet X-ray flux by this model, we must have at least one observation where the jet's effects were negligible. In that observation, the entire observed X-ray should be contributed only from the inflowing matter of the accretion disk and CENBOL. However, for V404 Cygni, a strong jet was present in all the observations. Thus we were not able to separate the jet X-ray contribution from the total X-ray. Random variation in normalization may be due to a fluctuating magnetic field or unsettling disk, leading to the source's flaring activity.

6. Summary

The first epoch of the 2015 outburst of the Galactic black hole V404 Cygni was unusual and violent. It did not behave like other typical outbursts of Galactic transient black hole candidates. Rapid variations were observed in both spectral and timing properties in very short time scales, ranging from a few minutes to hours. We have used 0.5–150 keV combined *Swift*/XRT and *Swift*/BAT data to study the accretion flow properties of the source. Spectral analysis was done using the TCAF model-based *fits* file in XSPEC. The model fitted/derived flow parameters allowed us to understand the evolution of this violent outburst's accretion flow parameters. We have also calculated the equipartition magnetic field for the outburst. No break is found in the power density spectra, which indicates that the Keplerian disk rate was very low. This is also confirmed from the spectral analysis. The presence of white noise in higher frequencies in the power density spectra indicates a highly turbulent disk. The strong magnetic field could be the reason behind it. It is also responsible for the flares. We find that the Compton cooling process is much faster than the synchrotron cooling process. Since the resonance condition between cooling and infall time scales inside the CENBOL is not satisfied, we did not expect any sharp low-frequency QPO. Indeed the object did not show any signature of prominent type-C QPO.

Author Contributions: Conceptualization, S.K.C.; Data curation, A.J.; Formal analysis, A.J.; Methodology, A.J.; Software, D.D. and S.K.C.; Supervision, D.D., S.K.C. and H.-K.C.; Writing—original draft, Arghajit Jana; Writing—review & editing, J.-R.S., D.D., S.K.C., D.C. and H.-K.C. All authors have read and agreed to the published version of the manuscript.

Funding: This research received no external funding.

Institutional Review Board Statement: Not applicable.

Informed Consent Statement: Not applicable.

Data Availability Statement: All data used here, are publicly available.

Acknowledgments: This work made use of XRT and BAT data supplied by the UK *Swift* Science Data Centre at the University of Leicester. We acknowledge anonymous referee for his kind suggestion to improve the quality of the paper. A.J. and D.D. acknowledge support from DST/GITA sponsored

India-Taiwan collaborative project (GITA/DST/TWN/P-76/2017) fund. Research of D.D. and S.K.C. is supported in part by the Higher Education Dept. of the Govt. of West Bengal, India. D.D. also acknowledges the ISRO sponsored RESPOND project (ISRO/RES/2/418/17-18) fund. D.C. and D.D. acknowledge support from DST/SERB sponsored Extra Mural Research project (EMR/2016/003918) fund. J.-R.S., and H.-K.C. are supported by MOST of Taiwan under grants MOST/106-2923-M-007-002-MY3 and MOST/107-2119-M-007-012.

Conflicts of Interest: The authors declare no conflict of interest.

Abbreviations

TCAF	Two components advective flow
CENBOL	CENtrifugal pressure supported BOundary Layer
QPO	Quasi periodic oscillation
XRT	X-ray telescope
BAT	Burst alert telescope

References

1. Shakura, N.I.; Sunyaev, R.A. Black holes in binary systems: Observational appearance. *Astron. Astrophys.* **1973**, *24*, 337–355.
2. Sunyaev, R.A.; Titarchuk, L.G. Comptonization of X-rays in plasma clouds. Typical radiation spectra. *Astron. Astrophys.* **1980**, *86*, 121–138.
3. Sunyaev, R.A.; Titarchuk, L.G. Comptonization of low-frequency radiation in accretion disks Angular distribution and polarization of hard radiation. *Astron. Astrophys.* **1985**, *143*, 374–388. [CrossRef]
4. Chakrabarti, S.K.; Titarchuk, L.G. Spectral Properties of Accretion Disks around Galactic and Extragalactic Black Holes. *Astrophys. J.* **1995**, *455*, 623. [CrossRef]
5. Chakrabarti, S.K. Spectral Properties of Accretion Disks around Black Holes. II. Sub-Keplerian Flows with and without Shocks. *Astrophys. J.* **1997**, *484*, 313. [CrossRef]
6. Zdziarski, A.A.; Zycki, P.T.; Svensson, R.; Boldt, E. On Compton Reflection in the Sources of the Cosmic X-Ray Background. *Astrophys. J.* **1993**, *405*, 125–129. [CrossRef]
7. Haardt, F.; Maraschi, L. X-ray Spectra from Two-Phase Accretion Disks. *Astrophys. J.* **1993**, *413*, 507–517. [CrossRef]
8. Esin, A.A.; McClintock, J.E.; Narayan, R. Advection-Dominated Accretion and the Spectral States of Black Hole X-Ray Binaries: Application to Nova Muscae 1991. *Astrophys. J.* **1997**, *489*, 865. [CrossRef]
9. Ritcher, G.A. PG 0818+513—An Eclipsing Binary. *Inf. Bul. Var. Stars* **1989**, *3362*, 1.
10. Barthelmy, S.D.; DÁi, A.; DÁvanzo, P.; Krimm, H.A.; Lien, A.Y.; Marshall, F.E.; Maselli, A.; Siegel, M.H. Swift trigger 643949 is V404 Cyg. *GCN Circ.* **2015**, 17929.
11. Casares, J.; Charles, P.A.; Naylor, T. A 6.5-day periodicity in the recurrent nova V404 Cygni implying the presence of a black hole. *Nature* **1992**, *355*, 614. [CrossRef]
12. Shahbaz, T.; Ringwald, F.A.; Bunn, J.C.; Naylor, T.; Charles, P.A.; Casares, J. The mass of the black hole in V404 Cygni. *MNRAS* **1994**, *271*, L10. [CrossRef]
13. Khargharia, J.; Froning, C.S.; Robinson, E.L. Near-infrared Spectroscopy of Low-mass X-ray Binaries: Accretion Disk Contamination and Compact Object Mass Determination in V404 Cyg and Cen X-4. *Astrophys. J.* **2010**, *716*, 1105. [CrossRef]
14. Miller-Jones, J.C.A.; Jonker, P.G.; Dhawan, V.; Brisken, W.; Rupen, M.P.; Nelemans, G.; Gallo, E. The First Accurate Parallax Distance to a Black Hole. *Astrophys. J.* **2009**, *706*, L230. [CrossRef]
15. Walton, D.J.; Mooley, K.; King, A.L.; Tomsick, J.A.; Miller, J.M.; Dauser, T.; García, J.A.; Bachetti, M.; Brightman, M.; Fabian, A.C.; et al. Living on a Flare: Relativistic Reflection in V404 Cyg Observed by NuSTAR during Its Summer 2015 Outburst. *Astrophys. J.* **2017**, *839*, 110. [CrossRef]
16. Negoro, H.; Matsumitsu, T.; Mihara, T.; Serino, M.; Matsuoka, M.; Nakahira, S.; Ueno, S.; Tomida, H.; Kimura, M.; Ichikawa, M.; et al. *MAXI/GSC* detection of a new outburst from the Galactic black hole candidate GS 2023+338 (V* V404 Cyg). *ATel* **2015**, *7646*, 1.
17. Mooley, K.; Fender, R.; Anderson, G.; Staley, T.; Kuulkers, E.; Rumsey, C. Bright radio flaring from V404 Cyg detected by AMI-LA. *Astron. Telegr.* **2015**, *7658*, 1.
18. Gazeas, K.; Vasilopoulos, G.; Petropoulou, M.; Sapountzis, K. Optical follow up of V404 Cyg during the current enhanced activity. *Astron. Telegr.* **2015**, *7650*, 1.
19. Rodriguez, J.; Cadolle Bel, M.; Alfonso-Garzón, J.; Siegert, T.; Zhang, X.-L.; Grinberg, V.; Savchenko, V.; Tomsick, J.A.; Chenevez, J.; Clavel, M.; et al. Correlated optical, X-ray, and γ-ray flaring activity seen with INTEGRAL during the 2015 outburst of V404 Cygni. *Astron. Astrophys.* **2015**, *L581*, 9. [CrossRef]
20. Radhika, D.; Nandi, A.; Agrawal, V.K.; Mandal, S. SWIFT view of the 2015 outburst of GS 2023+338 (V404 Cyg): Complex evolution of spectral and temporal characteristics. *Mon. Not. R. Astron. Soc.* **2016**, *462*, 1834. [CrossRef]

21. Tetarenko, A.J.; Sivakoff, G.R.; Miller-Jones, J.C.A.; Rosolowsky, E.W.; Petitpas, G.; Gurwell, M.; Wouterloot, J.; Fender, R.; Heinz, S.; Maitra, D.; et al. Extreme jet ejections from the black hole X-ray binary V404 Cygni. *Mon. Not. R. Astron. Soc.* **2017**, *469*, 3141–3162. [CrossRef]

22. Tetarenko, A.J.; Sivakoff, G.R.; Miller-Jones, J.C.A.; Bremer, M.; Mooley, K.P.; Fender, R.P.; Rumsey, C.; Bahramian, A.; Altamirano, D.; Heinz, S.; et al. Tracking the variable jets of V404 Cygni during its 2015 outburst. *Mon. Not. R. Astron. Soc.* **2019**, *482*, 2950. [CrossRef]

23. Miller-Jones, J.C.A.; Tetarenko, A.J.; Sivakoff, G.R. A rapidly changing jet orientation in the stellar-mass black-hole system V404 Cygni. *Nature* **2019**, *569*, 374. [CrossRef] [PubMed]

24. Motta, S.E.; Kajava, J.J.E.; Sanchez-Fernandez, C.; Beardmore, A.P.; Sanna, A.; Page, K.L.; Fender, R.; Altamirano, D.; Charles, P.; Giustini, M.; et al. Swift observations of V404 Cyg during the 2015 outburst: X-ray outflows from super-Eddington accretion. *Mon. Not. R. Astron. Soc.* **2017**, *471*, 1797. [CrossRef]

25. Siegert, T.; Diehl, R.; Greiner, J.; Krause, M.G.H.; Beloborodov, A.M.; Bel, M.C.; Guglielmetti, F.; Rodriguez, J.; Strong, A.W.; Zhang, X. Positron annihilation signatures associated with the outburst of the microquasar V404 Cygni. *Nature* **2016**, *531*, 341. [CrossRef]

26. Loh, A.; Corbel, S.; Dubus, G.; Rodriguez, J.; Grenier, I.; Hovatta, T.; Pearson, T.; Readhead, A.; Fender, R.; Mooley, K. High-energy gamma-ray observations of the accreting black hole V404 Cygni during its 2015 June outburst. *Mon. Not. R. Astron. Soc.* **2016**, *462*, L111. [CrossRef]

27. King, A.L.; Miller, J.M.; Raymond, J.; Reynolds, M.T.; Morningstar, W. High-resolution Chandra HETG Spectroscopy of V404 Cygni in Outburst. *Astrophys. J.* **2015**, *813*, L37. [CrossRef]

28. Chakrabarti, S.K. *Theory of Transonic Astrophysical Flows*; World Scientific: Singapore, 1990.

29. Chakrabarti, S.K. Standing shocks in isothermal rotating winds and accretion. *Mon. Not. R. Astron. Soc.* **1989**, *240*, 7–21. [CrossRef]

30. Chakrabarti, S.K. Estimation and effects of the mass outflow from shock compressed flow around compact objects. *Astron. Astrophys.* **1999**, *351*, 185–191.

31. Chakrabarti, S.K.; D'Silva, S. Magnetic Activity in Thick Accretion Disks and Associated Observable Phenomena. I. Flux Expulsion. *Astrophys. J.* **1994**, *424*, 138. [CrossRef]

32. D'Silva, S.; Chakrabarti, S.K. Magnetic Activity in Thick Accretion Disks and Associated Observable Phenomena. II. Flux Storage. *Astrophys. J.* **1994**, *424*, 149. [CrossRef]

33. Molteni, D.; Sponholz, H.; Chakrabarti, S.K. Resonance Oscillation of Radiative Shock Waves in Accretion Disks around Compact Objects. *Astrophys. J.* **1996**, *457*, 805. [CrossRef]

34. Ryu, D.; Chakrabarti, S.K.; Molteni, D. Zero-Energy Rotating Accretion Flows near a Black Hole. *Astrophys. J.* **1997**, *474*, 378. [CrossRef]

35. Chakrabarti, S.K.; Mondal, S.; Debnath, D. Resonance condition and low-frequency quasi-periodic oscillations of the outbursting source H1743-322. *Mon. Not. R. Astron. Soc.* **2015**, *452*, 3451–3456. [CrossRef]

36. Remillard, R.A.; McClintock, J.E. X-ray Properties of Black-Hole Binaries. *Annu. Rev. Astron. Astrophys* **2006**, *44*, 49. [CrossRef]

37. Jana, A.; Jaisawal, G.K.; Naik, S.; Kumari, N.; Chhotaray, B.; Altamirano, D.; Remillard, R.A.; Gendreau, K.C. NICER observations of the black hole candidate MAXI J0637–430 during the 2019–2020 Outburst. *Mon. Not. R. Astron. Soc.* **2021**, *504*, 4793–4805. [CrossRef]

38. Chakrabarti, S.K. *MG14 Conf. Pro.*; Ruffini, R., Jantzen, R., Bianchi, M., Eds.; World Scientific Press: Singapore, 2018; p. 369.

39. Jana, A.; Debnath, D.; Chakrabarti, S.K.; Mondal, S.; Molla, A.A. Accretion Flow Dynamics of MAXI J1836-194 During Its 2011 Outburst from TCAF Solution. *Astrophys. J.* **2016**, *819*, 107. [CrossRef]

40. Mondal, S.; Chakrabarti, S.K.; Nagarkoti, S.; Arévalo, P. Possible Range of Viscosity Parameters to Trigger Black Hole Candidates to Exhibit Different States of Outbursts. *Astrophys. J.* **2017**, *850*, 47. [CrossRef]

41. Garain, S.K.; Ghosh, H.; Chakrabarti, S.K. Effects of Compton Cooling on Outflow in a Two-component Accretion Flow around a Black Hole: Results of a Coupled Monte Carlo Total Variation Diminishing Simulation. *Astrophys. J.* **2012**, *758*, 114. [CrossRef]

42. Arnaud, K.A. XSPEC: The First Ten Years. *ASP Conf. Ser. Astron. Data Anal. Softw. Syst. V Ed. G.H. Jacoby J. Barnes* **1996**, *101*, 17.

43. Debnath, D.; Mondal, S.; Chakrabarti, S.K. Implementation of two-component advective flow solution XSPEC. *Mon. Not. R. Astron. Soc.* **2014**, *440*, L121. [CrossRef]

44. Debnath, D.; Mondal, S.; Chakrabarti, S.K. Characterization of GX 339-4 outburst of 2010-11: Analysis by XSPEC using two component advective flow model. *Mon. Not. R. Astron. Soc.* **2015**, *447*, 1984–1995. [CrossRef]

45. Debnath, D.; Jana, A.; Chakrabarti, S.K.; Chatterjee, D.; Mondal, S. Accretion Flow Properties of Swift J1753.5-0127 during Its 2005 Outburst. *Astrophys. J.* **2017**, *850*, 92. [CrossRef]

46. Jana, A.; Debnath, D.; Chatterjee, D.; Chatterjee, K.; Chakrabarti, S.K.; Naik, S.; Bhowmick, R.; Kumari, N. Accretion Flow Evolution of a New Black Hole Candidate MAXI J1348-630 during the 2019 Outburst. *Astrophys. J.* **2020**, *897*, 3. [CrossRef]

47. Shang, J.-R.; Debnath, D.; Chatterjee, D.; Jana, A.; Chakrabarti, S.K.; Chang, H.-K.; Yap, Y.-X.; Chiu, C.-L. Evolution of X-Ray Properties of MAXI J1535-571: Analysis with the TCAF Solution. *Astrophys. J.* **2019**, *875*, 4. [CrossRef]

48. Chatterjee, K.; Debnath, D.; Chatterjee, D.; Jana, A.; Chakrabarti, S.K. Inference on accretion flow properties of XTE J1752-223 during its 2009-10 outburst. *Mon. Not. R. Astron. Soc.* **2020**, *493*, 2452–2462. [CrossRef]

49. Chatterjee, D.; Debnath, D.; Chakrabarti, S.K.; Mondal, S.; Jana, A. Accretion Flow Properties of MAXI J1543-564 during 2011 Outburst from the TCAF Solution. *Astrophys. J.* **2016**, *827*, 88. [CrossRef]

50. Molla, A.A.; Debnath, D.; Chakrabarti, S.K.; Monda, S. Estimation of the mass of the black hole candidate MAXI J1659-152 using TCAF and POS models. *Mon. Not. R. Astron. Soc.* **2016**, *460*, 3163–3169. [CrossRef]
51. Jana, A.; Jaisawal, G.K.; Naik, S.; Kumari, N.; Chatterjee, D.; Chatterjee, K.; Bhowmick, R.; Chakrabarti, S.K.; Chang, H.-K.; Debnath, D. Accretion Properties of MAXI J1813-095 during its Failed Outburst in 2018. *Res. Astron. Astrophys.* **2021**, *21*, 125.
52. Jana, A.; Chakrabarti, S.K.; Debnath, D. Properties of X-Ray Flux of Jets during the 2005 Outburst of Swift J1753.5-0127 Using the TCAF Solution. *Astrophys. J.* **2017**, *850*, 91. [CrossRef]
53. Jana, A.; Debnath, D.; Chakrabarti, S.K.; Chatterjee, D. Inference on disk-jet connection of MAXI J1836-194 from spectral analysis with the TCAF solution. *Res. Astron. Astrophys.* **2020**, *20*, 28. [CrossRef]
54. Chatterjee, D.; Debnath, D.; Jana, A.; Chakrabarti, S.K. Properties of the black hole candidate XTE J1118+480 with the TCAF solution during its jet activity induced 2000 outburst. *Astrophys. Space Sci.* **2019**, *364*, 14. [CrossRef]
55. Koljonen, K.I.I.; Tomsick, J.A. The obscured X-ray binaries V404 Cyg, Cyg X–3, V4641 Sgr, and GRS 1915+105. *Astron. Astrophys.* **2020**, *639*, 13. [CrossRef]
56. Dallilar, Y.; Eikenberry, S.S.; Garner, A.; Stelter, R.D.; Gottlieb, A.; Gandhi, P.; Casella, P.; Dhillon, V.S.; Marsh, T.R.; et al. A precise measurement of the magnetic field in the corona of the black hole binary V404 Cygni. *Science* **2017**, *358*, 1299–1302. [CrossRef]
57. Rybicki, G.B.; Lightman, A.P. *Radiative processes in Astrophysics*; Wiley: New York, NY, USA, 1979.
58. Huppenkothen, D.; Younes, G.; Ingram, A.; Kouveliotou, C.; Gogus, E.; Bachetti, M.; Sánchez-Fernández, C.; Chenevez, J.; Motta, S.; van der Klis, M.; et al. Detection of Very Low-frequency, Quasi-periodic Oscillations in the 2015 Outburst of V404 Cygni. *Astrophys. J.* **2017**, *834*, 90. [CrossRef]
59. Vasilopoulos, G.; Petropoulou, M. The X-ray dust-scattered rings of the black hole low-mass binary V404 Cyg. *Mon. Not. R. Astron. Soc.* **2016**, *455*, 4426–4441. [CrossRef]
60. Magdziarz, P.; Zdziarski, A.A. Angle-dependent Compton reflection of X-rays and gamma-rays. *Mon. Not. R. Astron. Soc.* **1995**, *273*, 837–848. [CrossRef]
61. Natalucci, L.; Fiocchi, M.; Bazzano, A.; Ubertini, P.; Roques, J.-P.; Jourdain, E. High Energy Spectral Evolution of V404 Cygni during the 2015 June Outburst as Observed by INTEGRAL. *Astrophys. J.* **2015**, *813*, L21. [CrossRef]
62. Roques, J.-P.; Jourdain, E.; Bazzano, A.; Fiocchi, M.; Natalucci, L.; Ubertini, P. First INTEGRAL Observations of V404 Cygni during the 2015 Outburst: Spectral Behavior in the 20–650 keV Energy Range. *Astrophys. J.* **2015**, *813*, L22. [CrossRef]
63. Chakrabarti, S.K.; Nagarkoti, S.; Debnath, D. Delayed outburst of H 1743-322 in 2003 and relation with its other outbursts. *Adv. Space Res.* **2019**, *63*, 3749–3759. [CrossRef]
64. Gandhi, P.; Littlefair, S.P.; Hardy, L.K.; Dhillon, V.S.; Marsh, T.R.; Shaw, A.W.; Altamirano, D.; Caballero-Garcia, M.D.; Casares, J.; Casella, P.; et al. Furiously fast and red: Sub-second optical flaring in V404 Cyg during the 2015 outburst peak. *Mon. Not. R. Astron. Soc.* **2016**, *459*, 554–572. [CrossRef]
65. Kajava, J.J.E.; Sánchez-Fernández, C.; Alfonso-Garzón, J.; Motta, S.E.; Veledina, A. Rapid spectral transition of the black hole binary V404 Cygni. *Astron. Astrophys.* **2020**, *634*, 94. [CrossRef]

Article

Properties of Faint X-ray Activity of XTE J1908+094 in 2019

Debjit Chatterjee [1,2,*], Arghajit Jana [3], Kaushik Chatterjee [2], Riya Bhowmick [2], Sujoy Kumar Nath [2], Sandip K. Chakrabarti [2] , A. Mangalam [1] and Dipak Debnath [2]

[1] Indian Institute of Astrophysics, Koramangala, Bangalore 560034, India; mangalam@iiap.res.in
[2] Indian Centre for Space Physics, 43 Chalantika, Garia St. Rd., Kolkata 700084, India;
mails.kc.physics@gmail.com (K.C.); riyabhowmickmalda@gmail.com (R.B.);
sujoynath0007@gmail.com (S.K.N.); sandipchakrabarti9@gmail.com (S.K.C.); dipakcsp@gmail.com (D.D.)
[3] Physical Research Laboratory, Navrangpura, Ahmedabad 380009, India; argha0004@gmail.com
* Correspondence: debjit.chatterjee@iiap.res.in

Abstract: We study the properties of the faint X-ray activity of Galactic transient black hole candidate XTE J1908+094 during its 2019 outburst. Here, we report the results of detailed spectral and temporal analysis during this outburst using observations from *Nuclear Spectroscopic Telescope Array (NuSTAR)*. We have not observed any quasi-periodic-oscillations (QPOs) in the power density spectrum (PDS). The spectral study suggests that the source remained in the softer (more precisely, in the soft–intermediate) spectral state during this short period of X-ray activity. We notice a faint but broad Fe Kα emission line at around 6.5 keV. We also estimate the probable mass of the black hole to be $6.5^{+0.5}_{-0.7}\ M_\odot$, with 90% confidence.

Keywords: X-Rays:binaries—stars individual: (XTE J1908+094)—stars:black holes—accretion; accretion disks—shock waves—radiation:dynamics

Citation: Chatterjee, D.; Jana, A.; Chatterjee, K.; Bhowmick, R.; Nath, S.K.; Chakrabarti, S.K.; Mangalam, A.; Debnath, D. Properties of Faint X-ray Activity of XTE J1908+094 in 2019. *Galaxies* **2021**, *9*, 25. https://doi.org/10.3390/galaxies9020025

Academic Editor: Fulai Guo

Received: 27 March 2021
Accepted: 13 April 2021
Published: 16 April 2021

Publisher's Note: MDPI stays neutral with regard to jurisdictional claims in published maps and institutional affiliations.

1. Introduction

Black hole transients (BHTs) are fascinating objects to study. After a long period of quiescence, they show a sudden outburst. The sudden enhancement of viscosity at the piling radius could trigger the outburst [1–3]. The spectral and temporal properties of the source change during the outburst and evolves through the hard state (HS), hard intermediate state (HIMS), soft intermediate state (SIMS), and soft state (SS) [4–6]. Evolution of the state can be seen through hardness intensity diagram (HID) or "q" diagram [7,8] and accretion rate ratio intensity diagram (ARRID; [9,10]). A "failed" outburst is also a commonly known event where the source does not enter the softer spectral states [11–13]. In case of "failed" outbursts, sources do not follow the standard state transition or "q" diagram. The spectral shape varies in different spectral states, mainly due to the relative contribution of the thermal component [14,15] and non-thermal component [16,17].

In Two-Component Advective Flow (TCAF) solution [18,19], the accretion flow consists of two components: high viscous Keplerian flow with high angular momentum and low viscous sub-Keplerian flow with low angular momentum. The sub-Keplerian flow moves towards the black hole almost radially, and it almost stops at the centrifugal barrier and forms an axisymmetric shock [20]. The matter is puffed up beyond the shock and creates a hot electron cloud or Compton corona. This corona is called the CENtrifugal -pressure-supported BOundary Layer (CENBOL). This region intercepts the soft photons coming from the Keplerian disk and emits high-energy photons through inverse-Comptonization. In this way, TCAF can self-consistently explain the accretion dynamics around an accreting black hole. The oscillation of the same shock can explain the observed low-frequency quasi-periodic-oscillations (LFQPOs) [21,22]. The CENBOL is also considered to be the base of jets and outflows [23]. To get an estimation of the physical flow parameters directly from the spectral fit, TCAF model was implemented as an additive table model in XSPEC [24,25]. From spectral fit with the model, we obtain two accretion rate parameters, namely, the Keplerian disk rate ($\dot{m}_d$) in Eddington rate ($\dot{M}_{Edd}$), the sub-Keplerian halo rate ($\dot{m}_h$) in Eddington rate ($\dot{M}_{Edd}$); two shock or Compton cloud parameters' shock location (X_s) in Schwarzschild radius (r_s), compression ratio (R), which is the ratio between post-shock and pre-shock matter densities ($R = \rho_+/\rho_-$). One can also obtain the best-fitted value of the mass of the black hole (M_{BH} in $M_\odot$) and a normalization parameter from each spectral fit. If the mass of the black hole is well known, it could be kept constant during the spectral fitting.

The Galactic black hole (GBH) candidate XTE J1908+094 was discovered on 21 February 2002 by the Proportional Counter Array (PCA) on-board Rossi X-ray Timing Explorer (*RXTE*) [26]. The source spectrum was fitted with an absorbed power-law with photon index (Γ) of 1.55. The power density spectrum did not show any pulsations and exhibited a flat spectrum between 1 mHz and 0.1 Hz. A broad quasi-periodic-oscillation peak at 1 Hz was observed with a power-law break which continued to 4 Hz. A high-energy cutoff at $\sim$100 keV was also observed using *BeppoSAX* [27]. The source was suggested to be a black hole (BH) candidate from its spectral and timing properties [26–28]. High interstellar absorption (column density, $N_H \sim 2.5 \times 10^{22}$ cm^{-2}) was reported while fitting the spectrum with a multi-color blackbody, a Comptonization and a broad emission line [28]. The dimensionless spin parameter was also measured to be 0.75 from the broadening of Fe Kα line [29].

A radio counterpart was discovered at R.A. = $19^h08^m53.^s07$, DEC. = $+09°23'05''.0$ by *Very Large Array (VLA)* [30] which was also consistent with the Chandra observation [31]. Two possible near-infrared (NIR) counterparts were detected [32,33]—one of them is indicated as an intermediate/late-type (A-K) main-sequence companion, while the other is suggested to be a late-type (later than K) main-sequence secondary star [33]. After two similar outbursts in 2002 and early 2003, XTE J1908+094 went through another outburst on 2013 October 26 [34–38]. The 2013 outburst was well-studied in X-rays by *Swift*, *NuSTAR* [39,40]. Although a relativistic broadening of Fe-Kα line was observed, the spin of the BH could not be constrained due to data quality. A disk reflection contribution was also observed. The source was in the high/soft state during the *NuSTAR* observation. A flare was observed during the studied period. The flare was suggested to be related to the relativistic jet activity. Changes in the corona could be the reason for the flare. Multi-frequency radio and X-ray observation, and radio polarimetry with VLA and AMI-LA during the entire 2013 outburst was done [41]. The source followed the standard hardness-intensity diagram during the outburst. The common behavior of radio jets was also observed, which changes from compact to discrete as the state transits from hard to soft. From the VLBI monitoring of XTE J1908+094, a lateral expansion of resolved, asymmetric jet knots was noticed, which was ejected following the hard- to soft-state transition [42]. The knots are suggested to be the working surface where the ejected materials interacted with the surrounding dense interstellar medium. An external shock formed in this region causes the acceleration of particles, which subsequently diffused outwards over time.

XTE J1908+094 recently showed a "faint" X-ray activity on 2019 April 1 [43]. The source spectrum was fitted with a power-law model with photon index, Γ = 2.3. AMI-LA 15.5 GHz observation on 2019 April 5, detected the radio counterpart of XTE J1908+094 [44]. The obtained spectra on 2019 April 4 from the photon-counting mode of Swift also yield a soft spectrum with high absorption [45]. The source was suggested to be in the soft spectral state while using NICER data on 2019 April 6 and 9 [46].

In this *paper*, we have studied the timing and spectral properties of XTE J1908+094 during its X-ray activity on 2019 April 10 using *NuSTAR* observation of $\sim$40 ks. The *paper* is organized in the following way. In Section 2, we discuss the observations and the data analysis procedure. In Section 3, we present the temporal and spectral results of our analysis. In Section 4, we carry out the discussion based on our results.

2. Observation and Data Analysis

We processed the *NuSTAR* observation Id 90501317002 (https://heasarc.gsfc.nasa.gov/cgi-bin/W3Browse/w3browse.pl (accessed on 10 April 2019)) using the `nupipeline` command of NuSTAR Data Analysis Software (NuSTARDAS) version 1.8.0 (https://heasarc.gsfc.nasa.gov/docs/nustar/analysis/ (accessed on 10 April 2019)) and with the Calibration database (CALDB) version 1.0.2 (http://heasarc.gsfc.nasa.gov/FTP/caldb/data/nustar/fpm/ (accessed on 10 April 2019)). The spectra and light curves of the source were extracted from the FPMA detectors using a 60″ circle centered at the position of XTE J1908+094. The background region was selected carefully, since the source could have contamination from the nearby bright source GRS 1915+105. The background was chosen as a 60″ radius circle as far away from the XTE J1908+094. The background count rate is less than 2% of the source count rate, indicating that the source is still dominant. We also divided the total ∼40 ks data in three segments of ∼16 ks, ∼10 ks, and ∼12 ks for detailed study. We used `XSELECT` command `filter time` for this. The spectra and light curves were then generated using `nuproducts`. The spectra were re-binned to have 20 counts/s using `grppha`. The light curves were binned with 100 s time resolution.

For spectral analysis, we used both phenomenological (combined `diskbb`, `powerlaw` models) and physical (TCAF based `fits` file as an additive table model) model in `XSPEC` version 12.10.1. The hydrogen column density (N_H) was fixed at 2.5×10^{22} cm^{-2} [28]. The multiplicative model, `Tbabs` was used as an absorption table model considering the vern scattering cross-sections [47] and `wilm` abundances [48].

3. Results

We studied the 2019 X-ray activity of BHC XTE J1908+094 using *NuSTAR* observation. The source is close to a bright BHC GRS 1915+105. To verify the presence of any contamination due to GRS 1915+105, we studied the source and background count rate variation of XTE J1908+094 (Figure 1). The background count rate was found to be less than 2% of the source count rate. Therefore, we conclude that the X-ray activity of XTE J1908+094 during April 2019 to be "faint" and, yet, the variation observed was inherent rather than due to the nearby sources.

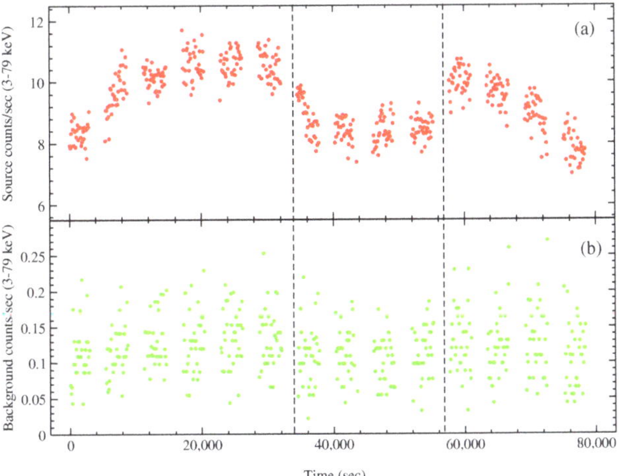

Figure 1. Variation of count rates of XTE J1908+094 from *NuSTAR* observation on 10 April 2019. The upper panel (**a**) shows the count rate for the source and the lower panel (**b**) shows the count rate of the background. The! background count is less than 2% of the source counts.

3.1. Variability Study from the Light Curve

We generated the light curve of 100 s time binning of the total 40 ks data to study the variability (Figure 1a). The light curve shows small variability during this period. We also generated the light curves of 14 individual orbits of 0.01 s time binning. The power density spectra (PDS) do not show any quasi-periodic-oscillations (QPOs). This observation contrasted with the 2002 observations of the object where a broad QPO was observed, and the spectrum was harder [49].

We generated light curves of the entire observation in the energy ranges of 3–10 keV and 10–79 keV. The variation in these light curves, along with the hardness ratio (HR; between 10–79 keV and 3–10 keV), are shown in Figure 2. We noticed that the soft photon (3–10 keV) counts always dominate the hard photon (10–79 keV) counts. Both the soft and hard count rates showed coherent periodic variation during the observation. Thus, the HR always remained at low values ($\sim$0.05 to $\sim$0.15). The soft and the hard count rates and the value of the HR indicate that the source was in the softer spectral regime during our studied period.

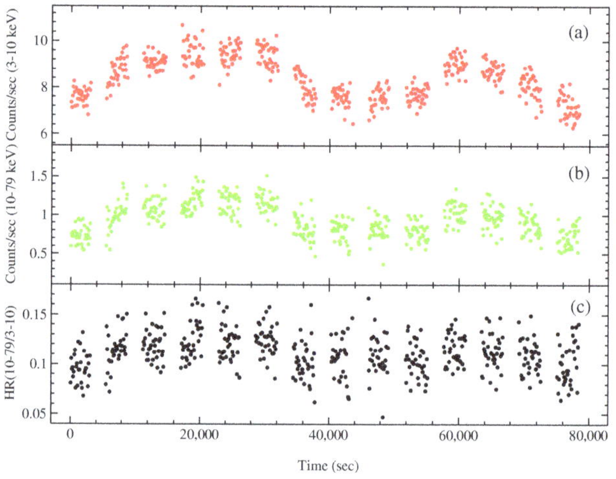

Figure 2. The variation of (**a**) 3–10 keV count rate, (**b**) 10–79 keV count rate, and (**c**) hardness ratio (HR) with time (in sec). The light curves are binned with a time resolution of 100 s.

3.2. Accretion Properties from Spectral Analysis

We studied the spectral properties with both phenomenological and physical models separately. We first fitted the spectrum of the full observation ($\sim$40 ks) using combined `diskbb` and `powerlaw` models (Figure 3a). The inner disk temperature (T_{in}) was obtained as 0.779 keV. The high photon index ($\Gamma\sim$2.085) indicates a softer spectral state. The fitted spectrum retained a χ^2_{red} value of 744/597$\sim$1.246. We noticed a small residual at around 6.5 keV (see Figure 3a). To check the contribution of any Fe-Kα line in the spectrum, we fitted the spectrum with `Tbabs(diskbb+powerlaw+Gaussian)` model (Figure 3b). The inner disk temperature (T_{in}) decreased very little (0.645 keV). The photon index also decreased to ($\Gamma\sim$2.02). We obtained 6.48 keV as the line energy and $\sim$1 keV as the line width (σ) from `Gaussian` model. The best fitted spectrum gave a χ^2_{red} value of 549/594$\sim$0.924. The fit improved with the inclusion of the `Gaussian` model for the Fe Kα emission line. We also fitted the spectrum with TCAF `fits` file along with `Gaussian` model (Figure 3c). The disk rate was high ($\dot{m}_d\sim$1.699) compared to the halo rate ($\dot{m}_h\sim$0.160). The shock location and compression value were obtained as $X_s\sim$49.83 r_s and $R\sim$1.1, respectively. These accretion rates and shock parameters, indicating a softer state, also justify the results obtained from the combined `diskbb+powerlaw+Gaussian` model fitting. The best-fitted spectrum retained a χ^2_{red} value of 562/592~0.949. The best-fitted spectra using these three different combinations of models are shown in Figure 3. The spectral parameters are listed in Table 1.

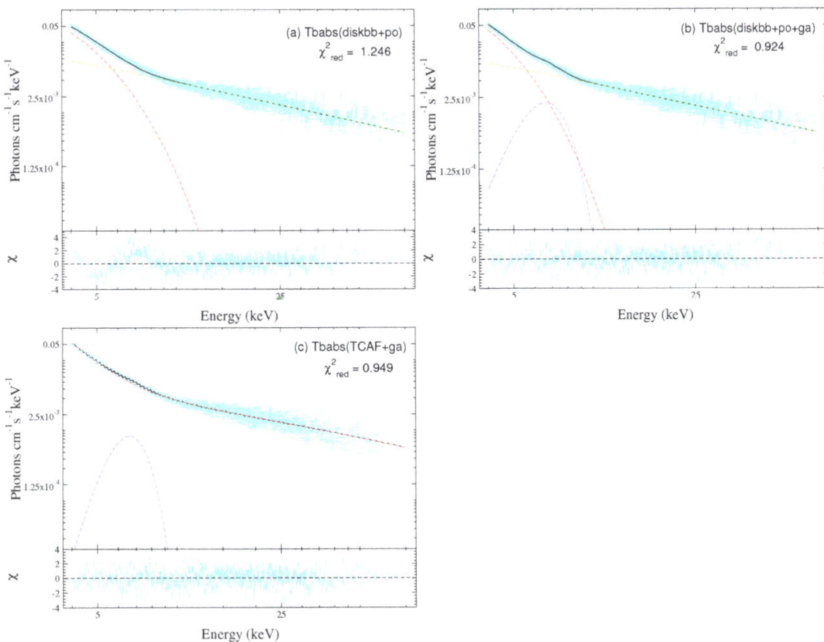

Figure 3. Unfolded spectra using (**a**) Tbabs(diskbb+powerlaw), (**b**) Tbabs(diskbb+powerlaw+Gaussian), and (**c**) Tbabs(TCAF+Gaussian) with their ratio. The red, green dashed lines are the blackbody and powerlaw model components. The solid black line refers to the combined best fitted spectra. The blue dashed lines are the Gaussian component.

Since a variation can be easily noticed in the light curve of the full observation (Figure 1a), we divided the light curve into three segments of ~16 ks, ~10 ks and ~12 ks and analyzed for detailed study. We fitted these three spectra using two combinations Tbabs(diskbb+powerlaw+Gaussian) and Tbabs(TCAF+Gaussian). The spectral parameters are listed in Table 2. Only TCAF plus Gaussian model fitted spectra are shown in Figure 4.

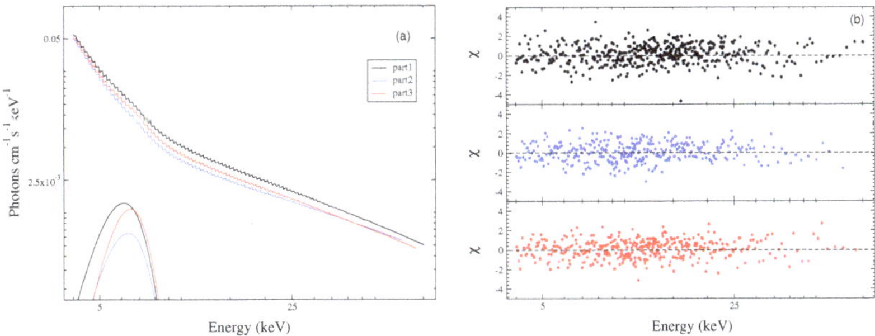

Figure 4. (**a**) **Best-fitted** spectra for the three segments of the *NuSTAR* observation. The right panel (**b**) shows the χ (i.e., (data − model)/error) variations of the three spectra in respective colors.

Table 1. Spectral Analysis Results obtained from *NuSTAR* Observation.

Model1	Diskbb	T_{in} (keV)	$0.779^{+0.013}_{-0.013}$
		norm	153^{+19}_{-16}
	Powerlaw	Γ	$2.085^{+0.028}_{-0.028}$
		norm	$0.057^{+0.004}_{-0.004}$
		χ^2/dof	$744/597 \sim 1.246$
Model2	Diskbb	T_{in} (keV)	$0.645^{+0.019}_{-0.019}$
		norm	493^{+95}_{-72}
	Powerlaw	Γ	$2.023^{+0.028}_{-0.028}$
		norm	$0.047^{+0.003}_{-0.003}$
	Gaussian	E (keV)	$6.483^{+0.152}_{-0.147}$
		σ (keV)	$1.000_{-0.041}$
		norm	$9 \times 10^{-4} \pm 1 \times 10^{-4}$
		χ^2/dof	$549/594 \sim 0.924$
Model3	TCAF	$\dot{m}_d$ ($\dot{M}_{Edd}$)	$1.699^{+0.03}_{-0.04}$
		$\dot{m}_h$ ($\dot{M}_{Edd}$)	$0.160^{+0.03}_{-0.04}$
		X_s (r_s)	49.831^{+9}_{-11}
		R	$1.100^{+0.15}_{-0.16}$
		M_{BH} ($M_\odot$)	$6.516^{+0.53}_{-0.76}$
		N_{TCAF}	$312^{+0.08}_{-0.11}$
	Gaussian	E (keV)	$6.521^{+0.183}_{-0.175}$
		σ (keV)	$0.814^{+0.045}_{-0.151}$
		norm	$5 \times 10^{-4} \pm 1 \times 10^{-4}$
		χ^2/dof	$562/592 \sim 0.949$

Table 2. TCAF model fitted spectral parameters for three segments of the *NuSTAR* observation.

	Obs.	Part1	Part2	Part3
	MJD	58,583.471	58,583.709	58,583.975
Diskbb	T_{in} (keV)	$0.652^{+0.033}_{-0.035}$	$0.701^{+0.034}_{-0.032}$	$0.668^{+0.034}_{-0.035}$
	norm	593^{+55}_{-99}	467^{+47}_{-96}	487^{+34}_{-51}
Powerlaw	Γ	$2.084^{+0.043}_{-0.043}$	$1.936^{+0.063}_{-0.063}$	$2.044^{+0.056}_{-0.055}$
	norm	$0.063^{+0.007}_{-0.006}$	$0.032^{+0.005}_{-0.005}$	$0.047^{+0.007}_{-0.006}$
Gaussian	E (keV)	$6.179^{+0.255}_{-0.245}$	$6.768^{+0.311}_{-0.338}$	$6.450^{+0.243}_{-0.227}$
	σ (keV)	$1.000_{-0.088}$	$1.000_{-0.137}$	$1.000_{-0.137}$
	norm	$9 \times 10^{-4} \pm 2 \times 10^{-4}$	$6 \times 10^{-4} \pm 1 \times 10^{-4}$	$9 \times 10^{-4} \pm 2 \times 10^{-4}$
	χ^2/dof	$483/439 \sim 1.100$	$316/332 \sim 0.952$	$352/368 \sim 0.956$
TCAF	$\dot{m}_d$ ($\dot{M}_{Edd}$)	$1.644^{+0.03}_{-0.04}$	$1.572^{+0.03}_{-0.04}$	$1.884^{+0.03}_{-0.04}$
	$\dot{m}_h$ ($\dot{M}_{Edd}$)	$0.137^{+0.03}_{-0.04}$	$0.165^{+0.03}_{-0.04}$	$0.139^{+0.03}_{-0.04}$
	X_s (r_s)	46.117^{+9}_{-11}	50.140^{+9}_{-11}	47.480^{+9}_{-11}
	R	$1.100^{+0.15}_{-0.16}$	$1.101^{+0.15}_{-0.16}$	$1.100^{+0.15}_{-0.16}$
	M_{BH} ($M_\odot$)	$6.380^{+0.73}_{-0.66}$	$6.702^{+0.73}_{-0.66}$	$6.721^{+0.53}_{-0.76}$
	N_{TCAF}	$326^{+0.08}_{-0.11}$	$227^{+0.08}_{-0.11}$	$330^{+0.08}_{-0.11}$
Gaussian	E (keV)	$5.959^{+0.297}_{-0.278}$	$6.600^{+0.310}_{-215}$	$6.348^{+0.195}_{-0.201}$
	σ (keV)	$1.000_{-0.145}$	$0.901^{+0.139}_{-0.220}$	$0.900^{+0.099}_{-0.131}$
	norm	$7 \times 10^{-4} \pm 2 \times 10^{-4}$	$3 \times 10^{-4} \pm 3 \times 10^{-5}$	$7 \times 10^{-4} \pm 3 \times 10^{-4}$
	χ^2/dof	$483/437 \sim 1.105$	$321/330 \sim 0.973$	$350/366 \sim 0.956$

3.3. Estimation of Inner Disk Radius and Prediction of Mass

We estimated the mass for XTE J1908-094 from the spectral analysis. From the obtained χ^2_{red} values of model 1 and model 2 in Table 1, we suggested model 2 is the best fitted phenomenological model. We considered the fitted disk-normalization (N_{disk}) to estimate the mass of the BHC XTE J1908+094. We average out the N_{disk} values obtained from fittings of the three segments of the whole observation (Table 2). The inner radius (r_{in}) of the disk and the `diskbb` normalization (N_{disk}) are related, as in

$$N_{disk} = \left(\frac{r_{in}}{D/10 \text{ kpc}} \right)^2 \cos\theta. \tag{1}$$

D is the distance of the system in kpc and r_{in} is in km. θ is the inclination of the disk in degree. This estimated inner radius from the above equation is subjected to some errors (see [50,51]). The corrected inner radius (R_{in} (km)) is,

$$R_{in}(\text{km}) \simeq \kappa^2 \zeta r_{in}. \tag{2}$$

κ and ζ are the hardening factor [51] and inner boundary correction factor [50], respectively. Since the inclination angle of the system is not confirmed, we considered three guess values for the inclination as: $\theta \sim 30°$, $50°$, and $80°$. The average value of N_{disk} was 516. We consider the κ and ζ are 1.7 and 0.41, respectively ([50,51]), and assume the distance of the system to be 10 kpc. We obtained the inner radius (R_{in}) for to be 29 km, 34 km, and 65 km, for the inclination angle $30°$, $50°$, and $80°$, respectively. The inner edge of the disk (R_{in}) is considered to be truncated at the inner most stable circular orbit (ISCO). For a Schwarzschild black hole it is $6GM_{BH}/c^2$, where G, M_{BH} and c are the Gravitational constant, mass of the black hole and speed of light at vacuum. We obtained the mass for this black hole from the obtained R_{in} values to be 3.2, 3.7, and 7.2 $M_\odot$, respectively.

In TCAF, the mass of the BH is an important input parameter. If it is not well known from dynamical or other methods, one can obtain the best-fitted value of the M_{BH} from each spectral fit. Here, we obtained the mass of XTE J1908+094 as $\sim$6.5 $M_\odot$ from the combined TCAF, and Gaussian model fit when the entire duration of the observation was considered. (Table 1).

4. Conclusions

We studied the spectral and temporal properties of BHC XTE J1908+094 during its renewed X-ray activity using one *NuSTAR* observation in 10 April 2019. The X-ray activity was very faint and mostly dominated by soft photons. No quasi-periodic oscillations (QPOs) were observed in its power density spectra (PDS). The light curves generated in the soft (3–10 keV) and hard (10–79 keV) energy range showed coherent periodic variation. The low value of hardness ratio suggests that the source was in a softer spectral state during the observation.

We analyzed the spectrum using both phenomenological and physical models separately. The high power-law photon index ($\Gamma \sim 2.23$) indicates that the source was either in the soft–intermediate or in soft state. The disk normalization was also relatively high compared to the power-law normalization. Due to the small residual around 6.5 keV, we refitted the spectrum, adding a `Gaussian` for the contribution of Fe Kα line emission. A broad `Gaussian` was obtained with line energy $\sim$6.48 keV and line width (σ) $\sim$1 keV. The photon index (Γ) decreased to 2.02, suggesting that the source probably was in the soft intermediate state. The presence of a broad width ($\sigma \geq 1$) Gaussian line also signifies the spectral state as soft intermediate. While fitting the three segments of the whole observation divided due to the periodic variability, we noticed that the second segment (see, Table 2) showed a slightly harder spectrum than the other two segments (part1 and part3).

We refitted the spectrum using TCAF plus Gaussian model. We noticed an unusually high disk rate ($\dot{m}_d \sim 1.699\ \dot{M}_{Edd}$) over the halo rate ($\dot{m}_h \sim 0.160\ \dot{M}_{Edd}$). This high amount of hard photon contribution is also noticed in the variation in light curves (see, Figure 2).

The obtained shock location ($X_s \sim 49.83\ r_s$) and compression ratio ($R \sim 1.1$) also refer to soft intermediate spectral state. The spectral parameters obtained from the whole observation's partial fitting also suggest that during the mid-segment, when both soft and hard count decreased, the spectrum became harder but remained in the same spectral state. This faint X-ray activity of XTE J1908+094 in 2019 could be due to the sudden enhancement of viscosity and supply of matter from the pile-up radius. This type of behavior was observed in the case of H1743-322 [2] and GX 339-4 [52]. As the supply of residual matter from the previous outburst (2012–2013) was exhausted, the very short-term X-ray activity was also faded.

We also estimated the possible value of the mass of the black hole (M_{BH}) from the spectral analysis. During our studied period, the spectra are dominated by the soft photons coming from the disk. The inner radius can be constrained in the soft spectral state from the `diskbb` normalization (N_{disk}). Since the other two unknown variables (distance and inclination) are not confirmed for this source, we consider the distance to be 10 kpc and inclination as $30°$, $50°$ and $80°$. The corrected inner radius (R_{in}) was obtained as 29 km, 34 km, and 65 km. As the source is found in the soft-intermediate state, the inner radius must be located very close to the ISCO [53]. Considering a Schwarzschild black hole the ISCO is at $\sim \frac{6GM_{BH}}{c^2}$ or $3\ r_s$. We obtained the black hole mass of 3.2, 3.7 and 7.2 $M_\odot$ for those three R_{in}. This result would only be valid for a Schwarzschild black hole. For a Kerr black hole, the inner radius would be located at less than $3\ r_s$, affecting the prediction of mass from the disk normalization. We could suggest that the system is located at $\sim$8–10 kpc and is highly inclined ($>70°$). From the TCAF model fitted spectral fit, we also obtained the mass of the black hole to be $\sim$6.5 $M_\odot$.

5. Summary

We studied the faint outburst of XTE J1908+094 in 2019 using archival *NuSTAR* data. From the timing and spectral study, we conclude that:

(i)   No quasi-periodic oscillation (QPO) was found in the PDS;
(ii)   The source was in SIMS during our studied period;
(iii)   We also estimated the most probable mass of the black hole to be $\sim$6.5 $M_\odot$;

Author Contributions: Conceptualization, D.C.; Data curation, D.C.; Formal analysis, D.C.; Methodology, D.C. and A.J.; Software, D.D. and S.K.C.; Supervision, D.D., S.K.C. and A.M.; Writing—original draft, D.C.; Writing—review and editing, D.C., D.D., A.J., K.C., R.B., S.K.N., S.K.C., A.M. All authors have read and agreed to the published version of the manuscript.

Funding: This research received no external funding.

Data Availability Statement: The data used here are publicly available. This research has made use of the NuSTAR Data Analysis Software (NuSTARDAS) jointly developed by the ASI Science Data Center (ASDC, Italy) and the California Institute of Technology (USA).

Acknowledgments: We would like to thank the anonymous referees for their constructive and insightful suggestions. D.C. acknowledges the support of the PDR fellowship, IIA, Bengaluru, Karnataka, India. A.J. acknowledges the support of the Post-Doctoral Fellowship from Physical Research Laboratory, Ahmedabad, India, funded by the Department of Space, Government of India. K.C. acknowledges support from DST/INSPIRE (IF170233) fellowship. R.B. acknowledges support from CSIR-UGC NET qualified UGC fellowship (June-2018, 527223). S.K.N., D.D. and S.K.C. acknowledge support ISRO sponsored RESPOND project (ISRO/RES/2/418/17-18) fund. D.D. and S.K.C. also acknowledge support from Govt. of West Bengal, India and DST/GITA sponsored India-Taiwan collaborative project (GITA/DST/TWN/P-76/2017) fund.

Conflicts of Interest: The authors declare no conflict of interest.

References

1. Chakrabarti, S.K. Grand Unification of Solutions of Accretion and Winds around Black Holes and Neutron Stars. *Astrophys. J.* **1996**, *464*, 664–683. [CrossRef]
2. Chakrabarti, S.K.; Debnath, D.; Nagarkoti, S. Delayed outburst of H 1743-322 in 2003 and relation with its other outbursts. *Adv. Space Res.* **2019**, *63*, 3749–3759. [CrossRef]
3. Mondal, S.; Chakrabarti, S.K.; Nagarkoti, S.; Arévalo, P. Possible Range of Viscosity Parameters to Trigger Black Hole Candidates to Exhibit Different States of Outbursts. *Astrophys. J.* **2017**, *850*, 47–55. [CrossRef]
4. Debnath, D.; Chakrabarti, S.K.; Nandi, A. Evolution of the temporal and the spectral properties in 2010 and 2011 outbursts of H 1743-322. *Adv. Space Res.* **2013**, *52*, 2143–2155. [CrossRef]
5. McClintock, J.E.; Remillard, R.A. *Compact Stellar X-ray Sources*; Cambridge University Press: Cambridge, UK, 2009; pp. 157–214.
6. Remillard, R.A.; McClintock, J.E. X-Ray Properties of Black-Hole Binaries. *Annu. Rev. Astron. Astrophys.* **2006**, *44*, 49–92. [CrossRef]
7. Belloni, T.; Homan, J.; Casella, P.; van der Klis, M.; Nespoli, E.; Lewin, W.H.G.; Miller, J.; Mèndez, M. The evolution of the timing properties of the black-hole transientGX 339-4 during its 2002/2003 outburst. *Astron. Astrophys.* **2005**, *440*, 207–222. [CrossRef]
8. Belloni, T.M. States and Transitions in Black Hole Binaries. In *The Jet Paradigm: From Microquasars to Quasars*; Belloni, T.M., Ed.; Springer: Berlin/Heidelberg, Germany, 2010; Volume 794, pp. 53–84.
9. Jana, A.; Debnath, D.; Chakrabarti, S.K.; Mondal, S.; Molla, A.A. Accretion Flow Dynamics of MAXI J1836-194 During Its 2011 Outburst from TCAF Solution. *Astrophys. J.* **2016**, *819*, 107–117. [CrossRef]
10. Chatterjee, K.; Debnath, D.; Chatterjee, D.; Jana, A.; Chakrabarti, S.K. Inference on accretion flow properties of XTE J1752-223 during its 2009-10 outburst. *Mon. Not. R. Astron. Soc.* **2020**, *493*, 2452–2462. [CrossRef]
11. Chatterjee, D.; Debnath, D.; Jana, A.; Chakrabarti, S.K. Properties of the black hole candidate XTE J1118+480 with the TCAF solution during its jet activity induced 2000 outburst. *Astrophys. Space Sci.* **2019**, *364*, 14. [CrossRef]
12. García, J.A.; Tomsick, J.A.; Sridhar, N.; Grinberg, V.; Connors, R.M.; Wang, J.; Fabian, A. The 2017 Failed Outburst of GX 339-4: Relativistic X-Ray Reflection near the Black Hole Revealed by NuSTAR and Swift Spectroscopy. *Astrophys. J.* **2019**, *885*, 48–59. [CrossRef]
13. Tetarenko, B.E.; Sivakoff, G.R.; Heinke, C.O.; Gladstone, J.C. WATCHDOG: A Comprehensive All-sky Database of Galactic Black Hole X-ray Binaries. *Astrophys. J. Suppl. Ser.* **2016**, *222*, 15–112. [CrossRef]
14. Novikov, I.D.; Thorne, K.S. *Black Holes*; DeWitt, C., DeWitt, B., Eds.; Gordon and Breach: New York, NY, USA, 1973.
15. Shakura, N.I.; Sunyaev, R.A. Black holes in binary systems. Observational appearance. *Astron. Astrophys.* **1973**, *24*, 337–355.
16. Sunyaev, R.A.; Titarchuk, L.G. Comptonization of X-rays in plasma clouds. Typical radiation spectra. *Astron. Astrophys.* **1980**, *86*, 121–138.
17. Sunyaev, R.A.; Titarchuk, L.G. Comptonization of low-frequency radiation in accretion disks Angular distribution and polarization of hard radiation. *Astron. Astrophys.* **1985**, *143*, 374–388. [CrossRef]
18. Chakrabarti, S.K.; Titarchuk, L.G. Spectral Properties of Accretion Disks around Galactic and Extragalactic Black Holes. *Astrophys. J.* **1995**, *455*, 623–639. [CrossRef]
19. Chakrabarti, S.K. Spectral Properties of Accretion Disks around Black Holes. II. Sub-Keplerian Flows with and without Shocks. *Astrophys. J.* **1997**, *484*, 313–322. [CrossRef]
20. Chakrabarti, S.K. *Theory of Transonic Astrophysical Flows*; World Scientific: Singapore, 1990.
21. Molteni, D.; Sponholz, H.; Chakrabarti, S.K. Resonance Oscillation of Radiative Shock Waves in Accretion Disks around Compact Objects. *Astrophys. J.* **1996**, *457*, 805–812. [CrossRef]
22. Nandi, A.; Debnath, D.; Mandal, S.; Chakrabarti, S.K. Accretion flow dynamics during the evolution of timing and spectral properties of GX 339-4 during its 2010-11 outburst. *Astron. Astrophys.* **2012**, *542*, 56–66. [CrossRef]
23. Chakrabarti, S.K. Estimation and effects of the mass outflowfrom shock compressed flow around compact objects. *Astron. Astrophys.* **1999**, *351*, 185–191.
24. Debnath, D.; Mondal, S.; Chakrabarti, S.K. Implementation of two-component advective flow solution in xspec. *Mon. Not. R. Astron. Soc. Lett.* **2014**, *440*, L121–L125. [CrossRef]
25. Debnath, D.; Mondal, S.; Chakrabarti, S.K. Characterization of GX 339-4 outburst of 2010-11: Analysis by XSPEC using two component advective flow model. *Mon. Not. R. Astron. Soc.* **2015**, *447*, 1984–1995. [CrossRef]
26. Woods, P.M.; Kouveliotou, C.; Finger, M.H.; Gogus, E.; Swank, J.; Markwardt, C.; Strohmayer, T. XTE J1908+094. *Int. Astron. Union Circ.* **2002**, *7856*, 1.
27. Feroci, M.; Reboa, L.; BEPPOSAX Team. XTE J1908+094. *Int. Astron. Union Circ.* **2002**, *7861*, 2.
28. Miller, J.M.; Oosterbroek, T.; Parmar, A.N. Broad-band X-ray measurements of the black hole candidateXTE J1908+094. *Astron. Astrophys.* **2002**, *394*, 553–560.
29. Miller, J.M.; Reynolds, C.S.; Fabian, A.C.; Miniutti, G.; Gallo, L.C. Stellar-mass black hole spin constraints from disk reflection and continuum modeling. *Astrophys. J.* **2009**, *697*, 900–912. [CrossRef]
30. Rupen, M.P.; Dhawan, V.; Mioduszewski, A.J. XTE J1908+094. *Int. Astron. Union Circ.* **2002**, *7874*, 1.
31. Jonker, P.G.; Gallo, E.; Dhawan, V.; Rupen, M.; Fender, R.P.; Dubus, G. Radio and X-ray observations during the outburst decay of the black holecandidate XTE J1908+094. *Mon. Not. R. Astron. Soc.* **2004**, *351*, 1359–1364. [CrossRef]

32. Chaty, S.; Mignani, R.P.; Israel, G.L. Discovery of the near-infrared counterpart of the X-ray transient XTE J1908+094. *Mon. Not. R. Astron. Soc.* **2002**, *337*, L23–L26. [CrossRef]
33. Chaty, S.; Mignani, R.P.; Israel, G.L. A closer look at the X-ray transient XTE J1908+094: Identification of two new near-infrared candidate counterparts. *Mon. Not. R. Astron. Soc.* **2006**, *365*, 1387–1391. [CrossRef]
34. Coriat, M.; Tzioumis, T.; Corbel, S.; Fender, R. Optically thin synchrotron emission from XTE J1908+094 observed by the ATCA. *Astron. Telegr.* **2013**, *5575*, 1.
35. Krimm, H.A.; Barthelmy, S.D.; Baumgartner, W.; Cummings, J.; Gehrels, N.; Lien, A.Y.; Markwardt, C.B.; Palmer, D.; Sakamoto, T.; Stamatikos, M.; et al. Swift/BAT detects a new outburst from the HXMB/BHC XTE J1908+094. *Astron. Telegr.* **2013**, *5523*, 1.
36. Miller-Jones, J.C.A.; Sivakoff, G.R.; Krimm, H.A. VLA detection of radio emission from the new outburst of XTE J1908+094. *Astron. Telegr.* **2013**, *5530*, 1.
37. Negoro, H.; Suzuki, K.; Ueno, J.S.S.; Tomida, H.; Kimura, M.; Nakahira, S.; Ishikawa, M.; Nakagawa, Y.E.; Mihara, T.; Sugizaki, S.; et al. MAXI/GSC detection of a hard-to-soft state transition of XTE J1908+094. *ATel* **2013**, *5549*, 1.
38. Rushton, A.P.; Fender, R.; Anderson, G.; Staley, T.; Rumsey, C.; Titterington, D. AMI detection of 2 cm radio emission from the XRB BHC J1908+094. *Astron. Telegr.* **2013**, *5532*, 1.
39. Tao L.; Tomsick, J.A.; Walton, D.J.; Fürst, F.; Kennea, J.; Miller, J.M.; Boggs, S.E.; Christensen, F.E.; Craig, W.W.; Gandhi, P.; et al. NuSTAR and Swift Observations of the Black Hole Candidate XTE J1908+094 during its 2013 Outburst. *Astrophys. J.* **2015**, *811*, 51–59. [CrossRef]
40. Zhang, L.; Chen, L.; Qu, J.-L.; Bu, Q.-C.; Zhang, W. The 2013-2014 Outburst of XTE J1908+094 Observed with Swift and NuSTAR: Spectral Evolution and Black Hole Spin Constraint. *Astrophys. J.* **2015**, *813*, 90–102. [CrossRef]
41. Curran, P.A.; Miller-Jones, J.C.; Rushton, A.P.; Pawar, D.D.; Anderson, G.E.; Altamirano, D.; Tudose, V. Radio polarimetry as a probe of unresolved jets: the 2013 outburstof XTE J1908+094. *Mon. Not. R. Astron. Soc.* **2015**, *451*, 3975–3985. [CrossRef]
42. Rushton, A.P.; Miller-Jones, J.C.A.; Curran, P.A.; Sivakoff, G.R.; Rupen, M.P.; Paragi, Z.; Spencer, R.E.; Yang, J.; Altamirano, D.; Belloni, T.; et al. Resolved, expanding jets in the Galactic black hole candidate XTE J1908+094. *Mon. Not. R. Astron. Soc.* **2017**, *468*, 2788–2802. [CrossRef]
43. Rodriguez, J.; Mereminskiy, I.; Grebenev, S.A.; Cangemi, F.; Clavel, M.; Coleiro, A.; Egron, E.; Grinberg, V.; Pottschmidt, K.; Remillard, R.; et al. INTEGRAL detects renewed activity from the microquasar XTE J1908+094. *Astron. Telegr.* **2019**, *12628*, 1.
44. Williams, D.; Motta, S.; Fender, R.; Titterington, D.; Green, D.; Perrott, Y. AMI-LA 15.5 GHz observation of the Black Hole candidate XTE J1908+094. *Astron. Telegr.* **2019**, *12620*, 1.
45. Miller, J.M.; Reynolds, M.; Tetarenko, B.; Ali, S.; Balakrishnan, M.; Chen, J.; Vozza, D. A Neil Gehrels Swift Observatory Snapshot of the Black Hole Candidate XTE J1908+094. *Astron. Telegr.* **2019**, *12632*, 1.
46. Ludlam, R.M.; Remillard, R.; Homan, J.; Strohmayer, T.E.; Gendreau, K.; Arzoumanian, Z. NICER Observations of the new outburst from XTE J1908+094. *Astron. Telegr.* **2019**, *12652*, 1.
47. Verner, D.A.; Ferland, G.J.; Korista, K.T.; Yakovlev, D.G. Atomic Data for Astrophysics. II. New Analytic FITS for Photoionization Cross Sections of Atoms and Ions. *Astrophys. J.* **1996**, *465*, 487–498. [CrossRef]
48. Wilms, J.; Allen, A.; McCray, R. On the Absorption of X-Rays in the Interstellar Medium. *Astrophys. J.* **2000**, *542*, 914–924. [CrossRef]
49. Göğüş, E.; Finger, M.H.; Kouveliotou, C.; Woods, P.M.; Patel, S.K.; Rupen, M.; Van Der Klis, M. Long-term spectral and timing behavior of the black hole candidate XTE J1908+094. *Astrophys. J.* **2004**, *609*, 977–987. [CrossRef]
50. Kubota, A.; Tanaka, Y.; Makishima, K.; Ueda, Y.; Dotani, T.; Inoue, H.; Yamaoka, K. Evidence for a Black Hole in the X-ray Transient GRS 1009-45. *Publ. Astron. Soc. Jpn.* **1998**, *50*, 667–673. [CrossRef]
51. Shimura, T.; Takahara, F. On the Spectral Hardening Factor of the X-Ray Emission from Accretion Disks in Black Hole Candidates. *Astrophys. J.* **1995**, *445*, 780–788. [CrossRef]
52. Bhowmick, R.; Debnath, D.; Chatterjee, K.; Nagarkoti, S.; Chakrabarti, S.K.; Sarkar, R.; Jana, A. Relation Between Quiescence and Outbursting Properties of GX 339-4. *Astrophys. J.* **2021**, *910*, 138. [CrossRef]
53. Done, C.; Gierlinski, M.; Kubota, A. Modelling the behaviour of accretion flows in X-ray binaries: Everything you always wanted to know about accretion but were afraid to ask. *Astron. Astrophys. R* **2007**, *15*, 1–66. [CrossRef]

Article

Changing Accretion Geometry of Seyfert 1 Mrk 335 with *NuSTAR*: A Comparative Study

Santanu Mondal * and C. S. Stalin

Indian Institute of Astrophysics, II Block, Koramangala, Bangalore 560034, India; stalin@iiap.res.in
* Correspondence: santanu.mondal@iiap.res.in

Abstract: We present a detailed spectral study of the narrow-line Seyfert 1 galaxy, Markarian 335, using eight epoch observations made between 2013 and 2020 with the *Nuclear Spectroscopic Telescope Array*. The source was variable during this period both in spectral flux and flow geometry. We estimated the height of the Compton cloud from the model fitted parameters for the whole observation period. This allowed us to investigate the underlying physical processes that drive the variability in X-rays. Our model fitted mass varies in a narrow range, between $(2.44 \pm 0.45 - 3.04 \pm 0.56) \times 10^7 M_\odot$, however, given the large error bars, it is consistent with being constant and is in agreement with that known from optical reverberation mapping observations. The disk mass accretion rate reached a maximum of 10% of the Eddington rate during June 2013. Our study sheds light on mass outflows from the system and also compares different aspects of accretion with X-ray binaries.

Keywords: galaxies: active; galaxies: jets; galaxies: nuclei; radiative transfer; Seyfert 1 objects: individual: Mrk 335

Citation: Mondal, S.; Stalin, C.S. Changing Accretion Geometry of Seyfert 1 Mrk 335 with *NuSTAR*: A Comparative Study. *Galaxies* **2021**, *9*, 21. https://doi.org/10.3390/galaxies9020021

Received: 19 February 2021
Accepted: 23 March 2021
Published: 25 March 2021

Publisher's Note: MDPI stays neutral with regard to jurisdictional claims in published maps and institutional affiliations.

1. Introduction

Active galactic nuclei (AGN) host an accretion disk with an energetic X-ray-producing Compton cloud (so-called corona) around a supermassive black hole at the center. The emitted X-ray radiation also acts as one of the most direct probes of accretion onto AGN. AGN show rapid flux and polarization variations, which indicate that the observed high energy radiation mainly originates from the inner few to tens of Schwarzschild radii ($r_s = 2GM_{BH}/c^2$; M_{BH}, G, and c are the mass of the black hole, gravitational constant, and the speed of light, respectively) from the event horizon of the black hole [1]. The power-law (PL) component is widely accepted as the effect of inverse Compton scattering (ICS) of thermally produced soft photons from the accretion disk by a corona of hot electrons at the inner edge of the disk [2,3]. However, the geometry and size of the corona, as well as the physical mechanisms governing the energy transfer between the disk and the corona, are not well understood, which necessitates a physical model to better understand the observed signatures in them.

There are proposed models in the literature [4] that consider the corona to be the region between the truncation radius and the innermost stable circular orbit. However, it is not clear why the disk truncates at a certain radius from the central black hole and how the truncation radius is connected with the corona. Furthermore, most of the models in the literature [5,6] use optical depth, the coronal temperature, or the spectral index as a parameter to compute the spectrum from the corona. None of these are the basic physical quantities of accretion. According to the two-component advective flow (TCAF) model [7,8], at a certain distance from the black hole (BH) where the gravitational force balances with the centrifugal force, shock forms by the low angular momentum, hot, sub-Keplerian halo, and satisfies Rankine–Hugoniot conditions, which is the region of the truncation of the disk. Apart from that, as the two forces balance each other, accretion of matter virtually stops and its velocity decreases, which also increases the optical depth of the corona region. As the gravitational force dominates closer to the BH, velocity again increases and the flow passes through the inner sonic point and falls onto the BH, therefore the flow is transonic [7]. Beyond this shocked region, both the disk and halo matter pile up to decide the optical depth and temperature of the corona region. The same region also upscatters the incoming soft radiation from the disk via ICS. A typical X-ray spectrum of AGN in the 2–10 keV region shows primarily the signature of a PL. Due to ICS, the corona cools down and its area decreases, leading to a low brightness level and a softer spectrum [9]. These physical properties and the dynamics of the flow make TCAF more favorable as a physical model than other existing models in the literature. In addition to the above, TCAF uses mass accretion rates and the mass of the BH as model parameters, which are the basic physical quantities of a flow. Furthermore, the same shock location, which can explain the spectral features, may also be able to explain the temporal features from its oscillation [10,11]. Therefore, among the various models available in the literature to understand the observed X-ray emission in AGN, we preferred to consider the TCAF model for our study.

The TCAF model has five parameters: (i) mass of the black hole (M_{BH}), (ii) disk accretion rate ($\dot{m}_d$), (iii) halo accretion rate ($\dot{m}_h$), (iv) location of the shock (X_s), and (v) shock compression ratio (R). Increasing $\dot{m}_d$ and keeping the other parameters fixed increases the number of soft photons from the Keplerian disk [12], making the spectrum softer. Higher $\dot{m}_h$ increases the corona temperature and hardens the spectrum. Similarly, increase of either X_s or R leads to hardening of the spectrum. However, in reality all parameters change in multidimensional space. Therefore, such variations may not sustain due to non-linear changes in optical depth and coronal temperature in different epochs. A detailed parameter study has been made in [13]. In recent years, the TCAF model has been implemented in XSPEC and has been used widely to study both X-ray binaries (XRBs) [14–16] and AGN [17–19]. However, there are physical processes that are not yet incorporated in the current TCAF model, e.g., jets, bulk motion Comptonization, and the spin effect of the BH.

Apart from the above components, many AGN exhibit extra features in their X-ray spectra that likely originate from the accretion disk [20]. The so-called X-ray reflection features produced by an accretion disk might be because of its illumination from the central compact object, such as a magnetically dominated corona residing above the surface of the disk (e.g., [2,21,22]). A prominent feature in the X-ray spectrum is the Fe Kα emission line at around 6.4 keV, which is due to fluorescence in a dense and relatively cold medium [23–25]. The shape of the Fe Kα line is a powerful probe of the general relativistic effects close to the BH [26,27], leading to estimates of the spin of the BH [28–30]. Indeed, observations over the last decades by X-ray satellites have uncovered relativistically broadened Fe Kα lines in the spectra of several AGN [31–33] and black hole XRBs (e.g., [34]). Models of X-ray reflection spectra [5,35] show additional features that can collectively constrain the ionization state, heating, and cooling of the illuminated surface [36].

Mrk 335 is a bright, narrow-line Seyfert 1 (NLS1) galaxy at a redshift of z = 0.026 [37], with a black hole mass $M_{BH} = 2.6 \times 10^7 M_\odot$ [38], and shows ionized absorption in the X-ray and ultra-violet (UV) bands [39]. It has been observed to go into extremely low-flux states [40] where the soft X-ray flux dropped by a factor of up to $\sim$30, while the hard flux dropped only by a factor of $\sim$2 in 7 years. Using observations from XMM-Newton, Ref. [41] found signatures of absorption and high reflection fractions in the spectrum of Mrk 335 in its lowest flux state, whereas at the intermediate flux state the observed X-ray spectrum is explained well by a blurred reflection model without the requirement of variable absorption [42]. Ref. [43] estimated a high-frequency lag of $\sim$150 s between the continuum-dominated energy bands and the iron line and soft excess components for this source. This time lag suggests that the continuum source is located very close to the central black hole and that relativistic effects are worth considering. Using combined *Swift* and *NuSTAR* observations, Ref. [44] interpreted the X-ray flare during 2014 as arising from the vertical collimation and ejection of the X-ray-emitting corona at a mildly relativistic velocity, which causes the continuum emission to be relativistically beamed away from the disk. Ref. [45] discussed the effect of the geometry of the corona on the relativistically blurred X-ray reflection arising from the accretion disk, which can also explain the variability between low- and high-flux states. Ref. [46] performed structure function analysis using long-term *Swift* optical/UV and X-ray data for Mrk 335 and showed that the X-ray low-flux state could be due to the physical changes in the corona or absorption, whereas the variability in the optical/UV band is more consistent with the thermal and dynamic timescales associated with the accretion disk. Ref. [47], studied the low-flux state data using *XMM-Newton* and observed absorption lines from a highly ionized outflowing wind.

Several studies on Mrk 335 clearly show that the system is highly variable, leaving a range of possibilities for its dynamical behavior and emission features. Mrk 335 is known to show changes in geometry [45], and these can be found from a disk-based model fit such as TCAF to data. It is not clear to date if there is any change in the mass accretion of the source, and if so, what effects it can have on the geometry of the corona in the source. To investigate the above, in this work we chose this object and carried out an analysis of the X-ray data on the source acquired by *NuSTAR* during the period 2013–2020. This paper is organized as follows: in the next section, we describe the observation details and the analysis procedures. In §3, we explain our results obtained from model fits to the data. We summarize our conclusions in the final section.

2. Observation and Data Analysis

In the present manuscript, we analyzed archival data (https://heasarc.gsfc.nasa.gov/docs/cgro/db-perl/W3Browse/w3browse.pl (accessed on 10 June 2020)) from *NuSTAR* observations of Mrk 335 made during 2013–2020. During this time interval Mrk 335 was observed nine times; for one observation data quality was not good, so we considered the remaining eight observations. The details of the observations are listed in Table 1. The *NuSTAR* data in the energy range of 3.0–30 keV (as the data are noisy above 30 keV), were extracted using the standard NUSTARDAS v1.3.1 (https://heasarc.gsfc.nasa.gov/docs/nustar/analysis/ (accessed on 10 June 2020)) software. We ran the NUPIPELINE task to produce cleaned event lists and NUPRODUCTS for generation of the spectra. We used a region of $80''$ for the source and $100''$ for the background using ds9. The data were grouped by the *grppha* command, with a minimum of 30 counts per energy bin. The same binning was used for all the observations. For spectral analysis of the data we used XSPEC (https://heasarc.gsfc.nasa.gov/xanadu/xspec/ (accessed on 10 June 2020)) [48] version 12.8.1. We used the absorption model TBABS [49] with the Galactic hydrogen column density fixed at 3.6×10^{20} atoms cm^{-2} [50] throughout the analysis.

Table 1. Log of observations.

Date	MJD	OBSID	Exposure (s)
13 June 2013	56,456	60001041002	21,299
13 June 2013	56,456	60001041003	21,525
25 June 2013	56,468	60001041005	93,028
20 September 2014	56,920	80001020002	68,908
10 July 2018	58,309	80201001002	82,257
06 June 2020	59,006	90602619004	30,156
07 June 2020	59,007	90602619006	30,495
08 June 2020	59,008	90602619008	22,452

3. Results and Discussion

We fit the *NuSTAR* observations of Mrk 335 taken between 2013 to 2020. During this observation period the source showed significant variability in spectral flux (see Figure 1). We performed spectral analyses using different models. First, we performed spectral fitting using a simple power-law (PL) model as TBABS(GAUSS+PL), where the PL index (Γ) varied from 0.72–1.84, a change by a factor greater than two, which is clearly an indication of significant changes in the flow behavior and corona properties. The model fitted parameters are provided in Table 2. Our model fits show that the line energy and width vary in a broad range, which point to complex geometry changes and gravity effects (more especially line broadening) in the source.

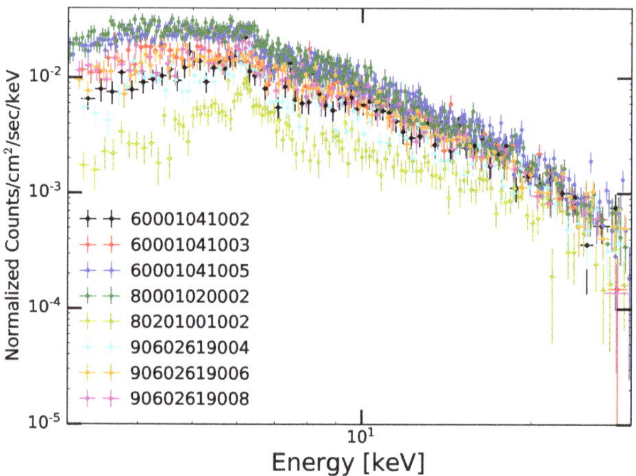

Figure 1. Spectra of all eight observations in the 3–30 keV energy range. Variability of the source is evident.

Table 2. Best-fitting parameters with the TBABS(GAUSS+PL) model.

OBSID	Γ_{PL}	$N_{PL}\,(\times 10^{-4})$ $ph./keV/cm^2/s$	E_g [keV]	σ_g [keV]	χ^2/dof
60001041002	1.09 ± 0.08	2.15 ± 0.46	5.47 ± 0.18	0.79 ± 0.21	71.91/65
60001041003	1.12 ± 0.10	2.79 ± 0.76	4.21 ± 0.45	1.52 ± 0.34	102.1/85
60001041005	1.66 ± 0.02	13.58 ± 0.54	6.00 ± 0.99	0.28 ± 0.06	452.94/316
80001020002	1.84 ± 0.02	20.92 ± 0.91	6.20 ± 0.13	0.34 ± 0.08	320.71/264
80201001002	0.72 ± 0.06	0.40 ± 0.06	6.07 ± 0.07	0.60 ± 0.08	136.14/124
90602619004	1.24 ± 0.06	2.50 ± 0.33	6.08 ± 0.07	0.41 ± 0.08	107.16/79
90602619006	1.17 ± 0.09	2.93 ± 0.71	4.83 ± 0.31	1.26 ± 0.30	82.39/106
90602619008	1.46 ± 0.05	6.07 ± 0.61	6.10 ± 0.06	0.13 ± 0.09	92.29/80

Γ_{PL} and N_{PL} are the PL index and normalization. E_g and σ_g are the Gaussian line energy and width, respectively.

In addition to the simple PL model, we also modeled the observed spectra using the complex phenomenological model PEXRAV [51] and the physical model TCAF [8]. The results of PEXRAV model fits are given in Table 3. Our PL model fits indicate that the source passed through hard and soft spectral states similar to XRBs, where the timescale for changing from one state to other is in the order of a few days to weeks, which is the viscous timescale ([52] and references therein). We found a similar timescale when the corona changed from its maximum size to its minimum, which is around 12 days. However, as the viscous timescale for AGN is in the order of a few hundred to a few Myr (scaled by the mass of the BH), it may not be possible to observe a similar timescale for other AGN. In future we aim to explore this aspect of accretion to estimate the viscosity of the flow.

From PEXRAV model fits, we found the reflection fraction to be high with $R_{ref} > 0.9$ for all the epochs. When the Fe abundance (A_{Fe}) was made a free parameter, A_{Fe} was found to vary from sub-solar to super-solar values during the epochs analyzed here, which is quite unphysical. It has been pointed out by [53] that reflection model fits to observed X-ray spectra are generally found to yield high A_{Fe} values. In addition, according to [54], the high A_{Fe} obtained from model fits to data could be due to some unknown physical effects being overlooked in current models. Therefore, for PEXRAV model fits, we froze A_{Fe} to solar value. For the observation day MJD = 56,468, we required one extra Gaussian line component at low energy $\sim$5 keV. From the line component and its width, it can be seen that both narrow and broad lines are present in this epoch. We kept the disk inclination ($i_{\rm inc}$) fixed to 45° throughout [55]. Next we applied the physical model TCAF, having the form TBABS(TCAF+GAUSS), to extract the accretion flow parameters. For all observations, goodness of fit is more or less similar to the PEXRAV model. The TCAF model fitted parameters are provided in Table 4. For some observations, better fits were achieved after adding some additive model components, which are detailed in the table. For the majority of the epochs, the χ^2/dof obtained from PEXRAV and TCAF model fits are in agreement, whereas for three epochs, the χ^2/dof from TCAF fits are marginally larger than those obtained from PEXRAV model fits. This might be due to differences in the spectral components between models, e.g., TCAF does not include line features, which may give relatively higher χ^2/dof. In Figure 2, we show different model fitted spectra for the OBSID 60001041002/3 in the left/right panels of the figure. Figure 3 shows the same spectral fitting but for the OBSID 60001041005 (left panel) and 80001020002 (right panel). The rest of the observational fits are shown in Appendix A.

Table 3. Best-fitting parameters with the TBABS(PEXRAV+GAUSS) model.

OBSID	Γ	$E_{\rm cut}$ [keV]	$R_{\rm ref}$	E_g [keV]	σ_g [keV]	χ^2/dof
60001041002	1.44 ± 0.03	31.55 ± 6.51	3.40 ± 1.17	5.66 ± 0.09	0.61 ± 0.21	$68.01/63$
60001041003	1.12 ± 0.02	16.56 ± 2.42	4.85 ± 1.07	3.40 ± 0.17	1.99 ± 0.37	$89.72/83$
60001041005 [a]	1.95 ± 0.10	96.30 ± 20.87	2.40 ± 0.41	6.20 ± 0.06	0.007 ± 0.003	$323.19/311$
80001020002	2.11 ± 0.09	107.90 ± 30.15	2.33 ± 0.92	5.50 ± 0.17	0.84 ± 0.17	$253.39/262$
80201001002	0.87 ± 0.03	20.40 ± 3.93	3.68 ± 0.81	6.09 ± 0.08	0.62 ± 0.08	$128.70/122$
90602619004	1.68 ± 0.03	54.63 ± 11.71	3.47 ± 1.12	6.11 ± 0.08	0.39 ± 0.09	$96.97/77$
90602619006	1.39 ± 0.02	69.19 ± 15.89	0.92 ± 0.17	5.20 ± 0.28	1.01 ± 0.36	$81.39/104$
90602619008	1.99 ± 0.04	32.69 ± 4.01	7.67 ± 1.07	6.12 ± 0.06	0.013 ± 0.003	$78.82/78$

Γ, $E_{\rm cut}$, and $R_{\rm ref}$ are model PL photon index, cutoff energy, and reflection scaling factor, respectively. E_g and σ_g are the Gaussian line energy and width. We kept the disk inclination fixed at 45° [55] and $A_{\rm Fe}$ at Solar abundance value. [a]: Gaussian component was added at 5.02 ± 0.18 keV with line width 0.64 ± 0.20 keV. The origin of this line is not clear yet, however, a similar line was observed for Mrk 335 earlier [56].

Table 4. Best-fitting parameters with the TBABS(TCAF+GAUSS).

OBSID	$M_{\rm BH} \times 10^7$ [$M_\odot$]	$\dot{m}_d$ [$\dot{M}_{\rm Edd}$]	$\dot{m}_h$ [$\dot{M}_{\rm Edd}$]	X_s [r_s]	R	E_g [keV]	σ_g [keV]	χ^2/dof
60001041002	2.94 ± 0.63	0.0024 ± 0.0008	2.000 ± 0.725	22.43 ± 9.96	2.45 ± 0.54	5.58 ± 0.19	0.69 ± 0.21	$72.45/61$
60001041003	3.02 ± 0.60	0.0026 ± 0.0010	1.158 ± 0.331	43.72 ± 12.97	5.78 ± 1.23	4.18 ± 0.61	1.53 ± 0.52	$101.75/81$
60001041005 [a]	2.44 ± 0.45	0.1001 ± 0.0433	1.781 ± 0.143	6.24 ± 0.94	5.21 ± 1.26	6.10 ± 0.05	0.22 ± 0.07	$333.54/309$
80001020002 [b]	2.93 ± 0.65	0.0059 ± 0.0001	3.117 ± 0.181	8.66 ± 1.66	1.49 ± 0.17	2.13 ± 0.52	2.27 ± 0.95	$244.18/256$
80201001002 [c]	2.79 ± 0.70	0.0145 ± 0.0011	3.017 ± 0.144	33.53 ± 7.62	4.24 ± 1.39	6.02 ± 0.13	0.66 ± 0.15	$130.29/117$
90602619004 [d]	3.02 ± 0.98	0.0059 ± 0.0014	2.945 ± 0.365	21.25 ± 3.19	2.79 ± 0.62	6.01 ± 0.08	0.51 ± 0.09	$87.65/72$
90602619006	3.04 ± 0.56	0.0024 ± 0.0007	1.978 ± 0.150	21.54 ± 2.57	2.20 ± 0.26	4.78 ± 0.62	1.29 ± 0.49	$82.35/102$
90602619008	2.87 ± 0.95	0.0045 ± 0.0012	1.540 ± 0.259	22.22 ± 2.49	1.20 ± 0.13	5.53 ± 0.19	0.70 ± 0.22	$95.42/76$

$\dot{m}_h$ and $\dot{m}_d$ represent TBABS(TCAF+GAUSS) model fitted sub-Keplerian (halo) and Keplerian (disk) rates, respectively. X_s and R are the model fitted shock location and shock compression ratio values, respectively. E_g and σ_g are the Gaussian line energy and width. [a]: Gaussian component was added at 3.44 ± 0.07 keV with line width 1.55 ± 0.21 keV. [b]: DISKLINE [26] component was added with line energy at 6.12 ± 0.10 keV, powerlaw of emissivity (β) = -2.98 ± 0.70, inner radius ($R_{\rm in}$) = 34.6 ± 24.4 $r_g(GM/c^2)$, and disk inclination was frozen at 45°. [c]: ZXIPCF [57] multiplicative model component was used for parameters: column density (N_H) = $68 \pm 10.91 \times 10^{22}$ cm^{-2}, covering fraction (f_c) = 0.45 ± 0.05, $log\xi$ = 1.2 ± 0.3. [d]: Gaussian component was added at 4.16 ± 0.11 keV with line width 0.29 ± 0.14 keV.

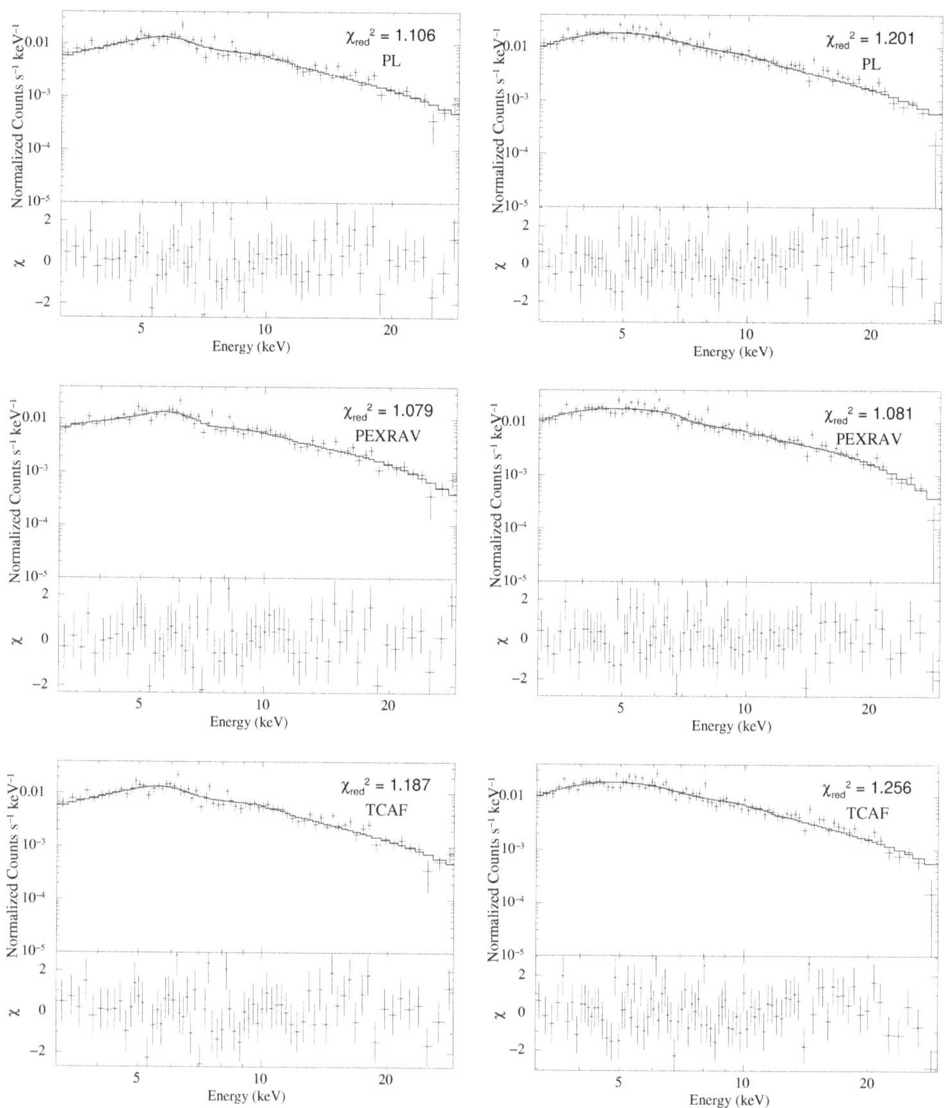

Figure 2. Sample fit of 3–30 keV spectra with PL, PEXRAV, and TCAF models for OBSIDs 60001041002 (**left panel**) and 60001041003 (**right panel**) along with the residuals.

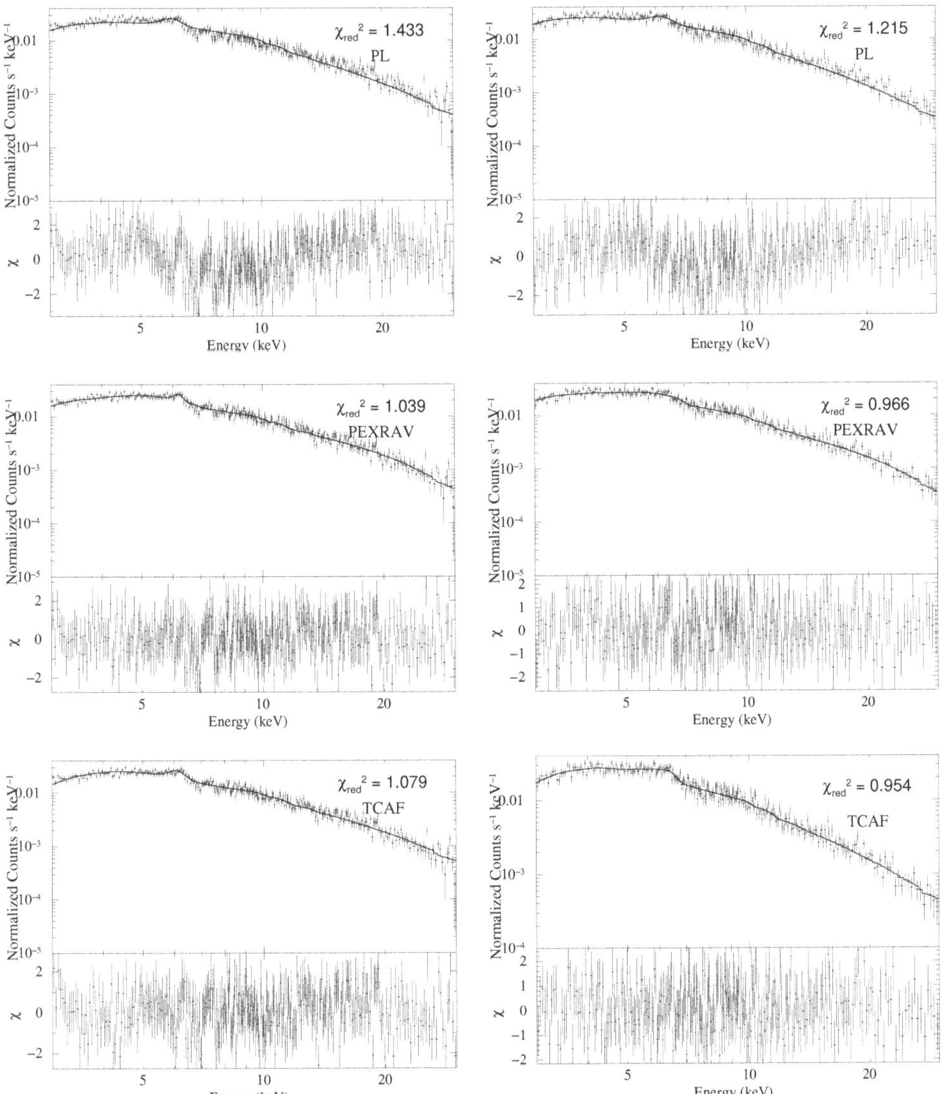

Figure 3. Same as Figure 2 but for the OBSIDs 60001041005 (**left panel**) and 80001020002 (**right panel**) along with the residuals.

We used TCAF model fitted parameters to estimate the geometrical height of the shock (H_{shk}). The height of the corona is basically the height of the shock, as the shock is the boundary layer of the corona. We estimated (H_{shk}) using the equation below (see also [14]):

$$H_{\text{shk}} = \left[\frac{\gamma(R-1)X_s^2}{R^2} \right]^{1/2},$$

(1)

where γ is the adiabatic index, X_s is the location of the shock or the boundary of the corona, and R is the shock compression ratio ($= \rho_+/\rho_-$), a ratio of the density of the downstream to the density of the upstream of the flow.

In Figure 4, we show the variation of various model parameters obtained from TCAF model fits as well as the variation of the hardness ratio (HR) with time. The halo rate was always higher than the disk rate by an order of magnitude or more, which implies that the flow was dominated by low angular momentum matter, which might be accreted from the wind or host galaxies. The $\dot{m}_d$ was found to vary by a factor of $\sim$40 from 0.24 to 10% of the Eddington rate and the $\dot{m}_h$ varied between 1.2–3.1 times the Eddington rate. The size of the shock location (X_s) or corona changed significantly and reached up to 6 r_s during 25 June 2013 (60001041005) and $8r_s$ during 2014 (80001020002). For OBSID 60001041005, we required an extra-broad (1.55 keV) Gaussian component at a low energy of 3.44 keV, which was also found to be broadened by strong gravitational effects [27]. For OBSID 80001020002, we required an additive DISKLINE [26] component at a line energy of 6.12 keV, with emissivity power-law index (β) $\sim$ -3.0 and $R_{in} = 24.4r_g(GM/c^2)$, as the current TCAF model does not include line emission due to gravitational effects in the fit. It is worth noting that the X_s is achieved for the highest $\dot{m}_d$, which might be due to the cooling effect inside the corona. As the accretion rate increased, more disk photons got intercepted by the corona. This cooled the corona, causing the shock to move inwards and thus causing a significant change in the size of the corona. Such change in the corona could also be due to the activity of the jet, which extracted a significant amount of thermal energy and contracted the corona [58]. Therefore, it is also possible to establish a jet–disk connection using TCAF model fits to AGN spectra similar to XRBs ([59–61] and references therein).

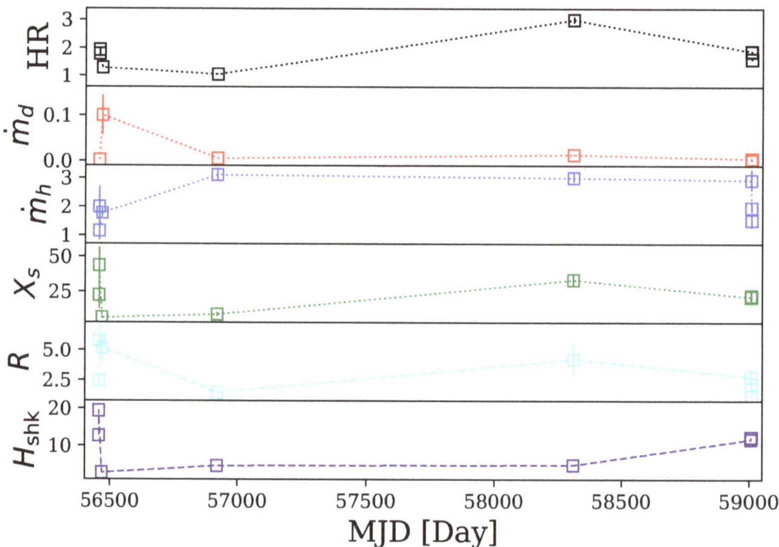

Figure 4. Two-component advective flow (TCAF) model fitted parameter evolutions with observation days are shown along with hardness ratio (HR) estimated from the power-law (PL) model fitted fluxes.

Such movement of the shock can give rise to the observed variability in spectral and temporal properties [62]. During 2014, $\dot{m}_h$ further increased and made the corona hotter and the shock receded slightly. During 2018 the shock moved further away and the corona became bigger, therefore intercepting more soft photons and generating high-energy photons, which can also ionize the disk atmosphere. For this particular observation, we required the partial covering fraction model (ZXIPCF), with a covering factor ($f_c =$ 0.45) and low ionization parameter ($log\xi = 1.2$). For OBSID 90602619004, during 2020, we required another narrow line (0.29 keV) component, when the shock again moved inward. The model-fitted shock compression ratio also varied in a broad range, which also dictates the strength of the shock and the optical depth of the corona. The 5th panel of

Figure 4 (from the top) shows the variation of this parameter. The bottom panel of Figure 4 shows the estimated height of the corona, which changes significantly for the whole observation period.

The mass of the BH is one of the parameters in TCAF. Therefore, keeping it as a free parameter in the TCAF model fits to the observed spectra will yield an M_{BH} value for Mrk 335. We kept M_{BH} as a free parameter during our fitting and we found M_{BH} to vary in a range between $(2.44 \pm 0.45$ to $3.04 \pm 0.56) \times 10^7 M_\odot$. Considering the large error bars, the derived BH mass is consistent with being constant. Our derived M_{BH} is consistent with that estimated by [38] using optical reverberation mapping observations, however, it is an order of magnitude larger than that obtained by [53]. Such a low mass found by [53] could be due to them not considering an expanded corona in their model fit to the data. Our results on M_{BH} also show that by fitting accretion-disk-based models such as TCAF to observed X-ray spectra, one can estimate the mass of BHs in AGN. We also found that freezing M_{BH} to some fixed value from the above range and redoing the fit has a negligible effect on other derived parameters, such as $\dot{m}_d$, $\dot{m}_h$, X_s, and R.

The viscous timescale in AGN is much longer compared to that of XRBs. At small timescales (in the order of a few days or months), significant changes in the accretion rate are not expected, similar to the case for heating and cooling rates. Thus, at such timescales, variability is expected to be less. This is in fact observed in the radio-quiet category of AGN and accretion-disk-based models can explain the observed flux variations in them. However, this might not be the case in the radio-loud category of AGN, as relativistic jets play a dominant role in the flux variations observed in them. Even in those sources where the ejection of jets extracts thermal energy from the corona, the change in the shock location can be drastic in them even though the cooling is less. Results on this will be reported elsewhere.

4. Conclusions

In the present paper, we analyzed *NuSTAR* data for Mrk 335, observed between 2013 and 2020, to understand the dynamics of the flow and the accretion behavior of the source. We found that the source was significantly variable both in spectral flux, accretion geometry, and reflection fraction. We also found the geometry of the corona to change between epochs. Such a change can affect the reflection fraction and the spectral energy distribution of the illuminating radiation, and consequently the ionization rate. From the TCAF model fits we obtained the dynamics of the flow along with the geometry change of the corona and accretion rates. In our TCAF model fits, the line profiles due to relativistic effects were taken into account using an additive DISKLINE model wherever needed rather than using it throughout for all observations. Among other models, even though the XILLVER model has relativistic and reflection effects incorporated, the accretion dynamics and the origin of corona are still lacking in the model. Therefore we did not take into account such models in our current study. Below we provide the key findings on the variable nature of the source.

1. During 2013, the corona at the inner region changed significantly from its elongated stage ($\sim r_s$) to destruction stage $\sim 6 r_s$, consistent with [44]. As the compression ratio changed and the corona contracted significantly, it could be possible that a jet/mass outflow was launched around 13 June 2013, therefore a significant amount of thermal energy was extracted by the jet/outflow and the corona contracted.
2. The observation during 2014 required a blurred reflection component along with TCAF to fit the spectra. This is quite natural as the corona contracted and the inner edge of the disk moved significantly inward, therefore the gravitational effect became dominant [26,27] and blurred the Fe Kα line. We also required a broad Gaussian line component at ~ 2 keV.
3. The HR was roughly constant during 2013, and was also similar to that obtained during 2014.

4. There was a significant change in Γ_{PL} between the 2013 and 2014 spectra, which may be due to a sudden change in accretion rates. During this period, disk accretion had increased by a factor of a few and also the size of the corona contracted significantly.

5. The steepening of the emissivity profile of Mrk 335 indicates that the corona is compact for this source [44]. In our study, we found that the size of the corona was indeed small and compact during 25 June 2013 and 20 September 2014 (see Table 4). It is also noticeable that the height of the corona reduced significantly during these two epochs.

6. During 2014, spectral flux between 3 and 30 keV changed/increased by a factor of $\sim$3 compared to 2013 and 2018. This could be due to an increase in accretion rates as well as the change in corona. As the accretion rate increased (in 25 June 2013 and 10 July 2018), the number of soft photons increased, thereby increasing the cooling rate, i.e., reduction of more energy from the corona by the seed photons from the Keplerian disk (see theoretical aspects in [9]). It should be noted that although the shock location changed significantly, the other parameters triggered that change, mainly the increase in disk accretion rate, and therefore the cooling rate. This also infers that not only the shock location but other parameters are equally important to explain the observed variability.

7. During 2018 and onward, the corona and H_{shk} again elongated. During this period the HR also increased. This also implies that there is a correlation between HR and the geometry of the corona.

8. During 2018, to take into account the disk ionization effects along with TCAF, we required partial covering and the ZXIPCF model to better fit the data. The model fit showed that an absorption column density with $N_H = 6.8 \times 10^{23}$ cm^{-2} is present. The fit also required a low ionization with $log\xi = 1.2$ erg cm s^{-1}, with a partial covering fraction of 0.45. This added component also indicates the presence of mass outflow from the system, which is evident in the monitoring observations of the source in optical/UV and X-rays ([63,64], and references therein).

9. The reflection fraction, R_{ref}, is measured as the ratio of the photon fluxes from the blurred reflection and power-law continuum model components. From PEXRAV model fitting we found this value was > 0.9.

10. The mass of the black hole, which was kept as a free parameter, was found to vary in a very narrow range $(2.44\text{--}3.04) \times 10^7 M_\odot$, and considering the error bars is consistent with a constant. This is in agreement with that of [38].

11. Earlier studies on observed X-ray variability inferred the origin of flux variations to changes in the primary power-law continuum, possibly exhibited through intrinsic variations in the corona, or possible changes in its size or location [42,44,65]. This is in agreement with our present findings.

Author Contributions: Conceptualization, methodology, formal analysis, writing—original draft preparation, S.M.; conceptualization, supervision, writing—review and editing, C.S.S. Both authors have read and agreed to the published version of the manuscript.

Funding: This research received no external funding.

Institutional Review Board Statement: Not applicable.

Informed Consent Statement: Not applicable.

Data Availability Statement: We used archival data for our analysis in this manuscript. Appropriate links are given in the manuscript. For the details of the data fitting, one can directly contact the first author.

Acknowledgments: We thank both referees for their constructive and insightful suggestions that improved the quality of the manuscript. SM acknowledges the Ramanujan Fellowship (# RJF/2020/000113) by SERB, Govt. of India. This research made use of the *NuSTAR* Data Analysis Software (NUSTARDAS) jointly developed by the ASI Science Data Center (ASDC), Italy, and the California Institute of Technology (Caltech), USA. This research also made use of data obtained through the High Energy Astrophysics Science Archive Research Center Online Service, provided by NASA/Goddard Space Flight Center.

Conflicts of Interest: The authors declare no conflicts of interest.

Appendix A

Here we show the spectral fitting plots for the rest of the observations using all three models.

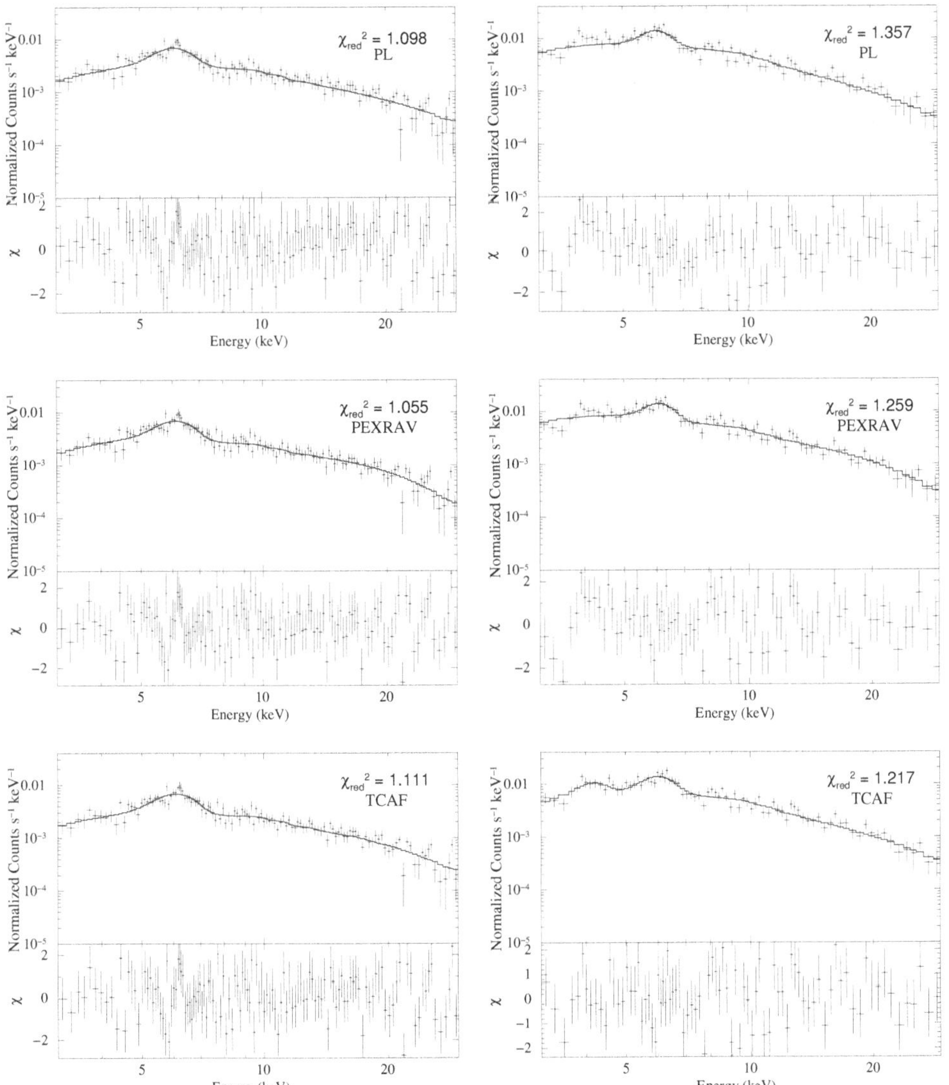

Figure A1. Same as Figure 2 but for the OBSIDs 80201001002 (**left panel**) and 90602619004 (**right panel**) along with the residuals.

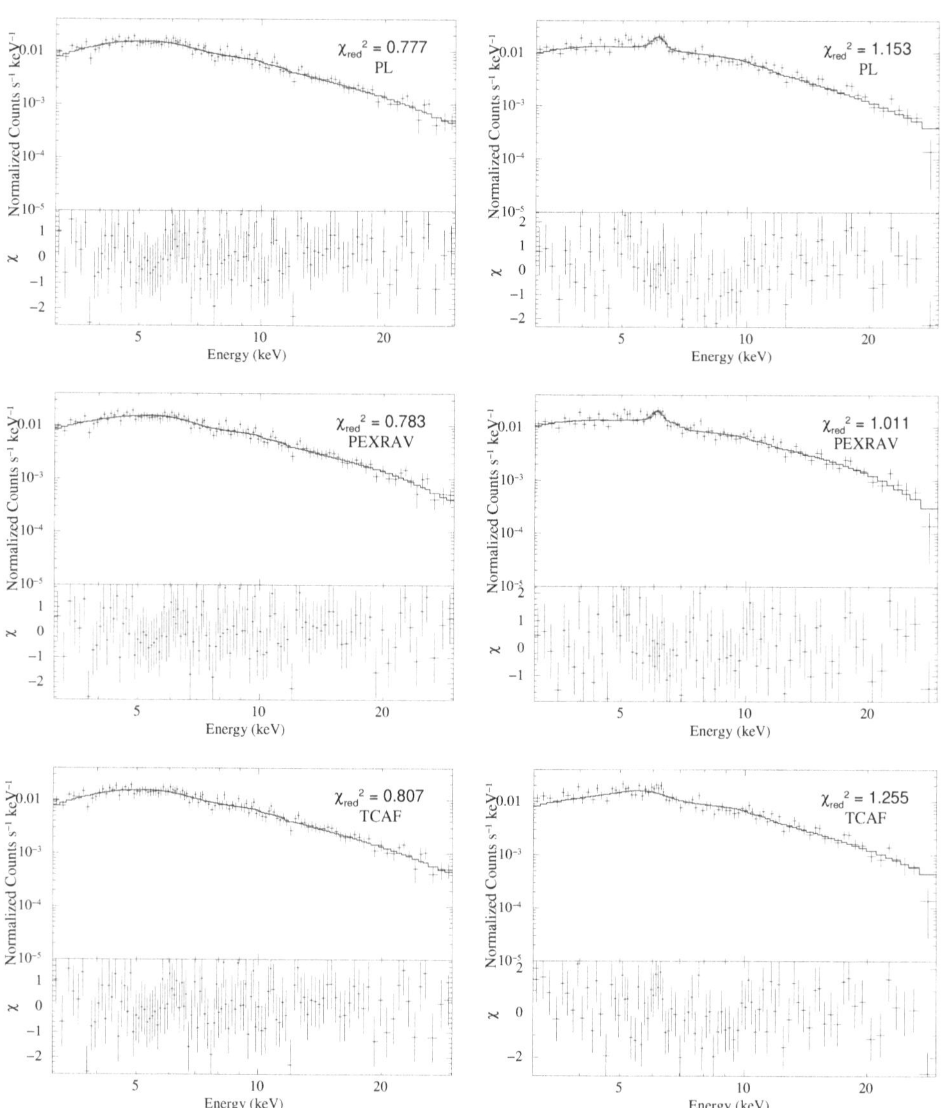

Figure A2. Same as Figure 2 but for the OBSIDs 90602619006 (**left panel**) and 90602619008 (**right panel**) along with the residuals.

References

1. McHardy, I.M.; Koerding, E.; Knigge, C.; Uttley, P.; Fender, R.P. Active galactic nuclei as scaled-up Galactic black holes. *Nature* **2006**, *444*, 730. [CrossRef] [PubMed]
2. Haardt, F.; Maraschi, L. A Two-Phase Model for the X-Ray Emission from Seyfert Galaxie. *Astrophys. J.* **1991**, *380*, L51. [CrossRef]
3. Zdziarski, A.A.; Poutanen, J.; Johnson, W.N. Observations of Seyfert Galaxies by OSSE and Parameters of Their X-ray/Gamma-Ray Sources. *Astrophys. J.* **2000**, *542*, 703. [CrossRef]
4. Esin, A.A.; McClintock, J.E.; Narayan, R. Advection-Dominated Accretion and the Spectral States of Black Hole X-ray Binaries: Application to Nova Muscae 1991. *Astrophys. J.* **1997**, *489*, 865. [CrossRef]

5. Garcia, J.; Dauser, T.; Reynolds, C.S.; Kallman, T.R.; McClintock, J.E.; Wilms, J.; Eikmann, W. X-Ray Reflected Spectra from Accretion Disk Models. III. A Complete Grid of Ionized Reflection Calculations. *Astrophys. J.* **2013**, *768*, 146. [CrossRef]
6. Titarchuk, L. Generalized Comptonization Models and Application to the Recent High-Energy Observations. *Astrophys. J.* **1994**, *434*, 570. [CrossRef]
7. Chakrabarti, S.K. Standing Rankine-Hugoniot Shocks in the Hybrid Model Flows of the Black Hole Accretion and Winds. *Astrophys. J.* **1989**, *347*, 365. [CrossRef]
8. Chakrabarti, S.; Titarchuk, L.G. Spectral Properties of Accretion Disks around Galactic and Extragalactic Black Holes. *Astrophys. J.* **1995**, *455*, 623. [CrossRef]
9. Mondal, S.; Chakrabarti, S.K. Spectral properties of two-component advective flows with standing shocks in the presence of Comptonization. *Mon. Not. R. Astron. Soc.* **2013**, *431*, 2716. [CrossRef]
10. Chakrabarti, S.K.; Mondal, S.; Debnath, D. Resonance condition and low-frequency quasi-periodic oscillations of the outbursting source H 1743- 322. *Mon. Not. R. Astron. Soc.* **2015**, *452*, 3451. [CrossRef]
11. Molteni, D.; Sponholtz, H.; Chakrabarti, S.K. Resonance oscillation of radiative shock waves in accretion disks around compact objects. *Astrophys. J.* **1996**, *457*, 805. [CrossRef]
12. Shakura, N.I.; Sunyaev, R.A. Black holes in binary systems. Observational appearance. *Astron. Astrophys.* **1973**, *24*, 337.
13. Chakrabarti, S.K. Spectral properties of accretion disks around black holes. II. Sub-Keplerian flows with and without shocks. *Astrophys. J.* **1997**, *484*, 313. [CrossRef]
14. Debnath, D.; Chakrabarti, S.K.; Mondal, S. Implementation of two-component advective flow solution in xspec. *Mon. Not. R. Astron. Soc.* **2014**, *440L*, 121. [CrossRef]
15. Debnath, D.; Mondal, S.; Chakrabarti, S.K. Characterization of GX 339-4 outburst of 2010-11: Analysis by XSPEC using two component advective flow model. *Mon. Not. R. Astron. Soc.* **2015**, *447*, 1984. [CrossRef]
16. Mondal, S.; Debnath, D.; Chakrabarti, S.K. Inference on Accretion Flow Dynamics Using TCAF Solution from the Analysis of Spectral Evolution of H 1743-322 during the 2010 Outburst. *Astrophys. J.* **2014**, *786*, 4. [CrossRef]
17. Mandal, S.; Chakrabarti, S.K. Spectrum of Two-Component Flows around a Supermassive Black Hole: An Application to M87. *Astrophys. J.* **2008**, *689*, 17. [CrossRef]
18. Mondal, S.; Rani, P.; Stalin, C.S. Flux and spectral variability of Mrk 421 during its low activity state using *NuSTAR*. *Mon. Not. R. Astron. Soc.* **2021**, submitted.
19. Nandi, P.; Chakrabarti, S.K.; Mondal, S. Spectral Properties of NGC 4151 and the Estimation of Black Hole Mass Using TCAF Solution. *Astrophys. J.* **2019**, *877*, 65. [CrossRef]
20. Fabian, A.C.; Ross, R.R. X-ray reflection. *SSRv* **2010**, *157*, 167.
21. Galeev, A.A.; Rosner, R.; Vaiana, G.S. Structured coronae of accretion disks. *Astrophys. J.* **1979**, *229*, 318. [CrossRef]
22. Merloni A.; Fabian A.C. Thunderclouds and accretion discs: a model for the spectral and temporal variability of Seyfert 1 galaxies. *Mon. Not. R. Astron. Soc.* **2001**, *328*, 958. [CrossRef]
23. Barr, P.; White, N.E.; Page, C.G. The discovery of low-level iron K line emission from CYG X-1. *Mon. Not. R. Astron. Soc.* **1985**, *216*, 65. [CrossRef]
24. Nandra, K.; Pounds, K.A.; Stewart, G.C.; Fabian, A.C.; Rees, M.J. Detection of iron features in the X-ray spectrum of the Seyfert I galaxy MCG -6-30-15. *Mon. Not. R. Astron. Soc.* **1989**, *236*, 39. [CrossRef]
25. Pounds, K.A.; Nandra, K.; Stewart, G.C.; George, I.M.; Fabian, A.C. X-ray reflection from cold matter in the nuclei of active galaxies. *Nature* **1990**, *344*, 132. [CrossRef]
26. Fabian, A.C.; Rees, M.J.; Stella, L.; White, N.E. X-ray fluorescence from the inner disc in Cygnus X-1. *Mon. Not. R. Astron. Soc.* **1989**, *238*, 729. [CrossRef]
27. Laor, A. Line Profiles from a Disk around a Rotating Black Hole. *Astrophys. J.* **1991**, *376*, 90. [CrossRef]
28. Brenneman, L.W.; Reynolds, C.S. Constraining Black Hole Spin via X-Ray Spectroscopy. *Astrophys. J.* **2006**, *652*, 1028. [CrossRef]
29. Reynolds, C.S. The spin of supermassive black holes. *Class. Quantum Gravity* **2013**, *30*, 244004. [CrossRef]
30. Reynolds, C.S.; Fabian, A.C. Broad Iron-K_α Emission Lines as a Diagnostic of Black Hole Spin. *Astrophys. J.* **2008**, *675*, 1048. [CrossRef]
31. Ballantyne, D.R.; Vaughan, S.; Fabian, A.C. A two-component ionized reflection model of MCG-6-30-15. *Mon. Not. R. Astron. Soc.* **2003**, *342*, 239. [CrossRef]
32. Iwasawa, K. The variable iron K emission line in MCG-6-30-15. *Mon. Not. R. Astron. Soc.* **1996**, *282*, 1038. [CrossRef]
33. Risaliti, G. A rapidly spinning supermassive black hole at the centre of NGC 1365. *Nature* **2013**, *494*, 449. [CrossRef]
34. Mondal, S.; Chakrabarti, S.K.; Debnath, D. Spectral study of GX 339-4 with TCAF using Swift and NuSTAR observation. *Astrophys. Space Sci.* **2016**, *361*, 309. [CrossRef]
35. Ross, R.R.; Fabian, A.C. A comprehensive range of X-ray ionized-reflection models. *Mon. Not. R. Astron. Soc.* **2005**, *358*, 211. [CrossRef]
36. Ross, R.R.; Fabian, A.C.; Young, A.J. X-ray reflection spectra from ionized slabs. *Mon. Not. R. Astron. Soc.* **1999**, *306*, 461. [CrossRef]
37. Huchra, J.P.; Vogeley, M.S.; Geller, M.J. The CFA Redshift Survey: Data for the South Galactic CAP. *Astrophys. J. Suppl.* **1999**, *121*, 287. [CrossRef]

38. Grier, C.J.; Peterson, B.M.; Pogge, R.W. A Reverberation Lag for the High-ionization Component of the Broad-line Region in the Narrow-line Seyfert 1 Mrk 335. *Astrophys. J.* **2012**, *744*, L4. [CrossRef]

39. Longinotti, A.L.; Krongold, Y.; Kriss, G.A.; Ely, J.; Pradhan, A. The Rise of an Ionized Wind in the Narrow-line Seyfert 1 Galaxy Mrk 335 Observed by XMM-Newton and HST. *Astrophys. J.* **2013**, *766*, 104. [CrossRef]

40. Grupe, D.; Komossa, S.; Gallo, L.C. Discovery of the Narrow-Line Seyfert 1 Galaxy Markarian 335 in a Historical Low X-Ray Flux State. *Astrophys. J.* **2007**, *668*, L111. [CrossRef]

41. Grupe, D. XMM-Newton Observations of the Narrow-Line Seyfert 1 Galaxy Mrk 335 in a Historical Low X-Ray Flux State. *Astrophys. J.* **2008**, *681*, 982. [CrossRef]

42. Gallo, L.C.; Fabian, A.C.; Grupe, D.; Bonson, K.; Komossa, S.; Longinotti, A.L.; Miniutti, G.; Walton, D.J.; Zoghbi, A.; Mathur, S. A blurred reflection interpretation for the intermediate flux state in Mrk 335. *Mon. Not. R. Astron. Soc.* **2013**, *428*, 1191. [CrossRef]

43. Kara, E.; Fabian, A.C.; Cackett, E.M.; Uttley, P.; Wilkins, D.R.; Zoghbi, A. Discovery of high-frequency iron K lags in Ark 564 and Mrk 335. *Mon. Not. R. Astron. Soc.* **2013**, *434*, 1129. [CrossRef]

44. Wilkins, D.R.; Gallo, L.C. Driving extreme variability: the evolving corona and evidence for jet launching in Markarian 335. *Mon. Not. R. Astron. Soc.* **2015**, *449*, 129. [CrossRef]

45. Wilkins, D.R.; Gallo, L.C.; Grupe, D.; Bonson, K.; Komossa, S.; Fabian, A.C. Flaring from the supermassive black hole in Mrk 335 studied with Swift and NuSTAR. *Mon. Not. R. Astron. Soc.* **2015**, *454*, 4440. [CrossRef]

46. Gallo, L.C.; Blue, D.M.; Grupe, D.; Komossa, S.; Wilkins, D.R. Eleven years of monitoring the Seyfert 1 Mrk 335 with Swift: Characterizing the X-ray and UV/optical variability. *Mon. Not. R. Astron. Soc.* **2018**, *478*, 2557. [CrossRef]

47. Gallo, L.C.; Gonzalez, A.G.; Waddell, S.G.H.; Ehler, H.J.S.; Wilkins, D.R.; Longinotti, A.L.; Grupe, D.; Komossa, S.; Kriss, G.A.; Pinto, C. Evidence for an emerging disc wind and collimated outflow during an X-ray flare in the narrow-line Seyfert 1 galaxy Mrk 335. *Mon. Not. R. Astron. Soc.* **2019**, *484*, 4287. [CrossRef]

48. Arnaud, K.A. XSPEC: The First Ten Years. *ASP Conf. Ser.* **1996**, *101*, 17.

49. Wilms J.; Allen A.; McCray R. On the Absorption of X-Rays in the Interstellar Medium. *Astrophys. J.* **2000**, *542*, 914. [CrossRef]

50. Kalberla, P.M.W.; Burton, W.B.; Hartmann, D.; Arnal, E.M.; Bajaja, E.; Morras, R.; Poppel, W.G.L. The Leiden/Argentine/Bonn (LAB) Survey of Galactic HI. Final data release of the combined LDS and IAR surveys with improved stray-radiation corrections. *Astron. Astrophys.* **2005**, *440*, 775. [CrossRef]

51. Magdziarz, P.; Zdziarski, A.A. Angle-dependent Compton reflection of X-rays and gamma-rays. *Mon. Not. R. Astron. Soc.* **1995**, *273*, 837. [CrossRef]

52. Mondal, S.; Chakrabarti, S.K.; Nagarkoti, S.; Arevalo, P. Possible range of viscosity parameter to trigger black hole candidates to exhibit different states of outbursts. *Astrophys. J.* **2017**, *850*, 47. [CrossRef]

53. Mastroserio, G.; Ingram, A.; van der Klis, M. Multi-timescale reverberation mapping of Mrk 335. *Mon. Not. R. Astron. Soc.* **2020**, *498*, 4971. [CrossRef]

54. Garcia, J.A.; Steiner, J.F.; Mcclintock, J.E.; Remillard, R.A.; Grinberg, V.; Dauser, T. X-Ray Reflection Spectroscopy of the Black Hole GX 339–4: Exploring the Hard State with Unprecedented Sensitivity. *Astrophys. J.* **2015**, *813*, 84. [CrossRef]

55. Chainakun, P.; Young, A.J. Simultaneous spectral and reverberation modelling of relativistic reflection in Mrk 335. *Mon. Not. R. Astron. Soc.* **2015**, *452*, 333. [CrossRef]

56. Ezhikode, S.H.; Dewangan, G.C.; Misra, R.; Philip, N.S. Correlation between relativistic reflection fraction and photon index in NuSTAR sample of Seyfert 1 AGN. *Mon. Not. R. Astron. Soc.* **2020**, *495*, 3373. [CrossRef]

57. Reeves, J.; Done, C.; Pounds, K.; Terashima, Y.; Hayashida, K.; Anabuki, N.; Uchino, M.; Turner, M. On why the Iron K-shell absorption in AGN is not a signature of the local Warm/Hot Intergalactic Medium. *Mon. Not. R. Astron. Soc.* **2008**, *385L*, 108. [CrossRef]

58. Chakrabarti, S.K. Estimation and effects of the mass outflow from shock compressed flow around compact objects. *Astron. Astrophys.* **1999**, *351*, 185.

59. Chakrabarti, S.K. Spectral signature of mass loss (and mass gain by) an accretion disk around a black hole. *Astrophys. J.* **2002**, *579L*, 21. [CrossRef]

60. Mondal, S.; Chakrabarti, S.K.; Debnath, D. Spectral signatures of dissipative standing shocks and mass outflow in presence of Comptonization around a black hole. *Astrophys. Space Sci.* **2014**, *353*, 223. [CrossRef]

61. Chatterjee, D.; Debnath, D.; Jana, A.; Chakrabarti, S.K. Properties of the black hole candidate XTE J1118+480 with the TCAF solution during its jet activity induced 2000 outburst. *Astrophys. Space Sci.* **2019**, *364*, 14. [CrossRef]

62. Chakrabarti, S.K.; Wiita, P.J. Spiral Shocks in Accretion Disks As a Contributor to Variability in Active Galactic Nuclei. *Astrophys. J.* **1993**, *411*, 602. [CrossRef]

63. Komossa, S. Lifting the curtain: The Seyfert galaxy Mrk 335 emerges from deep low-state in a sequence of rapid flare events. *Astron. Astrophys.* **2020**, *643L*, 7. [CrossRef]

64. Parker, M.L. The nuclear environment of the NLS1 Mrk 335: Obscuration of the X-ray line emission by a variable outflow. *Mon. Not. R. Astron. Soc.* **2019**, *490*, 683. [CrossRef]

65. Sarma, R.; Tripathi, S.; Misra, R.; Dewangan, G.; Pathak, A.; Sarma, J.K. Relationship between X-ray spectral index and X-ray Eddington ratio for Mrk 335 and Ark 564. *Mon. Not. R. Astron. Soc.* **2015**, *448*, 1541. [CrossRef]

Article

Powerful Jets from Radiatively Efficient Disks, a Decades-Old Unresolved Problem in High Energy Astrophysics

Chandra B. Singh [1,*,†], David Garofalo [2,†] and Benjamin Lang [3]

1   South-Western Institute for Astronomy Research, Yunnan University, University Town, Chenggong, Kunming 650500, China

2   Department of Physics, Kennesaw State University, Marietta, GA 30060, USA; dgarofal@kennesaw.edu

3   Department of Physics & Astronomy, California State University, Northridge, CA 91330, USA; benjamin.lang.447@my.csun.edu

*   Correspondence: chandrasingh@ynu.edu.cn

†   These authors contributed equally to this work.

Abstract: The discovery of 3C 273 in 1963, and the emergence of the Kerr solution shortly thereafter, precipitated the current era in astrophysics focused on using black holes to explain active galactic nuclei (AGN). But while partial success was achieved in separately explaining the bright nuclei of some AGN via thin disks, as well as powerful jets with thick disks, the combination of both powerful jets in an AGN with a bright nucleus, such as in 3C 273, remained elusive. Although numerical simulations have taken center stage in the last 25 years, they have struggled to produce the conditions that explain them. This is because radiatively efficient disks have proved a challenge to simulate. Radio quasars have thus been the least understood objects in high energy astrophysics. But recent simulations have begun to change this. We explore this milestone in light of scale-invariance and show that transitory jets, possibly related to the jets seen in these recent simulations, as some have proposed, cannot explain radio quasars. We then provide a road map for a resolution.

Keywords: black hole physics; rotating black holes; relativistic jets; active galactic nuclei; supermassive black holes; radio galaxies

Citation: Singh, C.B.; Garofalo, D.; Lang, B. Powerful Jets from Radiatively Efficient Disks, a Decades-Old Unresolved Problem in High Energy Astrophysics. *Galaxies* **2021**, *9*, 10. http://doi.org/10.3390/galaxies9010010

Received: 6 January 2021

Accepted: 23 January 2021

Published: 26 January 2021

Publisher's Note: MDPI stays neutral with regard to jurisdictional claims in published maps and institutional affiliations.

1. Introduction

Radio quasars are active galaxies that harbor strong, collimated jets, mostly of FRII radio morphology [1] , but that also shines brightly in the optical, suggesting they are accreting black holes in radiatively efficient mode [2]. They are not distributed randomly in space and time, with a greater probability of being found in isolated environments and at higher redshifts, at least compared to FRI radio galaxies, which are found in richer environments and comparably lower redshifts [3,4]. FRI radio galaxies differ from radio quasars not only in their jet morphology but also in their 'radio mode' accretion, as opposed to 'quasar mode', indicating the disk is radiatively inefficient.

Many X ray binaries cycle through states with radio-emitting jets present during so-called 'hard' states, but absent during so-called 'soft' states [5,6]. The consensus is that the former is related to radiatively inefficient accretion, similar to jet mode accretion in FRI radio galaxies, while the latter is associated with radiatively efficient accretion, similar to quasar mode ([7–10] and references therein). It is often found that as X-ray binaries transition from the hard state to the soft one, a transitory, ballistic, or more powerful and collimated jet, is generated [11,12]. It has been claimed that such a jet taps into black hole rotational energy [13] unlike the hard state jet [9]. Despite these general properties, a subclass of X-ray binaries does not suppress jets in soft states [14]. And when these neutron star X-ray binaries do suppress jets, their transitory jets tend to be weaker [15].

The observational evidence described above points to a picture in which accreting supermassive black holes combine radiatively efficient accretion with jets while stellar-mass black holes do not. In the past, and again recently via general relativistic magnetohydro-dynamic (GRMHD) simulations, resolving this issue hinges on the assumption that radio quasars are the large scale analog of transitory burst states in black hole X-ray binaries, during which X-ray binaries are approaching the soft state, so a stable radiatively efficient disk is in the process of formation, although it has not yet formed. The goal of this work is to highlight the incompatibility between well-accepted theory and the observations. We then produce a roadmap for GRMHD simulations designed for getting at a better solution, one that appeals to ideas that have been available for a decade.

In Section 2, we analyze the history of GRMHD simulations in their struggle to generate thin disks in an attempt to understand the disk radiative efficiency. The point here is to understand the conditions that produce jet suppression. Within the context of that history, we single out recent simulations by [16,17] that promise to take our understanding of jet formation in thin disks to a new level. This leads to an exploration of the jet-disk connection in light of scale invariance and the roadmap alluded above. We then conclude.

2. Disks and Jets in GRMHD

Within the last half-decade, GRMHD simulations have pushed the boundary of disk thickness to extremes, with values about an order of magnitude smaller than two decades ago. The height to radius ratio, or scaleheight, H/R, is found to be on average equal to 0.20–0.26 in the simulations dating back a decade and a half ago [18–24], while, a decade later, we see H/R decrease to 0.13 [25]. In-between we see a disk scaleheight of $H/R \sim 0.1$ associated with disk radiative efficiency of 6% that of the relativistic Novikov-Thorne model [26], followed up with disks as thin as $H/R \sim 0.06$ but up to 10% difference in efficiency with respect to the Novikov & Thorne disk [27,28]. Similar results were found in [29], with $H/R \sim 0.1$ and deviation from Novikov & Thorne disks of 5%. By 2019 we see $H/R \sim 0.03$ [16,17]. To model radio quasars, simulations need to produce disks with high radiative efficiency, as well as high jet efficiency. The quasar-like radiation from the nucleus appears to require a scaleheight of 0.01 from analytic models. Achieving this has been a challenge. The decrease in disk efficiency associated with an increase in jet efficiency was thought to be a good description of hard state jets in X-ray binaries and of radio galaxies, both observed to produce X-ray and radio jet signatures but weak emission lines. But radio quasars appear to upend this inverse trend between jet and disk efficiency. Ref. [30] have explored the blazar subclass of radio quasars (flat spectrum radio quasars (FSRQs), attempting to fit observations with simulations of disks in the context of moderately thin disks (those of [25]). We will explore this in the next section but the basic conclusion is the need for some missing ingredient that allows for both large jet and disk efficiency. Finally, Ref. [16] have produced the thinnest disks in GRMHD, with $H/R \sim 0.03$, with the absence of jet suppression. This is an important milestone for simulations. Ref. [16,17] also attempt to explain radio quasars, which they explore in the context of scale invariance, anchoring their arguments to transitory ballistic jets observed in X-ray binaries. We explore their analysis in that context, as well.

2.1. FSRQ Jets from Moderately Thin Disks

Flat-spectrum radio quasars (FSRQs) are the blazar subclass of active galactic nuclei (AGN) emitting the most relativistic jets along our line of sight with relatively high accretion rates. Using gamma-ray luminosity from the Fermi Large Area Telescope as a proxy for jet powers, and independent measurements of black hole mass, Ref. [30] produce the best fit correlation between jet power and black hole mass from observations of 154 FSRQs. The ideas are built on the work of [31] who uncovered a correlation between gamma-ray luminosity L_γ and total blazar jet power P_{jet} given by

$$log_{10}P_{jet} = 0.51 log_{10}L_\gamma + 21.2, \tag{1}$$

in units of erg/s. Soares & Nemmen (2020) compiled observational data for 154 FSRQs and fitted a relation between L_γ and black hole mass M, obtaining

$$log_{10}M = 0.37log_{10}L_\gamma - 8.95, \tag{2}$$

with M in terms of solar masses and L_γ in erg/s. Combining Equations (1) and (2), we obtain

$$log_{10}P_{jet} = 1.38log_{10}M + 33.53, \tag{3}$$

which produces constraints on models that connect jet power to black hole mass. Ref. [30] show that 97% of their FSRQ sample satisfies jet energetics as a function of black hole mass (Equation (3)) for moderately thin disks ($H/R = 0.13$). If the radiative efficiency in such moderately thin disks is insufficient to explain the quasar-like spectrum, the success of equation (3) to the sample becomes moot. In that case, it means the scaleheight must satisfy $H/R < 0.1$, which means the GRMHD simulations adopted by [30] fail to explain the sample. If, as mentioned, radiatively efficient disks satisfy $H/R = 0.01$, the mismatch is of course much worse. Ref. [30] report that it drops from 0.098 to 6×10^{-4}, which is 163 times smaller. The basic question here is whether or not the disk thickness that is compatible with jet energetics, is also compatible with the required radiative efficiency. What has been a constant staple in GRMHD simulations is the trend noted above. The thickness that seems to be needed to explain the high enough jet efficiency works against the thinness that is needed to explain the radiative output. A common way out of the above conundrum has been to assume that radio quasars do not have radiatively efficient disks into the inner regions and that while further out the radiation escapes, the energy is retained in the inner regions and the disk scaleheight puffs up there as a result. And this thick disk would be needed to explain the formation of a jet. This, it was thought for decades, would explain a quasar-like spectrum due to the outer cool, radiatively efficient disk, coupled to a powerful jet produced by a thick inner disk. Soares & Nemmen (2019) also mention this possibility. Several recent observations show that this idea fails ([32–35] and references therein).

While magnetically arrested GRMHD disk simulations struggle from a numerical perspective to accurately evolve strongly magnetized regions, radiation is arguably a more difficult problem. As a result, it is treated with a variety of prescriptions from ad hoc cooling functions to keep the disk as thin as possible (e.g., [25]) to more realistic ones that include radiative transport [36]. Given the differences in how the disk is cooled, the radiative nature of the disk can vary greatly among simulations, giving little confidence that GRMHD is providing insights into radiatively efficient disk physics. In [25], for example, the efficiency is twice that of Novikov & Thorne, but it is all emitted within the very central region. This appears difficult to square with typical quasar spectra.

A promising result is obtained in the GRMHD simulations of [16], with the GPU-accelerated code H-AMR [37], which simulates the thinnest possible tilted disk with good enough resolution to resolve the magnetorotational instability with $H/R = 0.03$ around a black hole of spin 0.9375. This simulation also provided a demonstration within a turbulent MHD accretion disk of the Bardeen-Petterson alignment of the disk and black hole. The total efficiency of the disk was found to be higher than that of the Novikov & Thorne prediction by a factor of 3. Perhaps most importantly, a powerful jet was present.

2.2. Jet and Disk Efficiency in GRMHD

Attempts at understanding the compatibility and/or difference between GRMHD and their ability to model thin disks and their compatibility with jets to model radio quasars are obscured in that some GRMHD simulations incorporate or produce physics that is completely different from others. For example, some simulations argue that the Novikov & Thorne solution is not too different from what the disk produces in GRMHD (e.g., [29,38] while other more recent simulations with magnetic fields flooding the system (e.g., [25,39–41] conclude that almost all the radiation is emitted near or inside of the innermost stable circular orbit (ISCO). Other GRMHD simulations suggest that strong magnetic fields might be generated in situ via dynamo action [42]. What is the origin of such vastly different physical pictures? One clear candidate is the variety of different mechanisms incorporated in simulations to treat radiation, with some ad-hoc while others are more faithful to the physics. For example, Ref. [36] performed 3D GRMHD simulations taking into account the time-dependent radiative transfer equations of accretion disks with $H/R \sim 0.1$ around a black hole with a spin of 0.5 using the HARMRAD code [43]. The radiative efficiency of the disk was found to be slightly lower than that of the Novikov & Thorne prediction while in the case of [25], the efficiency was twice the efficiency of Novikov & Thorne when the effects of scattering and absorption of radiation were not taken into account. These different processes change the efficiency but also the location where the radiation is emitted, as already discussed. Ref. [44] performed a radiative GRMHD simulation of a geometrically thin, sub-Eddington accreting disk ($H/R \sim 0.15$) around a zero spin black hole using the KORAL code [45,46]. Although a significant amount of dissipation was found inside the marginally stable orbit as in [25] simulations, the efficiency of the disk was found to be very close to that of Novikov & Thorne. All the results from available GRMHD simulations have been summarized in Table 1.

Uncertainty very much still dominates our understanding of numerical simulations of accretion onto black holes, but a bottom line is beginning to emerge: Strong disk magnetic fields seem to be a by-product of strong black hole threading magnetic fields but strong disk magnetic fields do not allow disks to remain thin [47–49]. And disks that are not thin might not be able to radiate efficiently. It seems there is possibly another factor that can allow the compatibility between the thinness of the disk and the condition of jet formation.

2.3. Scale Invariance and the Jet-Disk Connection

Let us assume that jet efficiency is sufficient in GRMHD to explain radio quasars. We will now show that arguments explaining radio quasars using these simulations nonetheless run into problems. As discussed by [16], radiatively efficient disks appear not to produce relativistic jets, which suggests to them that they are uncovering a transitional process that manifests itself in black hole X-ray binaries as the disk evolves from the hard state to the soft state [7]. The first issue here is the need to show in GRMHD that as the disk collapses from the hard state and transitions toward the soft state, the jet power depends on black hole spin, whereas in the state that precedes it (the hard state), the jet power is less dependent on black hole spin. In other words, it is crucial that this time-dependent scenario must be simulated. These conclusions are based on the observation that jet power does not correlate with a spin in the hard state [9], while it may correlate with it in the transitory state [13,50]. Second, and more fundamentally, the time evolution implicit in the above scenario does not lend itself to scaling up this transitory jet and applying it to radioquasars [51,52]. This crucial point has by and large gone unnoticed. The problem is that radio quasars distribute themselves on average at higher redshift compared to FRI radio galaxies and such a distribution cannot emerge from transitions that take hard state jets into transitory ballistic jets, which reverse the time sequence. The scale-invariant approach, thus, predicts radio quasars occupying lower redshifts compared to FRI radio galaxies. If we incorporate the full cyclical behavior of X-ray binaries in a scale-invariant application to radio quasars, then we would conclude that radio quasars and FRI radio galaxies do not show any redshift dependent difference, but that best-case scenario is not observed. In short, using the time evolution of X-ray binaries in trying to understand radio quasars predicts either that radio quasars are distributed at relatively lower redshifts or that they are not distributed differently in redshift compared to FRI radio galaxies. But the observations are not compatible with these scenarios.

Table 1. Results from some representative general relativistic magnetohydrodynamic (GRMHD) simulation works showing black hole spin (a/M), scale height (H/R), disk radiative efficiency (η_r), jet efficiency (η_j), total efficiency (η_T), and cooling method implemented. Here, η_{NT} means efficiency predicted by the Novikov Thorne (1973) model, and NA means not available information.

References	a/M	H/R	η_r	η_j	η_T	Cooling
Penna et al. (2010)	0–0.98	0.07–0.3	Less than 4.5% deviation from η_{NT}	NA	Close to η_{NT}	Ad hoc as in Shafee et al. (2008)
McKinney, Tchekhovskoy & Blandford (2012)	−0.9375–0.99	0.2–1	NA	Up to 50 times higher than η_{NT}	Up to 120 times higher than η_{NT}	No
Avara, McKinney & Reynolds (2016)	0.5	0.05–1	15% (almost 2 times higher than η_{NT} = 8.2%	1% (almost an order less than η_{NT})	20% (around 2.5 times higher than η_{NT})	Ad hoc as in Noble et al. (2010)
Sadowski (2016)	0	0.15	5.5 ± 0.5% (very close to η_{NT} = 5.7%)	NA	NA	Radiative transfer
Morales Texeira, Avara & McKinney (2018)	0.5	0.1	2.9% (less than half of η_{NT} = 8.2%)	4.3% (around half of η_{NT})	18.6% (more than twice of η_{NT})	Radiative transfer
Liska et al. (2019)	0.9375	0.03	18% (close to η_{NT} = 17.9%)	20–50% (Up to 2.5 times higher than η_{NT})	60–80% (3–4 times higher than η_{NT})	Ad hoc as in Noble et al. (2010)

These issues—and many others—have been resolved in semi-analytic models [4] via the introduction of a key feature for accreting black holes, namely counterrotation between the disk and the black hole, a window on accretion than many have found fruitful (a subset of these are [53–68]. In counter-rotating disk configurations, the process that suppresses the jet weakens as the black hole spin increases, thereby allowing the conditions that lead to strong, collimated jets, to couple to bright disk states. The radio-loud/radio-quiet dichotomy in this picture is based on a high black hole spin for both families of accreting black holes. Whereas radio quasars are high spinning counter-rotating accretion configurations, jetless quasars are high spinning, prograde accreting black holes. This is due to the small value of the inner disk boundary for high prograde configurations. Because the disk reaches deep into the gravitational potential of the black hole for high prograde spins, a greater amount of energy is generated from the disk at all disk locations, and the disk is radiatively dominated or quasar-like. Recently, we have been able to explain the lack of symmetry in the radio loud/radio quiet dichotomy with jetless (or radio quiet) AGN dominating the distribution at about 80% [33]. These ideas have allowed for theory to be compatible with scale invariance and with the observed distribution of radio-loud and radio-quiet quasars across cosmic time. Our roadmap for GRMHD simulations involves, therefore, exploring the nature of the jet-disk connection for corotating, as well as counterrotating accreting black holes. Whereas the jet efficiency was found to be slightly higher in the former, we suspect that, with better treatment of the radiation processes, the results will start to shift, and counterrotation will emerge as a key ingredient.

3. Conclusions

The existence of a subset of active galaxies with high observed radiative efficiency and powerful jets—radio quasars—has remained a major unsolved problem in high energy astrophysics for decades. In this work, we have explored the state of the art in numerical simulations of accretion onto black holes and find that where the comparison to observation is possible, tensions arise. We find a variety of very different physical scenarios that fail to produce a coherent picture for radio quasars and their FSRQ subclass from GRMHD. And we have singled out what appears to be the problem: the strong disk magnetic fields that accompany the strong black hole-threading field work to support greater disk thickness, which in turn decreases the radiative efficiency of the disk. How can we simulate accreting black holes capable of sustaining strong black hole horizon magnetic fields despite weak magnetic fields in their disk? The simulations of [16,17] constitute a milestone in this respect since they produce strong jets despite unprecedented small H/R values. Much work is required to single out the physical processes that would allow such disks to still drag the sufficiently strong magnetic field onto the black hole despite the disk thinness. And, this would need to be understood and juxtaposed to simulations where increased disk thinness has the opposite effect on the magnetic field threading the black hole, or on some other quantity associated with jet suppression. Whatever scale-invariant processes will be identified, it is clear that they cannot be associated with transitory ballistic jets in X-ray binaries. We have then made contact with the idea that counterrotating black holes may resolve such issues and have thus encouraged the GRMHD community to go back and explore the jet-disk connection in that context with these new simulations.

Author Contributions: Conceptualization, D.G. and C.B.S.; methodology, D.G.; validation, D.G., C.B.S., and B.L.; investigation, D.G., C.B.S., and B.L.; writing—original draft preparation, D.G. and C.B.S.; writing—review and editing, D.G. and C.B.S.; supervision, D.G.; project administration, D.G.; funding acquisition, D.G.and C.B.S. All authors have read and agreed to the published version of the manuscript.

Funding: This research was funded by the National Natural Science Foundation of China grant number 12073021.

Data Availability Statement: No data was generated during this work.

Conflicts of Interest: The authors declare no conflict of interest.

References

1. Fanaroff, B.L.; Riley, J.M. The morphology of extragalactic radio sources of high and low luminosity. *Mon. Not. R. Astron. Soc.* **1974**, *167*, 31P–36P. [CrossRef]
2. Shakura, N.I.; Sunyaev, R.A. Black holes in binary systems. Observational appearance. *Astron. Astrophys.* **1973**, *24*, 337–355.
3. Hardcastle, M.J.; Evans, D.A.; Croston, J.H. Hot and cold gas accretion and feedback in radio-loud active galaxies. *Mon. Not. R. Astron. Soc.* **2007**, *376*, 1849–1856. [CrossRef]
4. Garofalo, D.; Evans, D.A.; Sambruna, R.M. The evolution of radio-loud active galactic nuclei as a function of black hole spin. *Mon. Not. R. Astron. Soc.* **2010**, *406*, 975–986. [CrossRef]
5. Neilsen, J.; Lee, J. Accretion disk winds as the jet suppression mechanism in the microquasar GRS 1915+ 105. *Nature* **2009**, *458*, 481–484. [CrossRef] [PubMed]
6. Ponti, G.; Fender, R.P.; Begelman, M.C.; Dunn, R.J.H.; Neilsen, J.; Coriat, M. Ubiquitous equatorial accretion disc winds in black hole soft states. *Mon. Not. R. Astron. Soc. Lett.* **2012**, *422*, L11–L15. [CrossRef]
7. Fender, R.P.; Belloni, T.M.; Gallo, E. Towards a unified model for black hole X-ray binary jets. *Mon. Not. R. Astron. Soc.* **2004**, *355*, 1105–1118. [CrossRef]
8. Remillard, R.A.; McClintock, J.E. X-ray properties of black-hole binaries. *Annu. Rev. Astron. Astrophys.* **2006**, *44*, 49–92. [CrossRef]
9. Fender, R.P.; Gallo, E.; Russell, D. No evidence for black hole spin powering of jets in X-ray binaries. *Mon. Not. R. Astron. Soc.* **2010**, *406*, 1425–1434. [CrossRef]
10. Heckman, T.M.; Best, P.N. The coevolution of galaxies and supermassive black holes: Insights from surveys of the contemporary universe. *Annu. Rev. Astron. Astrophys.* **2014**, *52*, 589–660 [CrossRef]
11. Mirabel, I.F.; Rodriguez, L.F. A superluminal source in the Galaxy. *Nature* **1994**, *371*, 46–48. [CrossRef]
12. Hannikainen, D.; Campbell-Wilson, D.; Hunstead, R.; McIntyre, V.; Lovell, J.; Reynolds, J.; Wu, K. XTE J1550–564: A superluminal ejection during the September 1998 outburst. *Microquasars* **2001**, *276*, 45–50.
13. Narayan, R.; McClintock, J.E. Observational evidence for a correlation between jet power and black hole spin. *Mon. Not. R. Astron. Soc. Lett.* **2012**, *419*, L69–L73. [CrossRef]
14. Migliari, S.; Fender, R.P.; Rupen, M.; Wachter, S.; Jonker, P.G.; Homan, J.; Klis, M.V.D. Radio detections of the neutron star X-ray binaries 4U 1820–1830 and Ser X-1 in soft X-ray states. *Mon. Not. R. Astron. Soc.* **2004**, *351*, 186–192. [CrossRef]
15. Miller-Jones, J.C.A.; Sivakoff, G.R.; Altamirano, D.; Tudose, V.; Migliari, S.; Dhawan, V.; Spencer, R.E. Evolution of the radio-X-ray coupling throughout an entire outburst of Aquila X-1. *Astrophys. J. Lett.* **2010**, *716*, L109. [CrossRef]
16. Liska, M.; Tchekhovskoy, A.; Ingram, A.; van der Klis, M. Bardeen—Petterson alignment, jets, and magnetic truncation in GRMHD simulations of tilted thin accretion discs. *Mon. Not. R. Astron. Soc.* **2019**, *487*, 550–561. [CrossRef]
17. Chatterjee, K.; Liska, M.; Tchekhovskoy, A.; Markoff, S.B. Accelerating AGN jets to parsec scales using general relativistic MHD simulations. *Mon. Not. R. Astron. Soc.* **2019**, *490*, 2200–2218. [CrossRef]
18. De Villiers, J.P.; Hawley, J.F.; Krolik, J.H. Magnetically driven accretion flows in the Kerr metric. I. Models and overall structure. *Astrophys. J.* **2003**, *599*, 1238. [CrossRef]
19. Hirose, S.; Krolik, J.H.; De Villiers, J.P.; Hawley, J.H. Magnetically driven accretion flows in the Kerr metric. II. Structure of the magnetic field. *Astrophys. J.* **2004**, *606*, 1083. [CrossRef]
20. McKinney, J.C.; Gammie, C.F. A measurement of the electromagnetic luminosity of a Kerr black hole. *Astrophys. J.* **2004**, *611*, 977. [CrossRef]
21. De Villiers, J.P.; Hawley, J.F.; Krolik, J.H.; Hirose, S. Magnetically driven accretion in the Kerr metric. III. Unbound outflows. *Astrophys. J.* **2005**, *620*, 878. [CrossRef]
22. Krolik, J.H.; Hawley, J.F.; Hirose, S. Magnetically driven accretion flows in the Kerr metric. IV. Dynamical properties of the inner disk. *Astrophys. J.* **2005**, *622*, 1008. [CrossRef]
23. McKinney, J.C. General relativistic magnetohydrodynamic simulations of the jet formation and large-scale propagation from black hole accretion systems. *Mon. Not. R. Astron. Soc.* **2006**, *368*, 1561–1582. [CrossRef]
24. Hawley, J.F.; Krolik, J.H. Magnetically driven jets in the Kerr metric. *Astrophys. J.* **2006**, *641*, 103. [CrossRef]
25. Avara, M.J.; McKinney, J.C.; Reynolds, C.S. Efficiency of thin magnetically arrested discs around black holes. *Mon. Not. R. Astron. Soc.* **2016**, *462*, 636–648. [CrossRef]
26. Novikov, I.D.; Thorne, K.S. Black Holes. In *Les Astres Occlus*; De Witt, C., De Witt, B.S., Eds.; Gordon and Breach: New York, NY, USA, 1973; p. 343.
27. Noble, S.C.; Krolik, J.H.; Hawley, J.F. Direct calculation of the radiative efficiency of an accretion disk around a black hole. *Astrophys. J.* **2009**, *692*, 411. [CrossRef]
28. Noble, S.C.; Krolik, J.H.; Hawley, J.F. Dependence of inner accretion disk stress on parameters: The schwarzschild case. *Astrophys. J.* **2010**, *711*, 959. [CrossRef]
29. Penna, R.F.; McKinney, J.C.; Narayan, R.; Tchekhovskoy, A.; Shafee, R.; McClintock, J.E. Simulations of magnetized discs around black holes: Effects of black hole spin, disc thickness and magnetic field geometry. *Mon. Not. R. Astron. Soc.* **2010**, *408*, 752–782. [CrossRef]
30. Soares, G.; Nemmen, R. Jet efficiencies and black hole spins in jetted quasars. *Mon. Not. R. Astron. Soc.* **2020**, *495*, 981–991. [CrossRef]

31. Nemmen, R.S.; Georganopoulos, M.; Guiriec, S.; Meyer, E.T.; Gehrels, N.; Sambruna R.M. A universal scaling for the energetics of relativistic jets from black hole systems. *Science* **2012**, *338*, 1445–1448. [CrossRef] [PubMed]

32. Garofalo, D. Resolving the Radio-loud/Radio-quiet Dichotomy without Thick Disks. *Astrophys. J. Lett.* **2019**, *876*, L20. [CrossRef]

33. Garofalo, D.; North, M.; Belga, L.; Waddell, K. Why radio quiet quasars are preferred over radio loud quasars regardless of environment and redshift. *Astrophys. J.* **2020**, *890*, 144. [CrossRef]

34. Garofalo, D.; Webster, B.; Bishop, K. Merger Signatures in Radio Loud and Radio Quiet Quasars. *Acta Astron.* **2020**, *70*, 75–85.

35. Garofalo, D.; Bishop, K. Evidence for radio loud to radio quiet evolution from red and blue quasars. *Publ. Astron. Soc. Pac.* **2020**, *132*, 114103. [CrossRef]

36. Morales Texeira, D.; Avara, M.J.; McKinney, J.C. General relativistic radiation magnetohydrodynamic simulations of thin magnetically arrested discs. *Mon. Not. R. Astron. Soc.* **2018**, *480*, 3547–3561. [CrossRef]

37. Liska, M.; Hesp, C.; Tchekhovskoy, A.; Ingram, A.; van der Klis, M.; Markoff, S. Formation of precessing jets by tilted black hole discs in 3D general relativistic MHD simulations. *Mon. Not. R. Astron. Soc. Lett.* **2018**, *474*, L81–L85. [CrossRef]

38. Zhu, Y.; Davis S.W.; Narayan R.; Kulkarni, A.K.; Penna, R.F.; McClintock, J.E. The eye of the storm: Light from the inner plunging region of black hole accretion discs. *Mon. Not. R. Astron. Soc.* **2012**, *424*, 2504–2521. [CrossRef]

39. Beckwith, K.; Hawley, J.F.; Krolik, J.H. The influence of magnetic field geometry on the evolution of black hole accretion flows: Similar disks, drastically different jets. *Astrophys. J.* **2008**, *678*, 1180. [CrossRef]

40. Tchekhovskoy, A.; Narayan, R.; McKinney, J.C. Efficient generation of jets from magnetically arrested accretion on a rapidly spinning black hole. *Mon. Not. R. Astron. Soc. Lett.* **2011**, *418*, L79–L83. [CrossRef]

41. McKinney, J.C.; Tchekhovskoy, A.; Blandford, R.D. General relativistic magnetohydrodynamic simulations of magnetically choked accretion flows around black holes. *Mon. Not. R. Astron. Soc.* **2012**, *423*, 3083–3117. [CrossRef]

42. Liska, M.; Tchekhovskoy, A.; Quataert, E. Large-scale poloidal magnetic field dynamo leads to powerful jets in GRMHD simulations of black hole accretion with toroidal field. *Mon. Not. R. Astron. Soc.* **2020**, *494*, 3656–3662. [CrossRef]

43. McKinney, J.C.; Tchekhovskoy, A.; Sadowski, A.; Narayan, R. Three-dimensional general relativistic radiation magnetohydrodynamical simulation of super-Eddington accretion, using a new code HARMRAD with M1 closure. *Mon. Not. R. Astron. Soc.* **2014**, *441*, 3177–3208. [CrossRef]

44. Sadowski, A. Thin accretion discs are stabilized by a strong magnetic field. *Mon. Not. R. Astron. Soc.* **2016**, *459*, 4397–4407. [CrossRef]

45. Sadowski, A.; Narayan, R.; Tchekhovskoy, A.; Zhu, Y. Semi-implicit scheme for treating radiation under M1 closure in general relativistic conservative fluid dynamics codes. *Mon. Not. R. Astron. Soc.* **2013**, *429*, 3533–3550. [CrossRef]

46. Sadowski, A.; Narayan, R.; McKinney, J.C. Tchekhovskoy A.Numerical simulations of super-critical black hole accretion flows in general relativity. *Mon. Not. R. Astron. Soc.* **2014**, *439*, 503–520. [CrossRef]

47. Begelman M.C.; Pringle J.E. Accretion discs with strong toroidal magnetic fields. *Mon. Not. R. Astron. Soc.* **2007**, *375*, 1070–1076. [CrossRef]

48. Begelman, M.C.; Silk, J. Magnetically elevated accretion discs in active galactic nuclei: Broad emission-line regions and associated star formation. *Mon. Not. R. Astron. Soc.* **2017**, *464*, 2311–2317. [CrossRef]

49. Dexter, J.; Begelman, M.C. Extreme AGN variability: Evidence of magnetically elevated accretion? *Mon. Not. R. Astron. Soc. Lett.* **2019**, *483*, L17–L21. [CrossRef]

50. Garofalo, D.; Kim, M.I.; Christian, D.J. Constraints on the radio-loud/radio-quiet dichotomy from the Fundamental Plane. *Mon. Not. R. Astron. Soc.* **2014**, *442*, 3097–3104. [CrossRef]

51. Garofalo, D.; Singh, C.B. Scale-invariant jet suppression across the black hole mass scale. *Astrophys. Space Sci.* **2016**, *361*, 97. [CrossRef]

52. Garofalo, D. The jet–disc connection: Evidence for a reinterpretation in radio loud and radio quiet active galactic nuclei. *Mon. Not. R. Astron. Soc.* **2013**, *434*, 3196–3201. [CrossRef]

53. Rusinek, K.; Sikora, M.; Koziel-Wierzbowska, D.; Gupta, M. On the Diversity of Jet Production Efficiency in Swift/BAT AGNs. *Astrophys. J.* **2020**, *900*, 125. [CrossRef]

54. Miraghaei, H. The Effect of Environment on AGN Activity: The Properties of Radio and Optical AGN in Void, Isolated, and Group Galaxies. *Astron. J.* **2020**, *160*, 227. [CrossRef]

55. Piotrovich, M.Y.; Afanasiev, A.G.; Buliga, S.D.; Natsvlishvili, T.M. Determination of magnetic field strength on the event horizon of supermassive black holes in active galactic nuclei. *Mon. Not. R. Astron. Soc.* **2020**, *495*, 614–620. [CrossRef]

56. Baldi, R.D.; Williams, D.R.A.; McHardy, I.M.; Beswick, R.J.; Argo, M.K.; Dullo, B.T.; Westcott, J. LeMMINGs–I. The eMERLIN legacy survey of nearby galaxies. 1.5-GHz parsec-scale radio structures and cores. *Mon. Not. R. Astron. Soc.* **2018**, *476*, 3478–3522. [CrossRef]

57. Mikhailov, A.G.; Piotrovich, M.Y.; Gnedin, Y.N.; Natsvlishvili, T.M.; Buliga, S.D. Criteria for retrograde rotation of accreting black holes. *Mon. Not. R. Astron. Soc.* **2018**, *476*, 4872–4876. [CrossRef]

58. Afanasiev, V.L.; Gnedin, Y.N.; Piotrovich, M.Y.; Buliga, S.D.; Natsvlishvili, T.M.; Buliga, S.D. Determination of Supermassive Black Hole Spins Based on the Standard Shakura—Sunyaev Accretion Disk Model and Polarimetric Observations. *Astron. Lett.* **2018**, *44*, 362–369. [CrossRef]

59. Mondal, T.; Mukhopadhyay, B. Magnetized advective accretion flows: Formation of magnetic barriers in magnetically arrested discs. *Mon. Not. R. Astron. Soc.* **2018**, *476*, 2396–2409. [CrossRef]

60. Christodoulou, D.M.; Laycock, S.G.T.; Kazanas, D. Retrograde accretion discs in high-mass Be/X-ray binaries. *Mon. Not. R. Astron. Soc. Lett.* **2017**, *470*, L21–L24. [CrossRef]
61. Bhattacharya, D.; Parameswaran, S.; Mukhopadhyay, B.; Tomar, I. Does black hole spin play a key role in the FSRQ/BL Lac dichotomy? *Res. Astron. Astrophys.* **2016**, *16*, 54. [CrossRef]
62. Bonning, E.W.; Shields, G.A.; Stevens, A.C.; Salviander, S. Accretion disk temperatures of QSOs: Constraints from the emission lines. *Astrophys. J.* **2013**, *770*, 30. [CrossRef]
63. Kalfountzou, E.; Jarvis, M.J.; Bonfield, D.G.; Hardcastle M.J. Star formation in high-redshift quasars: excess [O II] emission in the radio-loud population. *Mon. Not. R. Astron. Soc.* **2012**, *427*, 2401–2410. [CrossRef]
64. Meier, D.L. *Black Hole Astrophysics: The Engine Paradigm*; Springer: Berlin/Heidelberg, Germany, 2012.
65. Komissarov, S.S. Central engines: Acceleration, Collimation and Confinement of Jets. In *Relativistic Jets from Active Galactic Nuclei*; Böottcher, M., Harris, D.E., Krawczynski, H., Eds.; Wiley-VCH: Weinheim, Germany, 2012; pp. 81–114.
66. Sambruna, R.M.; Tombesi, F.; Reeves, J.N.; Braito, V.; Ballo, L.; Gliozzi, M.; Reynolds, C.S. The Suzaku view of 3C 382. *Astrophys. J.* **2011**, *734*, 105. [CrossRef]
67. Meyer, E.T.; Fossati, G.; Georganopoulos, M.; Lister, M.L. From the blazar sequence to the blazar envelope: Revisiting the relativistic jet dichotomy in radio-loud active galactic nuclei. *Astrophys. J.* **2011**, *740*, 98. [CrossRef]
68. McNamara, B.R.; Rohanizadegan, M.; Nulsen, P.E.J. Are radio active galactic nuclei powered by accretion or black hole spin? *Astrophys. J.* **2010**, *727*, 39. [CrossRef]

MDPI

St. Alban-Anlage 66

4052 Basel

Switzerland

Tel. +41 61 683 77 34

Fax +41 61 302 89 18

www.mdpi.com

Galaxies Editorial Office

E mail: galaxies@mdpi.com

www.mdpi.com/journal/galaxies

9 783036 556109